DIRECTING
TELEVISION
AND FILM

WADSWORTH SERIES IN TELEVISION AND FILM

DIRECTING
TELEVISION
AND FILM

SECOND EDITION

Alan A. Armer

California State University, Northridge

Wadsworth Publishing Company
Belmont, California
A Division of Wadsworth, Inc.

Senior Editor: Rebecca Hayden
Editorial Assistant: Tamiko Verkler
Production Editor: Sandra Craig
Cover and Text Designer: Andrew H. Ogus
Print Buyer: Karen Hunt
Copy Editor: Pat Tompkins
Compositor: Thompson Type
Cover and Part Opener Illustration: Judith Ogus/Random Arts
Other Illustrations: Philip Li

Printed in the United States of America   85

2 3 4 5 6 7 8 9 10—94 93 92 91 90

Library of Congress Cataloging-in-Publication Data
Armer, Alan A.
    Directing television and film/Alan A. Armer.—2nd ed.
        p.   cm.
    Bibliography: p.
    Includes index.
    ISBN 0-534-11616-7
    1. Television—Production and direction.   2. Motion pictures—Production and
direction.   I. Title
PN1992.75.A76   1990   89-32811
791.43'0233—dc20   CIP

For Elaine

PREFACE

This second edition of *Directing Television and Film* is at once more basic and more advanced than the first. Using input from half a dozen universities, I have simplified the more sophisticated chapters and added sophistication to the simpler ones. No major organizational upheavals, just a clearer, crisper, richer text.

Specifically, this edition has:

— Added material on visual composition that defines the space within a frame (Chapter 3)

— Defined Part 2 as basically a study of single-camera operation and Part 3 as largely a study of multiple-camera operation and added introductions to Parts 2 and 3, examining single-camera and multiple-camera operations

— Added material on the actor–director relationship with helpful hints on how to obtain a good performance (Chapter 5)

— Defined the importance of dramatic beats and added a short scene that tests the reader's understanding of beats (Chapters 5 and 6)

— Added a detailed examination of the Master Angle as the beginning point in covering a scene (Chapter 7)

— Included an essay on control room commands (in the introduction to Part 3)

— Updated the chapter on news, comparing an independent station operation with a network affiliate operation, and added "How They Got Started" histories of four news directors (Chapter 11)

— Added material on scoring (composing background music to dramatic material) (Chapter 13)

— Added Projects for Aspiring Directors at the end of each chapter

— Expanded the Glossary to almost double its original size

In addition, this new edition includes dozens of minor clarifications that should make the text more valuable to students.

Directing Television and Film is not a production manual. It won't help you disassemble an electronic switcher, determine the gamma rating of Eastman's newest color film, or design a more versatile character generator.

The book assumes you already know something about switchers, CGs, and color film. If you don't, don't panic. Ninety percent of the material that follows is self-explanatory. If you've studied television or film production, so much the better.

After making the transition from the entertainment industry to academia, I searched for an appropriate text for my directing courses. I quickly discovered that although there are many excellent books on the subject—describing what directors do, how they do it, the problems they encounter, and various facets of the fascinating world they work in—none of these books provides sufficient insight into why directors do the things they do. The thinking. The theory. This book is an attempt to provide the *why* of directing.

The volume embraces two directorial arenas: fiction and nonfiction. The two fields are closely related; many of the principles of fiction (drama) apply forcefully to nonfiction. Theories relating to the staging of performers and cameras become foundation stones for day-to-day television programming. Principles of emphasis apply as much to a newscast or a cooking show as they do to a hit movie. Therefore, I encourage you to read the fiction section before undertaking the nonfiction material.

This book has three sections. Part 1, "The Basics," examines what kind of creature a director is, the elements with which he or she fashions a show, and the principles of visual composition that help directors create dynamic pictures.

Part 2, discussing fiction (drama), refers almost entirely to the one-camera operation that has become known as *film technique* or *motion picture technique*. It explores such bedrock areas as the script and the elements necessary to bring a story to life; the magic that changes actors into dimensional characters; how directors create movement for actors, externalizing their thoughts and emotions; how cameras should photograph the action; and the surprising nature of dramatic suspense.

Part 3, which explores nonfiction, deals almost exclusively with the multiple-camera pattern associated with live television or the filming of some sitcoms. It discusses five prominent television genres from a director's perspective: news, commercials, interviews, demonstrations, and music shows.

In fine art or popular art, few arbitrary rules exist. No set of procedures or techniques can be proper one hundred percent of the time. Those described in this text are used by most professional directors most of the time. They have become established through practice and tradition. The aspiring director should become comfortable with them. Later, after a few years' experience, a few rave reviews, and with a studio contract assured, he or she is free to break those rules—or to establish new ones.

In preparing this book, I spoke with a number of directors to get their opinions in certain key professional areas: What is a director? Where does he or she come from? How does a person become a director? The most practical of their answers appear in Chapter 1 and Chapter 11. In seeking the definition of a director, I discovered that certain characteristics or abilities cropped up again and again. Among them: leadership, taste, sensitivity, dramatic or technical expertise, and organization.

This book cannot make you a leader. It probably won't do much for your taste or sensitivity either. But it will provide you with the theory necessary to stage a dramatic scene or to prepare a newscast. It will guide you through the selection of camera angles. It will help you counsel performers in a broad spectrum of program forms. It will offer insights into the subtleties of creating mood and atmosphere. It will give advice on how to handle yourself in the directorial arena.

Of the characteristics necessary in a director, organization is critical. Confidence, security, feeling good about yourself, all arise from *preparation*—doing your professional homework. Had Thomas Edison been a director (and he would have made a good one!), he would have declared, "Good directing is 98 percent preparation and 2 percent inspiration." As much as any other factor, meticulous preparation is responsible for the success of most directors.

Fact of life: No book in the world can transform you into a director. Not even this one—although I certainly try! Practical, in-the-trenches experience is essential. But this book does provide reassuring signposts that will guide you on the appropriate path.

I wish you well. And when cast and crew assemble and the dream is ready to come alive, I hope that it will be *your* voice uttering the magic word that makes it happen: "Action!"

ACKNOWLEDGMENTS

I would like to express my deep appreciation to the following television and film professionals for sharing their expertise, opinions, experiences, and personal anecdotes—and for helping me get my act together: John Allyn, Tony Asher,

Tom Burrows, George Eckstein, Dick Fleischer, Walter E. Grauman, Delbert Mann, Walter Mirisch, Wayne Parsons, Jay Roper, and Sidney Salkow.

My appreciation also to the executives at KNBC and KHJ-TV who helped in the preparation of television news material. At NBC: news director Tom Capra and director Gene Leong. At KHJ-TV: news director Stephany Brady, producer Bill Northrup, and director Chris Stegner.

My thanks to the good people at Wadsworth Publishing Company for their friendliness, their eminently sensible suggestions, and their professionalism. In particular, I'm grateful to senior editor Becky Hayden for her continuing support and guidance; to the book's designer, Andrew Ogus; and to the production editor, Sandra Craig.

My appreciation to those generous academicians who took the time to give critical comments on the first edition or to review the manuscript for the second edition, contributing far more than criticisms or assessments. Their questions made me dig for significant answers. Their proffered nuggets of knowledge enriched the text. Thanks to David Barker, Texas Christian University; Lilly Ann Boruszkowski, Southern Illinois University; Diane J. Cody, University of Michigan; James Fauvell, New York Institute of Technology; Jim Friedman, Northern Kentucky University; and Robert Musburger, University of Houston.

My gratitude and love to all of the students in all of my classes at California State University, Northridge. They brighten my life. Special thanks to Carol Chamberlin, Tim Sharman, Brian Shockley, Terry Spalding, and Venessia Valentino for allowing me to include their pictures here.

ABOUT THE AUTHOR

At the age of fifteen, Alan A. Armer won the title of "World's Fastest Talker" in an NBC radio contest, reciting Lewis Carroll's "The Walrus and the Carpenter" (613 words) in 57 seconds. He's slowed down considerably since then.

After graduating from Stanford University with a degree in speech and drama, Alan Armer tried his hand first in radio as a disk jockey for San Jose's KEEN and then in advertising, writing, staging, editing, narrating, even acting in early TV commercials. Because of what he learned about television, Armer was able to write and produce a weekly showcase for young professional actors, *Lights, Camera, Action,* for Los Angeles station KNBH.

When the show expired three years later, Armer became a stage manager for the station. Six months later he started directing; he was a staff director there for over four years. When the days of live programming were fading, 20th Century Fox hired Armer to produce some of their early TV films.

During the next fifteen years, Armer wrote, produced, or directed over 350 television programs for Fox, Desilu, Quinn Martin Productions, Paramount, and Universal. Many of his series were in the Top Ten. They included "The Fugitive," "The Untouchables," "Cannon," "The Invaders," "Broken Arrow," "Lancer," and "Name of the Game," plus a number of TV movies.

During his years in television, Armer won almost every major industry award, including the TV Academy's distinguished Emmy Award, Mystery Writers' Award, TV Guide Award, Western Writers' Award, and Sound Editors' Award. He also served as president of the Hollywood Chapter of the Television Academy, was a trustee to the National Academy, and served repeatedly as a member of the Board of Directors for the Producers Guild.

After twenty-five years in television, Armer moved into academia. To enrich his teaching career, he attended graduate school at UCLA and earned his M.A. Now, after nearly a decade, Armer continues as a professor at California State University, Northridge, where he teaches classes in directing and screenwriting.

He is listed in *Who's Who in Entertainment* and has published two books with Wadsworth: *Writing the Screenplay* and *Directing Television and Film.*

CONTENTS

PART ONE
THE BASICS

PART

1

THE BASICS

WHAT IS
A DIRECTOR?

You walk into a television studio or motion picture soundstage. Overhead, electricians are making final adjustments to lights. In one mirrored corner, makeup artists apply the finishing touches to performers' faces while hairdressers brush and spray the performers' hair. Near the lighted set, the camera crew scratches chalk lines on the floor, marking camera positions for the upcoming scene.

Within the set, a dozen crew members politely but relentlessly question an Authority Figure about props, script, costume, production schedule, retakes, and location. The Authority Figure answers quickly, waves them away, and then nods to someone offstage. As you watch, the assistant director brings the cast into the set. The Authority Figure whispers a few final instructions and then a bell clangs for silence. Outside the soundstage, a red light blinks a warning to passersby. Inside, a camera assistant holds a slate in front of the camera. Someone calls, "Speed." And then the Authority Figure speaks the two-syllable word that will bring the characters to life.

In television and film, such authority figures are the spine of any production pattern. They organize, assemble, interpret, and dramatize the elements of a show. Most conspicuously, they create order out of chaos. We call them **directors** because they direct the activities of a wide-ranging group of technical experts and

If they changed all the rules and they said, "You've got to stand in line and pay money to do this work," I would be the first guy in line.

William Friedkin[1]

craftspeople as well as the efforts of creative artists. (Boldfaced terms are defined in the Glossary.) To complicate matters, many so-called technicians are also artists; many artists are also technicians (for example, directors of photography, editors, art directors, and special effects experts). Therefore, because they straddle two diverse worlds, the technical and the artistic, directors may be considered artist-technicians.

Art has often been defined as *communication*. In television and film (popular art), directors stand at the center of the communication process, shaping and transmitting the message from its Sender (writer/originator) to its Receiver (audience). The director shapes a program or film message through the manipulation of technical and artistic elements, seeking to maximize their effect upon the audience. If the director interprets the message with skill and sensitivity, the audience will become emotionally or intellectually involved. Generally, the greater the degree of audience involvement, the more successful the film or program.

[1]Eric Sherman, ed., *Directing the Film: Film Directors on Their Art* (Boston: Little, Brown, 1976), p. 3.

Awareness of audience needs is seldom in the front of a director's mind. Directors cannot act on (or even be aware of) a thousand disparate spectator preferences or prejudices. They can only please themselves. When staging a scene, directors necessarily become its most critical audience. In relying on personal taste, they must hope that their dramatic judgments and directorial styles reflect today's world and today's audiences. When directors find that they are no longer in the mainstream of today's world, they must update their taste—or get out of the business.

This chapter will examine the chemistry of directors, the nature of their world, and how they function in that world—in other words, the thinking, the theory, and the logic behind a director's actions. This examination consists of seven primary categories:

— THE DIRECTOR EMERGES: a capsule history of the director's contribution to theater, film, and TV

— WHAT IS A DIRECTOR?: an examination of the artist, the technician, and the parent-psychiatrist

— THE DIRECTOR AT WORK: a study of two professional worlds—"live" or videotape (directing from a control booth or in the field) and film (its applications in motion pictures and TV)

— DIRECTORIAL STYLE: a comparison of the manipulative versus the invisible approach in directing

— CHARACTERISTICS OF GOOD DIRECTORS: professionals' opinions on the qualities necessary for directorial success

— THE APPEAL OF DIRECTING: reasons why so many students and entertainment industry professionals aspire to direct

— WHERE DIRECTORS COME FROM: an exploration of the paths that take directors to their chosen careers

THE DIRECTOR EMERGES

Anyone involved in the production aspect of television or film would find it difficult to imagine a show without a director. Production patterns have become so incredibly complex in recent years that a set such as the one described at the beginning of this chapter would instantly degenerate into chaos without an authority figure to provide leadership and structure. Consequently, most people believe that directors have always been with us. Not so. Through most of drama's long history, no one was minding the store.

> When the director did finally appear toward the end of the nineteenth century, he filled so pressing a need that he quickly pre-empted the hegemony that had rested for centuries with playwrights and actors. . . . The appearance of the director ushered in a new and original theatrical epoch. His experiments, his failures, and his triumphs set and sustained the stage.[2]

In the brief century of their existence, directors have dominated the shaping (or reshaping) of four major entertainment forms: the stage, motion pictures, radio, and television.

When the great director-innovators (Konstantin Stanislavski, Adolphe Appia, Max Reinhardt, and others) first appeared, the theater was shapeless and lacked a point of view.

> They insisted that if theater was to retrieve its unique, primitive, communal power, a director would have to impose a point of view that would integrate play, production, and spectators. By his interpretation a director would weld a harmonious art and a cohesive audience out of the disturbing diversity increasingly apparent in our urban, industrial mass society.[3]

In American theater, audiences became increasingly unwilling to suspend their disbelief; they demanded more and more realism. Directors attempted to satisfy this need by doing away with flimsy canvas flats and constructing a lifelike architecturally solid stage world. At the turn of the century, herds of sheep, locomotive engines, and portions of restaurants appeared on the American stage, but audiences were still dissatisfied. Motion pictures provided the reality they were looking for—but stage traditions did not die easily.

Early film directors, unsure of their new medium, borrowed theatrical techniques. The stationary camera watched proceedings from an audience position, as if through a proscenium arch. Actors moved laterally across a stage, at a distance. Their gestures and makeup were exaggerated, as if for the benefit of spectators at the rear of the house.

But innovative directors soon broke out of the mold, experimenting, defining new forms, and creating a cinematic language. Edwin Porter, D. W. Griffith, Sergei Eisenstein, and others moved the camera closer, higher, lower, panning

---

[2]Toby Cole and Helen Krich Chinoy, eds., *Directors on Directing* (New York: Bobbs-Merrill, 1976), p. 3. Used with permission.

[3]Cole and Chinoy, *Directors on Directing*, p. 4. Used with permission.

and tilting, dollying and trucking, making it a participant in dramatic action rather than a spectator. They established the motion picture as an art form in its own right, evoking mood from lighting, creating new dimensions in character and plot, and evolving editing techniques still in use today.

In the late 1920s, the introduction of sound fascinated and perplexed a new generation of directors. Radio directors discovered the technique of creating pictures in their listeners' imaginations solely through the use of sound. Silent film directors, unused to actors who actually spoke dialog, stepped aside for stage directors, who were acknowledged experts with the spoken word. Stage directors tried their hand briefly at film but often failed because they lacked experience in cinematic techniques. They didn't understand the raw power of the visual image and returned audiences to the lukewarm custard of "proscenium arch cinema." Stage-trained directors soon gave way to a dynamic new breed, including Ernst Lubitsch, Alfred Hitchcock, and Rouben Mamoulian, who were able to combine sound and image effectively to enrich each other, creating new cinematic dimensions.

When television emerged in the late 1940s, the medium demanded a new set of directors. Many came from radio, despite a background generally limited to sound. These directors, working in familiar surroundings, sensibly borrowed techniques from motion pictures. (Many early TV studios were converted radio studios.) Because live television cameras were extremely mobile, many of these early directors developed fluid movement in their shots, tracking actors as they moved about a set, gliding from two-shots into closeups into three-shots. They preferred moving their cameras to actors (or actors to cameras) rather than frequent cutting from angle to angle. (Film aficionados will recognize that such camera mobility was actually pioneered in film by Max Ophuls and others years before the arrival of TV.) When many of these live TV directors later made the transition into film, they carried those techniques with them, further enriching the medium.

WHAT IS A DIRECTOR?

A portrait of a director displays many colors. It resembles certain Picasso paintings in which the spectator views a model's face simultaneously from two or three angles. Expressed another way, a director becomes many things to many people: parent, priest, psychiatrist, friend, writer, actor, photographer, costumer, electronics wizard, musician, graphics artist, and a dozen more.

Directors' work embraces two worlds, the artistic and the technical, which is why this chapter initially defined directors as artist-technicians. Let us begin our portrait with those two definitions, which are by no means mutually exclusive.

The Artist

In drama the script is usually a director's first artistic consideration. Good directors know that a fine picture is seldom possible without a fine script. Their taste, experience, and sensitivity enable them to recognize the potential that sometimes appears even in a mediocre script. When a script requires additional work—which is usually the case—directors seldom rewrite it themselves. Instead, they seek the services of the best writer available, who will shape their vision on paper. The ability to understand script structure, to analyze its progression, to recognize potholes, to build colorful, dimensional characters, and to help create pungent dialog all define the director as artist.

In nonfictional television or film also, the script marks a show's parameters for success and therefore merits the director's earliest attention. The director's artistic perceptions must combine with drawing board craftsmanship. Does the script accomplish its primary purpose? Can production values be improved? Does the show include humor? Does it build logically? Can the elements of showmanship be enhanced? (See Chapter 2.) If its purpose is educational, is the message understandable and vivid? The ability to conceptualize the transition from words to picture also defines the director as artist. The script contains the seeds of pictures that will take form first in the director's imagination and later on a soundstage.

The sensitive director has a salon photographer's understanding of pictorial composition. Such a director knows from which vantage point a scene should be shot, from a high angle or low, past foreground trees or with the horizon tilted. Such a director has a painter's sense of color, coordinating sets, costumes, and photographic tones to create specific emotional responses.

Directors of drama understand the artistry of acting, not just intellectually but on a deeply intuitive level. They may hate actors or love them, but directors surely understand them. In casting, directors search for more than acting ability and appropriate physical characteristics; they search for subtleties in style that will enrich a story, as well as chemistries between actors that will add dimension.

Recognizing actors' weaknesses, directors protect them. Recognizing their strengths, directors accept no less than their best efforts. (Surprisingly often, directors have themselves acted, extending their artistry to still another dimension.) But good directors understand performers of all types and talents: singers, dancers, comedians, mimes, and minstrels. They support them and bully them and inspire them to perform their best.

The director's artistry also extends to other facets of production. On one of television's earliest dramatic series, "Matinee Theater," the executive producer required all of his directors to design their own sets. His theory: Sets and their dressing are so closely related to a director's staging that both should be designed by the same person. Although such an assumption has some validity, it is undeniable that **art directors** make enormous contributions to TV and film. They design all sets. They are instrumental in planning the "look" of a show. The best art directors have learned to think like directors. Accordingly, they design settings that offer multiple options for effective staging.

Not surprisingly, directors of drama often make major contributions to a set's design. Because they have studied their characters, they usually understand them better than an art director does. And because a setting necessarily reflects the characters who inhabit it, the director is better prepared to supply this added dimension. Also, because directors understand their characters, they are prepared to evaluate the characters' tastes in furniture, books, paintings, and accessories—thus functioning in another artistic capacity, that of **set dresser**.

The director's artistic functions also include editing, special effects, costume design, makeup, hairdressing, as well as designing graphics for titles and credits.

The Technician

Successful directing requires more than just minimal understanding of the technical aspects of video and film. Artsy-craftsy directors who try to bluff their way through a project with neither technical know-how nor the desire to acquire it find that (1) they communicate ineffectively with experts on their crew, and (2) because of that communications breakdown, their project has suffered a loss of quality. Directors with an understanding of technical subtleties and current technological advances can discuss those subtleties with crew specialists, thereby providing an extra degree of polish for their show.

Whether in film or video, directors' technical expertise functions first at the planning stage, in choosing production equipment. Such decisions, usually discussed at length with crew members and finalized at a production meeting, include choice of camera and lenses, film, lighting options, sound equipment, special effects, crew size, crew members, switching needs, and microwave link.

In video, production patterns vary enormously, from a one- or two-person electronic news gathering (ENG) unit on location to a studio "special" with a crew of 100. Crew size depends on the director's technical needs, the scope of his or her production plans, and the nature of the program. Once production begins, a director's technical knowledge must encompass dozens of specialized areas, from color saturation to microphone characteristics, from computer-generated graphics to postproduction editing. In motion pictures, electronic aspects are replaced by a wide range of film-related technologies, beginning with camera and film characteristics and concluding with **automated dialog replacement (ADR)**, the dubbing of music and sound effects (integrating them with dialog into a composite track), and an analysis of answer print color values at the processing laboratory. An **answer print**, or **first trial composite**, represents the laboratory's first attempt at balancing color values from scene to scene in a film that contains the final dubbed sound track.

Production manuals can be useful, but aspiring directors would do well to take some training in film or television production classes. (See "Books That May Be Helpful" at the end of this text.)

The Parent-Psychiatrist

In any organization requiring the closely integrated services of skilled specialists operating in a pressure-cooker atmosphere, an authority figure is needed to prevent the pressure cooker from exploding. In film and video, where temperament frequently is part of the artistic equation and where enormous talent often conceals the most fragile of egos, such an authority figure must be a master of diplomacy.

Because skills and personalities are diverse, conflicts arise. Because problems are subjective, they elicit widely divergent solutions, again promoting conflict. Because the pressures of limited time and money foster insecurity, voices become unsteady, tempers flare, performers quit, and directors consider changing professions.

Successful directors control the divergent elements of a production company in the same way a mother or father raises a family—with discipline and love. Directors who are themselves well prepared set an example for others in their company. Directors who announce the rules up front, establishing a clearly defined production pattern, enable everyone to find some degree of security and comfort in that structure. Crew members and performers prefer to work on a relaxed, happy set. By their own demeanor directors establish the mood: A happy set reflects a secure, confident, businesslike director; a nervous set often reflects a tense, uneasy director.

Most of a director's discipline problems originate in the performers. Because actors must make the greatest ego commitment, exposing themselves to the possible ridicule of strangers, they are the most insecure. When temperamental conflicts arise, often there's an underlying reason. Maybe performers have personal problems. Perhaps they need more guidance or clarification—or love. After the director has provided these, if the performers continue to make imperious demands, they will lose the respect of their peers. Most performers pride themselves on their professionalism. Awareness of that pride is a director's strongest disciplinary weapon.

Directors don't need psychiatric training, but they do need practical common sense and insight—the ability to look inside a performer's mind to ascertain the real reason for unprofessional behavior. Sometimes television performers drag their heels because their roles don't seem as juicy as those of other series stars. They rebel because they feel their talents are being overlooked or (worse!) have been judged deficient.

In both motion pictures and television, performers sometimes behave inappropriately because of unresolved conflicts with the producer, perhaps over money or billing or the size of their dressing rooms. Script problems are also common, as are disagreements over a characterization. Both should be resolved, whenever possible, prior to production. Occasionally, personal problems—the loss of a parent, ill health, or a fight with a lover—interfere with concentration.

No book can provide easy answers to the myriad of personal or personality problems that disrupt a set or inhibit a performance. Wise directors begin with compassion and understanding, resolving problems that can be resolved and establishing priorities. The first priority: Turn off the outside world. Each performer and crew member must focus 100 percent attention within the dramatic arena. Such an edict is not always easy to enforce. The director may need the wisdom of Solomon, the kindness of St. Francis of Assisi, and the unyielding discipline of an army sergeant.

THE DIRECTOR AT WORK

When directing a live or taped *studio* program—usually a music or musical variety, drama, game or quiz, interview, demonstration, or news show—the director sits in a control booth surrounded by electronic wizardry and facing an array of video screens. Cameras are located some distance away, connected to the director through the umbilical cord of program lines. When directors want to communicate with talent, they must go through **floor managers**, who relay the information. Sequences—or entire programs—are edited as the show is performed.

Such a video production pattern also applies to directors of major programming away from the studio—for example, sports broadcasts, awards shows, or special events. Directors also edit these programs as the action takes place. When segments of a news or magazine format show are taped away from the studio, there is no need for a control booth. Now the director stands near the single lightweight camera, coordinating the action. These videotape elements usually are edited later to fit the space or time requirements of a broadcast.

In film, the director usually stands beside a single camera, a whisper away from performers. After a day's work has been completed, the film is processed, screened, and edited. Once the editing has been completed, weeks or months later, sound effects and music are added. One notable exception to this pattern is the filmed multicamera sitcom in which the director coordinates action from a control booth. (The procedures described here are those most commonly used; exceptions and modifications occur daily.)

Technologies differ widely between film and TV. Specific functions also differ. But the theory that guides directors remains exactly the same. The principles that determine whether film producers make $50 million at the box office or lose their shirts, whether a TV program garners a Nielsen rating of 50 or 5, whether audiences laugh, cry, or yawn are fundamentally the same in both media. Those principles relate to concepts of showmanship that were valid 2,000 years ago and will probably be equally valid in the year 4000.

Directing Live or Videotaped Television

The TV director works for many bosses: local stations, cable outlets, commercial or public broadcasting networks, plus independent production companies that create programming or commercials. Within that employment spectrum, program content ranges broadly from drama to commercials, from music to news, from kidvid to special events.

Two types of directors function within the still burgeoning television market: (1) the specialist who has refined his or her talents in a special program type and (2) the multifaceted director who can handle cooking shows and symphony orchestra broadcasts with equal grace and style. Multifaceted directors frequently work on the staffs of local television and cable stations. In time, because of unique talents or the caprices of destiny, these directors become more and more adept at a particular program type and make their transition into the ranks of the specialists.

Whether directors work from a control booth or beside a single field camera, their first consideration always must be what happens *in front of the camera*. Film school students and those in directing or production classes often tend to concentrate on hardware: control booth gadgetry, the camera, lighting, microphones, and other production paraphernalia. Such fascination is understandable; these are wondrous toys that smack of professionalism. On a TV stage, the size and raw power of the technology seem to dwarf performers.

Such misplaced emphasis immediately become apparent when you consider program material from the audience's perspective. Audiences generally remain unaware of microphone characteristics or subtleties of lighting. Most are (consciously) blind to cuts, oblivious of dissolves, unaware of camera moves. They become involved with *performers*. If they can see those performers and hear them clearly, all other aspects become secondary. Such cold reality is often traumatic for students who excel at television's technological aspects.

Multiple Camera Operation   Once TV directors have staged program action to their satisfaction, they then position their electronic cameras. Directors of drama follow an almost identical pattern. The action usually is meticulously rehearsed, first with actors (**dry**) and then with cameras. The director assigns specific shots to each of the camera operators, who may jot these down on a **shot sheet** or use shot sheets already prepared by the director.

In awards shows, rehearsal is necessarily limited because the show's outcome is unknown. The host and awards presenters receive a briefing on entrances and exits and what generally will be expected of them. Their spoken material is checked on cue cards or teleprompters, and final changes in copy are made. The director assigns specific shots or angles to each camera. Sometimes lettered cards substitute in rehearsal for nominated performers, giving camera operators the chance to know in advance where each star will sit. Lavish production numbers are carefully rehearsed, sometimes for many weeks prior to the air date.

In sports and special events, very little may be rehearsed. Each camera is assigned a position and a general area of coverage. Because they usually are complicated, technical aspects of such shows are double and triple checked. In special events, the director often works from an outline rather than a script. Frequently the planned sequence of events changes at the last minute, so directors must be able to replan their camera coverage, extemporizing to fit the redesigned program pattern.

Prior to airtime, the director takes his or her place in the control booth (often in a van), making sure that all program elements are ready and standing by. When the broadcast begins, the director cues performers and then edits the action as it occurs by cutting between cameras. The director selects those angles that most effectively reveal or dramatize the action, emphasizing the most significant aspects. Thus, in drama, the director might **cut** from a full shot on camera 2 to closeups of individual actors on cameras 1 and 3. In a live or taped commercial, the director might cut from a medium shot of a saleswoman to an extreme closeup of the product she is pitching. The director executes such cuts by cueing a **technical director** (**TD**), usually seated to the right. The TD uses a versatile electronic **switcher** to accomplish a wide variety of transitions between shots, such as cuts, dissolves, and wipes. The switcher also allows one picture to be superimposed upon or placed within another, that is, by **matte**, or by **keying**. In some stations, directors do their own switching.

Elsewhere in the control booth, usually isolated from the clamor by glass partitions, sits the audio control engineer. This important member of the production team manipulates the sound portions of a program: either music or dialog originating from the stage and picked up by microphones, or sound originating in the booth, such as recordings, cassettes, or cartridges ("carts"). The show's producer also usually sits in the control booth.

Single-Camera Operation—EFP   When drama is videotaped in the field away from the secure studio operation, it usually is recorded with a single camera, a technique called **electronic field production** (**EFP**). EFP resembles the traditional motion picture pattern because editing occurs at the conclusion of production rather than concurrently. The advantage of a single-camera technique over a multiple-camera production pattern is that each angle can be individually rehearsed for maximum effectiveness, ensuring clean sound, superior lighting, controlled photographic composition, and optimum performance. Most important of all: the director's proximity to actors makes for easy communication. The major disadvantage of EFP is that it is time consuming and therefore expensive. Editing also becomes a cost consideration because all of the angles (and there may be hundreds!) must be assembled with an eye to logical progression as well as dramatic effectiveness.

As in motion pictures, each scene is photographed angle by angle. Thus, if a director decides that a two-character scene must play in a master (wide) angle, two over-shoulder shots, and two closeups, the scene must be photographed five separate times: once for the master, twice for the over-shoulder shots (each favoring a different actor), and twice for the closeups. The videotape director may screen a "take" at its conclusion to make certain that the scene has the necessary impact. If not, he or she will record it again. (Chapters 6 and 7 cover the single-camera pattern in drama more extensively.)

EFP is not used solely for drama. Producers requiring high standards for their programs may take studio-quality camera and sound equipment into the field for any event that merits such special treatment. The term *EFP* generally implies greater quality than conventional news coverage. Remember this rule of thumb: The greater the quality demanded, the more sound, lighting, and camera equipment will be required.

Single-Camera Operation—ENG A far more economical use of the single-camera operation is **electronic news gathering (ENG)**, usually operated by two people. It is a compact, portable operation using a videotape recorder that may be taken anywhere instantly, with a minimum of equipment. Because of its great mobility, ENG is ideal for news operations. Its only drawback is that its picture and sound quality is lower than that of film or EFP.

For other than major news breaks, most local station ENG operations have no designated director. They are staged by a field producer or an associate producer or by the talent, usually a member of the station's news team. For aspiring directors, such a position affords excellent practical experience in a wide variety of entertainment arenas. ENG is also used to provide segments of programs other than news. Local station magazine format shows, in particular, utilize ENG, recording a broad cross section of feature stories for five, sometimes six, half-hour programs per week.

Directing Film

The film director works in both television and motion pictures. Television films primarily include half-hour or hour segments for continuing network series, TV movies ranging from ninety minutes to eight hours in length, commercials, animated children's shows (that is, animated films for children, not films for animated children!), and some documentaries. Motion picture films, intended for theatrical exhibition, usually run from 90 to 120 minutes. While they deal in both comedy and drama, movie producers tend to slant their story lines to audiences between the ages of 14 to 24.

There is relatively little difference in the mechanics of filming a motion picture feature and a television feature or even a TV series segment. The differences lie in the budgets and the subject matter.

Motion Pictures Because motion picture budgets usually are substantially larger than those provided for television films, movie directors may offer their audiences greater production scope. They generally have the luxury of top-of-the-line writers, actors, composers, editors, and camera crews, increased production scope, and more time. Increased production scope implies the use of elaborate sets, distant locations, special effects, animation—whatever will enhance the film's visual imagery. Although extra production time may not appear as significant as other budget considerations, almost any experienced director, if given a choice, would opt for it. There never seems to be enough time, either in preparation or production. Many TV series directors, impatient with assembly line production practices, eye the leisurely motion picture schedules wistfully.

The second major difference between television and feature films concerns subject matter. Attending a motion picture is not inexpensive. To coax spectators out of their homes, movie producers feel they must offer story material and behavior that patrons cannot receive free in their living rooms. Movie directors therefore usually work with more sexually explicit scenes than are possible in TV, with more overt violence than is permitted by network censors, and with stories that are (sometimes) more sophisticated than the lowest-common-denominator programming of network television. Such enticements have lost some of their value with the increased popularity of videocassettes.

Television Although it has become fashionable to deprecate series television, it nevertheless provides directors with basic training of incalculable value. After-school specials, Saturday morning children's shows, and filmed programming of every size, shape, or budget all offer valuable in-the-trenches experience, preparing directors for more challenging and more prestigious television or motion picture features. By working within limited hours and limited budgets, TV directors are forced to use their imaginations, to find more economical ways of planning and staging material. That is, they must limit the number of sets, cast and extras, and camera **setups** (camera positions and arrangements of lights) through resourceful staging of the actors' movements within a set. If a scene can be staged in two angles rather than three, for example, the director can probably save half an hour or more. On a tight schedule, an extra half hour is solid gold.

Three Phases of Directing Film

It might be illuminating to follow the activities of a television series director through film's three traditional phases: **preparation** (or **preproduction**), production, and **postproduction**. Because television's production pattern generally parallels that of motion pictures, such detail will (more or less) describe the entire film spectrum. Our prototypical director is a young woman named Sara.

Sara's odyssey begins when her agent makes a deal with the production company that packages a television series called "Buckeye." She will direct one episode: six days of preparation, seven days of filming. She must report to the studio on thus and such a date. If the company likes her work, it may exercise an option for two additional segments, dates to be arranged.

Preparation   A script arrives at Sara's home the day before she is due to report. She opens the cover. The title page states, "Rough Draft—Limited Distribution." As she reads the material, Sara understands why distribution is limited. The script is awkward and amateurish. Her heart sinks. This has been a quality series. Why are they sticking her with a second-rate script? She reads quickly, trying to get an overall impression. Not a bad concept, but it needs work. Lots of work. Can the script possibly be pulled together in the limited time remaining? She reads it again, more slowly this time, making notes as she goes, trying to visualize the characters.

The following morning she reports to the producer and discovers that he has already given the writer many of her suggested changes. He's bright and likeable, trying to soothe her concerns with assurances that the writer will complete all revisions in a couple of days. He asks if Sara is familiar with the show. Yes, she has watched it, but would like to screen a couple of episodes in order to more completely familiarize herself with the show's characters, style, and production pattern. Sara also wants to see how much camera **coverage** other directors provided—how many different angles in each scene—so that she can give the production company what it expects.

The first day Sara meets the casting director, a middle-aged woman of considerable experience. They discuss the characters at some length, suggesting possible actors for some of the key roles. The casting director will check salaries and availabilities. When she has answers—and a more complete list of choices—she will schedule a meeting. Perhaps tomorrow.

Sara also meets her **assistant director**, an intense young man, and the **unit manager,** a gruff, outspoken film veteran who shakes his head, grumbling that the "limited distribution" script is considerably over budget, totally unrealistic. He implies that it will probably never be filmed. The assistant director gives Sara a covert wink, as if to say, "Don't worry about him. He's paid to worry."

The assistant has broken down the rough draft script and put it "on the board." That is, he has converted the script into a series of cardboard strips, one strip for each scene. Each strip lists scene numbers, whether day or night, length of scene, which characters appear, probable number of extras, stunts, and special effects. The strips are then arranged on a **production board**, all scenes in the same locale grouped together. As much as possible, scenes for each character are also grouped together for economy; the more days an actor works, the more money it costs. Location days usually are placed at the front of the schedule so that bad weather will allow the company an escape hatch—to move indoors. (If location shots were placed at the end of the schedule, rain would leave the company nowhere to go because all soundstage material would already have been filmed.)

Sara goes over the board with her assistant, suggesting several changes for the benefit of actors and performance. Like most directors, Sara prefers to save the most demanding material for late in the schedule, to give actors a chance to absorb their characters, to "get into" their roles. For this reason, she prefers to shoot as much in sequence as possible.

The assistant has already looked at a number of possible sites for location sequences. To save Sara unnecessary work, he has eliminated those that were obviously wrong, narrowing choices to two or three for each location.

Sara visits each location with her assistant and the art director, selecting those that seem most appropriate to the script and would bring color or richness to the show (**production values**). For each day's work, she tries to find locations that are close to each other to avoid long (and therefore costly) moves. The assistant takes Polaroid photographs of each location so that the producer may subsequently approve or disapprove the director's choices. (When producers have time, they accompany directors in scouting locations.) Sara discusses with the art director how to conceal billboards or other advertising material that might cause sponsor or legal problems. She arranges for signs to be made, identifying locations per the script's requirements. Her happiest discovery is that her assistant is both perceptive and dependable. His location choices all were excellent. She will be able to delegate authority with confidence.

Sara's first real problem occurs the next day at the casting session. The casting director and the producer try to convince her to use a certain actress in the guest role. Sara is not a fan of this actress. She realizes that she has seen her act only once, in an inferior play on Broadway. But she knows that if this part is not cast superbly, the whole show could fall apart. So Sara asks to see film of other roles this actress has played. When the film is screened for her later, she's still not convinced that the actress is good enough. Sara asks if she will read for the part. The casting director is horrified. This actress is too important to come in for an office reading. Sara is adamant. Finally the casting director calls the actress's agent and pleads her case. The agent grudgingly agrees to ask the actress to read.

Later, after a lukewarm office reading, the producer grins ruefully and acknowledges that they can probably find someone better. Sara has been burned too often, taking someone's word about an actor. No matter where she works, if an actor is recommended whom she doesn't know, she doesn't pretend familiarity; she demands to see performances on tape or film.

The following day the script revisions arrive. Sara reads them with some apprehension. They're an improvement, but the script is still a long way from where it should be. The producer sends for the writer. With Sara, they go through the script, page by page—in some scenes, line by line—refining, suggesting additional rewrites, consolidating, trying to solve the script's creative problems as well as its budgetary ones.

When they are finished, the writer prepares to go home to complete his work. The producer grins, takes him by the arm, and ushers him to an office. He must complete the rewrites here. "*Here?*" the writer screams, complaining loudly that he has eight hours' work to do, maybe more. The producer puts an arm about his shoulders, promising to have dinner sent in—a thick juicy steak, french fries, even exotic dancers if that will make him happy—but the writer must finish here, tonight, so that the script can be speeded to the mimeograph department.

When the mimeographed scripts are delivered the next day, Sara breathes a sigh of relief. In spite of apparent mental anguish, the writer did an excellent job. For the first time, Sara begins to believe that this can be a good show.

The production pattern of most television series is that director A films one program while director B prepares the next. While director A is shooting, Sara visits the soundstage to meet the **cinematographer**—in charge of the camera crew—and the star. The latter, a handsome but relatively inexperienced actor, is cordial. Sara watches him work, realizing that he will require extra attention. He hasn't yet read the next script, so she gives the star a capsule description. The purpose of her visit is primarily social, so that when she begins her show two days hence, the star and crew will accept her as a member of the family.

After their agents have worked out financial and billing details, all cast members visit the wardrobe department, where they will meet Sara for the first time. Sara has already discussed with the wardrobe person the number of "changes" (of costume) each character will have. Now each wardrobe item is evaluated for fit, color, appropriateness to the character, and suitability to the scene's action. If a costume is scheduled for a fight (or other action) sequence, it probably will have to be "doubled," that is, several identical costumes must be on hand because one or more may become damaged.

As the production date looms closer, casting is completed, locations are confirmed, and set plans are okayed. The story editor or the producer gives the script a final polish. The unit manager calls a production meeting that all members of the production team attend. The cinematographer will get there if possible, but must give first priority to the show now being filmed. At the meeting, Sara goes through the script page by page, describing her specific needs. Although she has already discussed properties, set plans and dressing, transportation, stunts, and other production requirements with the people concerned, she now reinforces each of her needs, clarifying areas of confusion, making certain that each department head understands what is expected during production so that there will be no last-minute problems. (Some problems will inevitably surface later, but production meetings reduce their quantity and severity.)

On the eve of production, Sara studies each scene to be shot the next day. She makes rough sketches of contemplated action and business (see Chapter 6). She considers ways in which she can cover (photograph) the action effectively. Additionally, because she must shoot six location pages tomorrow, she estimates where she must be at each point during the day. A major scene is to be filmed in the morning; she must allow extra time for it. She will thus appear to be behind schedule—for a while. But she plans shortcuts for some unimportant scenes later in the day where she can sacrifice coverage (and thus gain time), thereby finishing the day on schedule. If she can finish coverage of her major scene by 3:00 P.M., she will be home free—for the day. She plans to inform her assistant of her plans while driving out to location.

Sara wants to work for this company again. She knows from experience that giving the producer an excellent picture is only part of the equation. If she goes over budget, spending company money recklessly, she will not be invited back.

(Occasionally, when directors appear either irresponsible or insufficiently prepared, they are replaced in the middle of filming a picture.) But if the picture is solid and finishes reasonably close to budget, everyone will be satisfied. A production manager once told Sara that most budgets contain a small amount of padding, a layer of protection for contingencies.

Production As soon as the crew arrives at location the next morning, Sara gives her first setup, or camera position, to the cinematographer and then tentatively walks it through with the cast, exploring the action for logic, determining if it seems comfortable for the actors. The star has not been scheduled until later; he worked late the previous night and is entitled to ten hours before his next call. This time between calls is known as **turnaround time**. The assistant director has tried to provide the star with additional time off because he carries a tremendous work load.

Sara is delighted with the crew. They are fast and efficient, giving her more time than she expected. The star is due to arrive at 10:00 A.M. At 10:30 he still hasn't arrived. His first scene (the major scene of the day) should be shot at 11:00. At 11:05 the star arrives. He looks terrible, has a throbbing headache and puffy eyes. Sara gives her assistant a reproving look; he never told her the star drinks. She welcomes the star warmly, gets him coffee, mentally restages the big scene. Now she decides to shoot her master angle from behind the star, saving his closeups until he feels better.

While the company waits for the star to finish his makeup, Sara rehearses with the rest of the cast. Then she goes over to talk it through with the star, quietly and privately. He has problems with a couple of lines. She offers alternatives that he accepts, and they decide to shoot the closeups both ways, as written and as revised, to give the producer a choice.

At 11:45 the star is ready for rehearsal. But by the time cast and crew are prepared to film the master angle, the crew must break for lunch. They've been on the job six hours. If they don't break, there will be a meal penalty, an extra and unnecessary expense. Sara makes a practical decision. It would ultimately cost more money to break for lunch, come back, rehearse the master angle again (the actors would be cold), and then shoot it. Better to pay the meal penalty and shoot the scene now. Concealing her apprehension, she tells the assistant her decision. He grins, good idea. They shoot the master scene. It's too slow. They shoot it again. Halfway through, the star blows a line. No matter. Rather than shoot the entire scene again, they pick it up, beginning just before the flubbed line. The scene goes beautifully. Sara is thrilled and they break for lunch.

Sara completes coverage for her master scene at 3:25 P.M. Her assistant had pleaded with her to sacrifice some of her intended coverage for that scene in order to make up lost time, but she had been adamant; the material was too important. She would be happy to sacrifice somewhere else. She is aware now that they will lose daylight at about 5:15. They may not finish the day's work. Sara gulps, retains her composure, and pushes on.

The company **wraps**—finishes for the day—at 5:15. One short sequence was not filmed. They will try to pick it up tomorrow. It will add to tomorrow's burden, but Sara knows she will be able to handle it. Most importantly, in spite of problems, she got through the first day, filming excellent material, missing only one short, unimportant sequence. When she pretends concern for the missing sequence, the assistant laughs. Who does she think she's kidding?

Postproduction   When they return from location at the end of the second day (yes, Sara picked up the missing sequence), Sara, her cinematographer, and her assistant grab sandwiches and run to a projection room to look at the **dailies**, the film shot the previous day. The editor watches with them although he ran the footage earlier with the producer. How did the producer like it? He was ecstatic.

As they watch, Sara discusses the film with the editor, preferring certain takes, suggesting certain patterns in putting scenes together. Later, after the editor has assembled the total footage into a **rough cut**, Sara will return to evaluate the assemblage, making sure that the editor has put the film together according to her vision. If not, she will suggest certain changes to be made before the producer views the assembled footage. The Directors Guild contract guarantees directors the right of "first cut" although, in television, directors are often so busy that they voluntarily relinquish the privilege.

Within a week after the producer views the assembled footage, the business affairs office contacts Sara's agent, exercising its option for her to direct two additional films. They also have a question: Would Sara be interested in directing other films for the same studio?

DIRECTORIAL STYLE

Aware that a show's success depends in great part on the degree of audience involvement, most directors try to remain invisible—that is, not allow an audience to become aware of their manipulations. Wise directors realize that audience involvement often is ruptured once viewers become aware of behind-the-scene elements such as camera angles, music, editing, or directing. But sometimes directors are compelled to reveal their presence.

The Manipulative Director

Occasionally in drama or documentaries that use dramatic elements, directors find that they need to create certain emotional effects, distorting space or time and venturing far from snapshot reality. With such an approach, directors inevitably reveal their presence, although usually only the most sophisticated spectators become aware that they are being manipulated. Sometimes labeled

expressionistic, such an approach uses camera, lighting, music, setting, actors, and special effects to shape the viewer's emotional response.

The opening sequence of Orson Welles's *Citizen Kane* provides an excellent example. It is night. Camera begins close on a sign, "No Trespassing," fastened to a chain-link fence. Music is ominous, foreboding. Camera moves up and over the fence and finds a hulking castle known as Xanadu. One window is lighted, high in the castle wall. We go inside the room to see Kane's lips in huge closeup whispering the word *rosebud*. A close shot of a glass globe in Kane's hand. Inside the globe: a charming snow scene. The hand relaxes suddenly. The globe drops and breaks. A shard of broken glass reflects the distorted low-angle view of a nurse hurrying into the room.

We have seen a bizarre succession of images. So intense was their effect that they linger in our minds, superimposed like a ghost image, coloring the remainder of the film. Watching this material, audiences are usually aware that the images bear little relationship to literal reality; the director created a sequence of "effect" shots that, together with music and other expressive elements, evoked powerful and disturbing emotional resonances.

Many beginning directors, especially in drama, deliberately remind spectators of their presence through use of flashy camera angles, distortion lenses, bizarre staging, or melodramatic lighting. Such dishonest manipulation is not motivated by subject matter; it creates a theatricality that inevitably intrudes when applied to nontheatrical subject matter. These directors call attention to themselves for reasons of career advancement or ego. Their obvious manipulation pulls attention away from the narrative, reminding audiences that the magical world they are watching has been contrived by artists and technicians. By contrast, when sensitive directors use expressionistic devices, it is for the purpose of enriching a project's emotional message, thereby deepening audience involvement.

The Invisible Approach

One hundred eighty degrees from the manipulative approach is the invisible approach. Now viewers remain unaware of any directorial presence. Most documentaries are shot with the invisible approach, the subject matter sharply outlined in the foreground, with style and technique apparently nonexistent.

Although most films and TV programs place content before form, with directorial manipulation virtually undetectable, no one can claim that manipulation does not take place. It is almost impossible for any director to record the content of a project without revealing some trace of personal attitude or bias. Even the camera's location in relation to a performer will subtly influence an audience. A high camera diminishes the performer; a low camera suggests strength. Choice of background creates inferences. Lighting can be kind, harsh, or somewhere in between. Because time is a factor, especially in TV news, editing becomes significant. What words are omitted? Do the omissions alter the message in any way?

Just as writers reveal something of themselves in everything they write, so do directors reveal in their work, consciously or unconsciously, some hint of their emotions or attitudes. If a director loathes a star, for example, it will be virtually impossible to keep some faint reflection of that loathing from appearing on tape or film. A director may treat a project in the sincere conviction that its presentation is totally objective. But until computers replace directors, complete objectivity is not possible.

CHARACTERISTICS OF GOOD DIRECTORS

Ask any group of directors—television or film, fiction or nonfiction—about the qualities necessary to become successful in their craft. Some will specify imagination or leadership. Others will cite showmanship or sensitivity or the ability to organize. All, of course, are prime attributes of the successful director. Some are inherent; others can be acquired, at least in some degree.

Taste

In preparing this book, I recorded conversations with a number of TV and film directors, asking each which qualities he or she felt are most necessary in a director. The answer voiced most often was *taste*. Most interviewees quickly assured me that taste is only a part of the total equation, but the brightest, most imaginative, most organized director without taste could not succeed.

How do aspiring directors acquire taste? By stimulating themselves through exposure to a variety of entertainment forms: by reading the works of outstanding authors, listening to fine music, viewing fine art, seeing well-crafted plays or movies. Whether in college or out, such exposure to the brightest creative minds of all time gives us the opportunity to enrich our imaginations—and our taste.

Exposure to the work of today's most creative talents also keeps our taste *current*. Every year styles change imperceptibly—in clothing, music, stories, and a dozen other areas. Because these changes occur slowly, we are only dimly aware of them. If we isolate ourselves, they pass us by; we become old-fashioned in our tastes and behavior. Remember a motion picture that you raved about ten years ago. When you view it again, perhaps on videocassette, it probably appears dated.

Expertise

Students frequently complain that many directors succeed because of luck. There can be no question that luck is a factor in achieving any goal. But most of us are given more than one opportunity to demonstrate our creative talents or abilities.

Yes, luck opens a door for us—but we must have the expertise that will take us past the threshold.

Expertise in the various arts and crafts comes primarily from three sources: from study, from curiosity, and from experience.

Study   Universities and film schools generally provide an excellent background in cameras, film or videotape characteristics, video control panels, and sound equipment. Many offer courses in acting, script writing, aesthetics, and directing. In 1984, over a thousand colleges and universities offered courses in film or television production techniques. And those departments seem to be proliferating. Books such as this provide additional insights.

Curiosity   A healthy curiosity frequently grows out of the need to succeed. It pushes aspiring directors beyond the knowledge acquired in school or in books. It provides them with the impetus to poke their noses into TV stations or motion picture sets, to visit companies shooting on location, to talk to cast and crew members, to wangle jobs as assistants to the assistants, or to seek out industry members when they speak in public. It takes them to little theater groups where, working for little or no money, they can gain insights into the subtleties of acting and the intricacies of directing on a legitimate stage. Curiosity prompts them to read scripts by fine screenwriters, learning structure and style.

Curiosity also sparks an interest in unusual theatrical films and television shows, thus encouraging aspiring directors to learn from the best teachers in the world, the directors themselves. Critical viewing may reveal how directors use their cameras and how they stage action. Students may discover the latest developments in special film or video effects. They can listen to dialog, trying to figure out what makes it effective, or to musical scores, to guess why the composer wrote specific music as a background for a specific scene. They can study acting, learning how fine actors build vivid and involving characters.

Here's a tip: When possible, see a show *twice*. The first time, in spite of good intentions, we all tend to watch as pure audience. We get involved in the story, worry for the hero, and completely forget our original purpose. But the second time, we can watch as students, critically observing how the director has manipulated the show's elements.

Videocassettes of well-directed films are also extremely helpful. You can stop the tape and repeat key sequences as often as necessary in order to study directorial technique.

Curiosity also drives many aspiring directors to create their own shows, on videotape or film, with or without synchronized sound. Such a project can be expensive, but it can also prove enormously enriching because it contributes experience.

Curiosity, of course, is not limited to aspiring directors. Professional directors ask questions relentlessly to expand their skills. They read trade and technical journals and attend seminars on new technologies. They attend private screenings by avant-garde directors to soak up new techniques. And they experiment constantly with their own techniques. One of the entertainment industry's most effective methods of generating audience excitement is through expansion into the deep space of new technologies. Directors must continue to educate themselves—or abandon their trade.

Experience NBC news director Jay Roper expressed the opinion that students should try to ally themselves with a TV or film operation *while* they're in school— even if the job is menial. "When students wait until after they graduate to interface with the world, it's too late. They have to be already working part time somewhere in the industry. They need to see what's going on." Getting a foot in the door early, Roper believes, provides students with an additional learning experience as well as with professional contacts that may prove valuable later.

One Director's Perspective

Although many believe that imagination and perseverance are necessary qualities in a director, Walter Grauman's definitions of them may provide additional insights. He is a director of "Murder, She Wrote" and numerous theatrical and TV movies.

Q. What specific qualities distinguish most good directors?

A. A number of things. Imagination. Even to the point, maybe, that some people say they're a little crazy. They don't see things quite the way the average person sees them. I think they see them filtered through a slightly bizarre or twisted perspective.

Q. What else?

A. Tenacity. Obstinate tenacity. To stick with whatever a director sees and senses or imagines in his own mind. The biggest treachery that a director faces is betraying himself because he listens to other people's opinions and advice. If you're wrong, you're wrong. But if you're right, you're right by yourself.

Educational Background

In addressing one of my directing classes, former Directors Guild of America president Delbert Mann cited the need for a well-rounded liberal arts background. He mentioned such areas as history, literature, drama, and political science as requirements for a successful director. Mann's views have been echoed by a number of other directors who valued such learning above technical or production know-how. In my own experience, the majority of good directors are well read. They know what's going on in the world; they're acquainted with Tennessee Williams and with Molière; they can cite batting averages in the National League. But they also know lens and film characteristics, lighting techniques, and the latest wrinkles in computerized special effects. Because, by definition, directors are both artists and technicians, both aspects must be nurtured.

Both Richard Fleischer and Delbert Mann attended Yale Drama School. Both feel that their experience as actors proved enormously valuable in their careers as dramatic directors. According to Fleischer,

> The most important advice that I ever give to students interested in being directors—not technicians, but directors—is to get theatrical training. If they had to choose between film school or drama school, I would strongly urge them to go to the drama school. The most important function that a director has is to get performances from actors. And getting performances means that you have to communicate with actors and understand them and know what their problems are and know how to relate to them and talk to them.
>
> I have great respect for actors. I think every director should act at some time in his career. Doesn't have to be a good actor. He just has to act, to put himself in the actor's situation and understand the problems that an actor has—and also face the experience of being directed, somebody telling you how to act.

A classic story from the early days of live television concerns an arrogant and relatively inexperienced director who was having difficulty in staging a dramatic scene. After a number of awkward trials and errors, the actors began to grow impatient. Trying to help, one of them suggested another way of staging the scene. Instead of being grateful, the director snapped: "Don't tell me how to direct! Do I tell you how to *act*?" Veteran directors guffaw at the story, aware that a dramatic director's most essential function is "telling performers how to act."

Fleischer added a postscript to his thoughts on acting. Knowledge of writing and dramatic structure is "equally important. Writing courses for directors are absolutely necessary. If a scene isn't working, you have to realize *why* the scene isn't working. If it's because of the dialog or if the structure is wrong, then you have to figure out answers and fix it."

THE APPEAL OF DIRECTING

Directing is what most people in the film or television business want to do. Most film editors, producers, assistant directors, writers, and network vice presidents in their secret hearts would instantly trade places with a director. Most students in most college production courses view directing as their ultimate goal.

There are several reasons for such wide appeal. Directing is enormously gratifying to the ego. As indicated at the beginning of this chapter, a director is the authority figure to whom everyone looks for answers. He or she is the prime mover on the set, the maker of miracles, wielder of power, source of creative inspiration, solver of problems, a Super Being who has stepped down from Mount Olympus to create Truth and Beauty.

In addition to its ego gratification, directing also offers the opportunity to exercise real creative control over a project. To the uninitiated, creative control doesn't appear to be a critical factor. After all, doesn't a writer have creative control? Or an actor? Or a producer? In fact, none of them does. In each case, someone higher on the executive ladder, someone with more "clout," is capable of wresting control away. To artists who genuinely care about the integrity of their work, such distortions can be worse than frustrating; they can be agony.

Finally, because directing requires a state of continual alertness and concentration, it is exhilarating. It is being 100 percent alive. Some have compared directing to mountain climbing. The sheer difficulty of the climb challenges and seduces you. Ignore the smallest detail and you hurtle to your death. And yet the view from the top is mind bending! Here is Delbert Mann again:

> I love the job of directing. I love the responsibility—the sense that it is on my shoulders to do. I'm one of the lucky people of the world to be able to make a living doing what I like to do best. I'm also awfully glad that none of my kids wanted to go into the film business.

WHERE DIRECTORS COME FROM

Directors seldom start out as directors. They approach their careers from many directions. In live television, floor managers normally "graduate" to become directors. In film, such in-line promotion seldom happens; when assistant directors move up the ladder, they usually become unit managers or production managers.

Where, then, do film directors come from? From any of a dozen affiliated arts and crafts. Because a director controls the expenditure of hundreds of thousands (and perhaps millions) of dollars as well as the fate of a project, untested directors are not selected lightly. Even in episodic television, normally cautious producers, anxious to protect the security of each episode, eager to gain the greatest quality in the least amount of production time, would never, never buy the services of an inexperienced director. Unless. . . .

Unless the director happened to be the producer's boss. Unless the director happened to be the producer. Unless the director happened to be one of the series stars, contractually guaranteed the right to direct an episode. Unless the director happened to be an excellent writer whose services the producer desperately needs. (The writer agrees to write three episodes in exchange for guess what?) Unless the originally scheduled director "falls out of the schedule" at the last minute, becoming sick or otherwise unavailable. Then the desperate producer, suddenly under the gun, must find a director wherever possible and as quickly as possible. Under such circumstances, producers sometimes are forced to take chances on newcomers.

Film editors sometimes become directors (such as Robert Wise). So do actors (D. W. Griffith, Charlie Chaplin, Orson Welles), cinematographers or artists from other visually oriented fields (Stanley Kubrick, Haskell Wexler), writers (Robert Towne), and art directors (Marvin Chomsky). And sometimes directors become directors. That is to say, small-time directors become big-time directors (Steven Spielberg, George Lucas). Directors with minimal experience catch fire on successful low-budget films and find that they are suddenly in demand.

With increasing frequency, directors seem to be emerging from college film departments. Steven Spielberg, director of *Jaws*, *Close Encounters of the Third Kind*, *E.T., the Extraterrestrial*, and other record-breaking films, attended the film department of California State University, Long Beach. A cinema buff, Spielberg had been producing, directing, and acting in his own films since childhood. While in college he directed a student film that was later screened by executives at Universal Studios. They were sufficiently impressed that they gave him the chance to direct some episodic television at Universal. Because his talent was extraordinary, Spielberg was soon directing TV movies. One of the most successful, *Duel*, was later released as a feature film in Europe. On the strength of *Duel* and other TV films, Zanuck-Brown offered Spielberg the opportunity to direct *Jaws*, which has grossed well over $100 million to date.

George Lucas, director of *American Graffiti* and creator, director, and/or producer of the *Star Wars* trilogy and the Indiana Jones series, came from the cinema department at the University of Southern California. His student film, *THX 1138*, later was expanded to feature length by Warner Brothers and became the launching pad for *American Graffiti*.

The executive producer of *American Graffiti*, Francis Ford Coppola, graduated from the film department at UCLA. The script he had written for a Master of Arts degree became his first feature, *You're a Big Boy Now*. Later he directed *Godfather I* and *II*, *Apocalypse Now*, and *The Conversation*. A number of other current directors also emerged from university film schools, among them Martin Scorsese, John Milius, Paul Schrader, and Brian DePalma.

CHAPTER HIGHLIGHTS

— Directors communicate concepts to an audience through the manipulation of creative and technological elements. When they interpret the message skillfully, audiences become emotionally or intellectually involved.

— Until a century ago, directors per se did not exist. They emerged toward the end of the nineteenth century, integrating play, production, and spectators, taking over the functions that previously had belonged to playwrights and actors. Early motion picture directors borrowed stage techniques. Edwin Porter, D. W. Griffith, and others discovered new cinematic techniques that gave film a form and style of its own. When television emerged, its directors borrowed techniques from motion picture production.

— Directors play many roles: artist, technician, and parent-psychiatrist. As an artist, the director must be writer, photographer, editor, actor, and set de-signer—able to function creatively in each of these essential areas. As a technician, the director must be able to communicate knowledgeably with experts and crew members regarding choice and use of equipment and sub-tleties in production techniques. As a parent-psychiatrist, the director must assume the role of a wise and loving authority figure, resolving personality conflicts and disciplining performers whose conduct becomes unprofes-sional.

— Directors of live or videotape programs work in two essential patterns. Sometimes they operate from a control booth in a studio or van, facing a bank of video screens. Then the director selects (edits) those pictures that dramatize the action most effectively, cueing a technical director to cut be-tween cameras. At other times, directors use a versatile ENG (electronic news gathering) unit in the field when recording on-the-spot interviews or fast-breaking news stories.

— Film directors work in two primary fields: TV and motion pictures. The differences lie primarily in budget and subject matter. Because movie thea-ters charge admission, producers usually offer patrons what they cannot receive free at home: lavish production, more explicit sex or violence, and greater story sophistication. A TV series director functions in three areas: preproduction, production, and postproduction. Preproduction, or prepara-tion, covers such essential bases as story and casting conferences, location scouting, wardrobe, set design, and a production meeting. Production con-cerns the actual physical shooting of the project, either on location or on a soundstage. Postproduction involves working with an editor, assembling the film into a "first cut."

— Directorial styles vary. Subject matter often dictates the appropriate treat-ment. Some directors distort time and space for emotional effect, thereby proclaiming their presence. Others prefer to remain invisible to avoid inter-fering with the audience's involvement.

— Certain characteristics are common to most good directors: taste, imagina-tion, leadership, sensitivity, showmanship, and tenacity. Some of these may be acquired through study or practical experience. Others are inherent. Many directors cite the need for a strong liberal arts background, giving such an education higher priority than technical training. For drama, experienced directors proclaim the importance of acting experience in order to empathize with actors. Knowledge of writing and dramatic structure are also important.

Directors emerge from a variety of fields. In live television, they often move up from the ranks. Many film directors previously worked as actors, cinematographers, film editors, or art directors. More and more, today's directors are graduating from film schools.

PROJECTS FOR ASPIRING DIRECTORS

1. Watch a TV soap opera. Force yourself to study the director's manipulations. How honest/believable are the performances and actions of actors? Are there similarities in staging from scene to scene? Do you feel that the director arbitrarily manipulates characters into position so that he or she can photograph them effectively?

2. Research a director of your choice in magazines, books, or trade journals. How did he or she become a director? Determine such career aspects as education, professional training, directorial strengths, weaknesses, and credits. How important a factor was luck in this career?

3. If possible, visit a local TV station or cable outlet. Watch the director at work. Make a list of the various functions that he or she performs while preparing and directing a program. What facets of the director's work created the most problems?

4. Write a short review of a book on directing other than this.

5. When watching television or movies, find examples of the invisible director and the director who loudly proclaims his or her presence. In what ways did the second type reveal that presence?

CHAPTER

2

ELEMENTS OF
ENTERTAINMENT

Certain facets of human nature have caused people to assemble in groups since our ancestors first sat by a campfire to hear a tribal wiseman narrate myths and legends. Those same elemental needs took audiences to see the ritual ballad dances that predate Greek drama, to watch gladiators fight for their lives in Roman circuses, to wonder in darkened parlors at hand-drawn lantern slides projected by candlelight onto a wall. Those needs for entertainment constitute the raw materials with which a director shapes a program.

Gregariousness may be a part of those audience needs, but it is not a requisite—although comedy shows certainly play more successfully to groups than to individuals. One reason why television sitcoms so frequently include a laugh track is to give viewers at home the comfortable reassurance of others also present, who join the viewers in savoring the program's humor.

In order for directors to stage a program effectively—to make intelligent decisions regarding script, camera, music, and other key production elements—they must understand the nature of that program's attraction for audiences. They must understand the nature of entertainment itself.

In the past century, entertainment forms have become so diverse that formulating any exact definition of *entertainment* is difficult. As cable channels proliferate and television audiences expand worldwide, such complexity grows even more

The film's job is to make the audience "help itself," not to "entertain"
it. To grip, not to amuse. To furnish the audience with cartridges, not to
dissipate the energies that it brought into the theater. "Entertainment"
is not really an innocuous term: beneath it
is a quite concrete, active process.

Sergei Eisenstein[1]

difficult to assess and define. Communications scholar Stefan Melnik describes a system of evaluating entertainment that he calls the "Uses and Gratifications" approach.[2] Melnik theorizes that all media functions must be evaluated in terms of *receiver needs*. Because the director stands at the center of the communication process, interpreting the Sender's message in such a way that the Receiver will be intellectually or emotionally gratified, Melnik's approach seems appropriate. Thus, an analysis of audience needs and expectations will begin our examination of the elements of showmanship.

Researchers have categorized audience needs into the following general classifications:

1. To increase aesthetic, pleasurable, and emotional experience

2. To gain information, knowledge, and understanding

---

[1]*Film Form—Essays in Film Theory* (New York: Meridian Books, 1957), p. 84.

[2]Heinz-Dietrich Fischer and Stefan Melnik, eds., *Entertainment—a Cross-Cultural Examination* (New York: Hastings House, 1979).

3. To decrease the pressure of personal, professional, or social problems

4. To strengthen contacts with family and friends[3]

Expressed in simpler terms, in the first three categories audiences seek *pleasure*, *information*, and *escape*. Because so many television viewers and radio listeners (for example, shut-ins or senior citizens) use the media to relieve loneliness, the fourth category must expand beyond the social experience of enjoying entertainment in the company of others and be considered a need for *companionship*.

Within each major category we will examine the program forms and entertainment elements most likely to be encountered by directors in traditional film and television. Inevitably, some elements overlap from category to category.

— PLEASURE: the most abundant category, contains the sensually pleasing elements of spectacle, survival, male-female relationship, order/symmetry, and surprise and humor

— INFORMATION: contains twin elements of morbid curiosity and healthy curiosity

— ESCAPE: compares the entertainment world with the real world

— COMPANIONSHIP: describes the relationship that viewers form with entertainers and its dramatic equivalent: identification

PLEASURE

Of the four major entertainment categories, *pleasure* contains the greatest number of components with which directors can deal directly. An understanding of the elements that provide pleasure enables directors to enrich their program material, thereby making it more appealing to audiences. Some combination of *spectacle*, *survival, male-female relationship, order/symmetry*, and *surprise and humor* appears in every successful TV program or motion picture.

Spectacle

One of my favorite high school teachers once told me the secret of the popularity of his chemistry classes. He prevented lectures from becoming dull or boring by performing illustrative experiments featuring fire and smoke, liquids that

---

[3]Adapted from Elihu Katz and Michael Gurevitch, "On the Use of Mass Media for Important Things," *American Sociological Review* 38, no. 2 (1973): 166–67.

changed colors, occasional small explosions, or terrible smells. Such spectacular effects invariably brought his students to life.

From infancy we're fascinated with sights and sounds that appeal primarily to the senses. Although the word **spectacle** literally refers only to the visual, I will use the term here in a wider interpretation—to describe all elements of extraordinary sensual appeal, primarily *sound*, *motion*, and *color*. Examine a baby's toys and notice how they incorporate these elements: the rattle (sound); the mobile with slowly rotating plastic animals (color, motion); music boxes (sound); wind-up animals (color, motion); celluloid whirligigs that spin in the wind (color, motion).

As we grow older, the same spectacular elements continue to attract us. We explode firecrackers on the Fourth of July and watch skyrockets spray colored plumes across the sky. We hold kaleidoscopes to our eyes, observing the myriad color forms in constantly changing patterns. We watch parades, fascinated by costumed marching bands, strutting drum majorettes, and floats festooned with flowers and pretty girls. We attend the circus to enjoy its clowns, trapeze artists, trained animal acts, and garish music: an extravaganza of sound, motion, and color.

Spectacle, as defined here, is not limited to the bright, the dazzling, the upbeat. It also contains darker elements of sensual appeal such as those in (1) the awesome natural forces of thunder and lightning, raging fires, floods, and windstorms and (2) the contravention of natural laws. The second category includes such fantasy elements as magic and illusion, witchcraft, demonology, and ghosts and spirits. Such traditional magician's fare as floating a figure in air, sawing a woman in half, or making an elephant disappear continue to fascinate us because they defy logic and the laws of nature. Such entertainment is part *spectacle* (contravention of nature) and part *puzzle*. We gasp at a feat of levitation and simultaneously wonder how it is done.

If the term *spectacle* describes elements of sensual appeal, why do we accept such spectacular creatures as witches, demons, and ghosts as appealing? Because in the entertainment world it is *safe* to be frightened by them. When twenty-foot sharks or poltergeists or demons terrorize us, it is a delicious fear that we experience. Why else would audiences flock to theaters to see so-called horror films? The apprehension, the fear, the shock become acceptable because the more practical side of our minds remains aware that we are spectators in a theater. When the terror grows too intense, a safety valve reminds us it is all make-believe. The terror is thus doubly enjoyed: We experience the thrill of an intense emotional experience, yet we are secure in our comfortable seats. We may reassure ourselves at any time merely by glancing around the theater at other audience members and smiling as they recoil from the screen's images.

Similarly, when we watch floods or hurricanes on the six o'clock news or tornadoes in a television movie, we may be awed by nature's spectacular displays while we remain isolated and indestructible in our warm living rooms. If anxiety should overwhelm us, we can always change the channel or turn off the set.

Spectacle adds luster to a motion picture or television program. Like other elements of audience appeal, it cannot compensate for content that is ill conceived, but it does tend to make such content more palatable. Let's examine a few different program types and determine how the element of spectacle fits into each.

Sports Almost nowhere is the element of spectacle more evident than in the field of sports. Its marching bands, cheerleaders, card stunts, electronic displays, blaring music, and uniforms are the apotheosis of sound, motion, and color. Directors once gave full coverage to such elements. In time, as sports audiences wearied of this peripheral color, directors adjusted their coverage, concentrating more on the event itself, filling half times with coach interviews and predictions by authorities.

The sports event contains strong spectacular elements, the most compelling of which is conflict. Fringe color enhances the contest, but should not infringe upon its drama. Astute sound engineers maintain viewers' awareness of cheers, band music, and shouted instruction on the playing field when cameras are directed elsewhere.

Newscasts In newscasts, the element of spectacle equates with *visuals*. Audiences want to see newsworthy events. Presenting a visual—videotape, film, or even a still photograph—is preferable to showing a stationary seated figure, a "talking head," merely describing the event. The former is dynamic, the latter static. The first creates a sense of physical presence at a news event; the second is little better than reading about it in a newspaper.

Conscious of the need for visuals, newscast producers frequently include stories for which film or tape is available and minimize or abandon those without. Such awareness of the importance of visuals makes a more entertaining newscast, but frequently distorts the relative values of news stories, failing to provide a balanced perspective on world or local happenings. If there's film, a story gets aired. Without film, it often doesn't.

> A modern version of Bishop Berkeley's dictum prevails. If a tree fell in the forest, he wrote, and no one was there to hear it, it would make no sound. So it is with television news. Events may take place—important events by any standard—and if there is no camera and correspondent present to record it, it is not "news" and at least 50% of our countrymen will never learn of it.[4]

The dilemma is an old one: integrity versus commerce. In terms of television news, the dilemma involves choosing a balanced perspective on world events or

---

[4]Frank Mankiewicz and Joel Swerdlow, *Remote Control* (New York: Ballantine Books, 1979), p. 99.

stories that assume disproportionate importance because of available tape or film footage. Because each TV station struggles to seduce viewers away from competing channels, commerce frequently wins. (See Chapter 11 for more on visuals in the news program.)

Drama Lavish spectacle has become an intrinsic part of most feature motion pictures and a factor in their soaring production costs. In decades past, Hollywood studios manufactured as a part of their total output a number of low-budget B pictures called "programmers" containing relatively few elements of spectacle. With the emergence of television in the 1940s, Hollywood relinquished that portion of its production package to its new competitor. What was left had to be powerful enough to attract a mass audience, offering theatergoers what they could not see on the tube: expensive production values, star names, explicit sex, and/or violence.

The term *production value* approximates *spectacle* and includes those elements of a motion picture or television show that combine to give a rich or handsome impression, implying that no production corners have been cut. The term relates to use of actual locations, elaborate sets, costumes, special effects, stunts, chases, perhaps even extravagant makeup or hairdressing.

In the early years of television, audiences grew rapidly. Drama, largely, was live; production values were minimal, which was fine with viewers. They were thrilled to be watching drama in their living rooms without paying admission. As live television was replaced by film, the use of minimal production values survived—for a while. Then, as audiences became more sophisticated, production values improved. "Naked City" and "The Untouchables" were the first television series to offer night-for-night photography, extensive location shooting, and movie-quality sets. (**Night-for-night photography** is just what its name implies: shooting night scenes at night rather than the less time-consuming practice of photographing such scenes in the daytime and making them appear to be shot at night through use of filters.) Other series quickly followed suit.

In recent years, the element of spectacle has probably been abused (or perhaps overused) more frequently in TV series than in any other entertainment form. Television directors or producers, perhaps compensating for the lack of substantial story or character values, have repeatedly resorted to such affordable sound, motion, and color elements as car chases, fights, stunts, and special effects. Emphasis on such elements often is fostered by pressure from networks in their continuing struggle to win Nielsen rating points.

Responding to TV's new expensive look, the fragmented motion picture industry expanded its horizons, reaching new heights of spectacle with science fiction films (the *Star Wars* and *Star Trek* trilogies, *Alien, Close Encounters of the Third Kind*), fantasy adventure (*Superman, Raiders of the Lost Ark, Indiana Jones and the Temple of Doom*), and horror epics (*Amityville Horror, Poltergeist*). These films explored new areas of spectacle in the awesomeness of deep space and alien life forms; they reached new frontiers of technical sophistication in computerized miniaturization and special effects photography.

From a director's point of view, spectacle usually enriches a project. Successful directors are like kids in a candy store; they want to cram their projects with as much richness as possible (including not just spectacle, but also good actors, crew, and writers). Production values are desirable when they legitimately relate to the script and its theme. However, if spectacular elements simply call attention to themselves, distorting the honesty of a dramatic situation, they must be arbitrarily rejected.

For example, if a script tells the intimate story of a love that emerges between two charming old people, *simplicity* would probably be the keynote for directorial treatment. Lighting, staging, costumes, locations, and music would all reflect the intimate mood. Soaring helicopter shots and tympany drums would seem jarringly out of place.

A noteworthy exception to the requirement for honesty is when a director works with a less-than-mediocre script or (alas) a hopelessly inadequate cast. Such a project appears doomed unless the director can add sufficient theatricality to make it attractive. In such a case, the director is justified in using whatever elements of sound, motion, or color he or she can pack into the project, creating an illusory sense of excitement.

Music Programs   In 1984, **music video** exploded onto television screens, exploiting the elements of color and motion (and certainly sound) as never before. Appealing primarily to young audiences, these enormously visual rock, country, or soul numbers include bizarre imagery, computer graphics, dancing, singing or dramatic action, and kaleidoscopic or hallucinatory special effects, with few limits set on the nature of subject matter except that it relate generally to the theme or flavor of the music or its lyric content. In some cases, the presentations lean so heavily on the element of spectacle that they wander far from the musical content, apparently exploiting effect for its own sake. More than almost any other entertainment form, music videos offer directors the opportunity to exercise their imaginations with few restrictions.

More traditional music programs cannot use the elements of sound, motion, and color so flamboyantly. Directors of symphony orchestra broadcasts find more muted elements of spectacle in the settings, in the impact of close angles on musical instruments, and in whatever drama the conductor may impart. Vocal numbers on musical variety shows sometimes include spectacle in their settings and in visual or special effects, such as smoke, flashing lights, overlapping images, and so on. (See Chapter 13 for more on music shows.)

Commercials   With the exception of a few major feature films, no "entertainment" form uses spectacle more consistently and more effectively than television commercials do. It has almost become a cliché that they frequently outshine the programs that surround them.

A couple of decades ago the advertising industry was agog when a canned soup company filmed a thirty-second commercial featuring Ann Miller and other dancers for a cost of approximately $250,000. Today, such a figure is not uncommon. The cost factor on handsomely mounted one-minute commercials can run between $300,000 and $500,000 apiece.

Why do advertisers spend such staggering sums of money on commercials? For two reasons. The first is obvious and relates directly to the chapter premise: to add sensual appeal and thereby attract audiences. Aware that television viewers often use commercial breaks for going to the bathroom or getting a beer from the refrigerator, advertisers must first attract before they can sell.

The second reason why advertisers spend such enormous sums is that rich production values "attach" to a product. Images of attractive, fun-loving teenagers playing ball, painting a sailboat, or enjoying a picnic, enhanced by a lively jingle, create an ambience of exhilaration, of warm acceptance in a joyous male-female relationship. In the audience's minds, those images become inextricably involved with the product itself. Ultimately, the ambience *becomes* the product; the two are inseparable. The sales message, by inference, becomes, "Buy our product and participate in this joyous life-style."

Survival

In *The Dragons of Eden*, Carl Sagan discusses a survey of the dreams of college students. The dream most frequently reported concerns a fear of falling.

> The fear of falling seems clearly connected with our arboreal origins and is a fear we apparently share with other primates. If you live in a tree, the easiest way to die is simply to forget the danger of falling.[5]

The second most frequent dream reported by college students is that of *being pursued or attacked*. Like the dream of falling, the second dream directly concerns survival. Steeped in race memory, it relates to a time when nightmarish reptiles warred with predatory mammals, when all of life was a carnivoral carnival: Kill or be killed, eat or be eaten.

The battle rages on in our subconscious, tormenting us yet in dreams, sublimating itself in games and contests. We have replaced tooth and claw with more acceptable forms of mayhem. We still participate in the life-and-death struggle but vicariously now, by extension. Watch the faces of fans at a boxing match, howling for blood.

---

[5]Carl Sagan, *The Dragons of Eden* (New York: Ballantine Books, 1977), p. 158.

Observe puppies at play. They fight with each other endlessly, growling, biting, attacking, lunging for the throat, retreating, and attacking again—but seldom hurting each other. For them, as for humans and other animals, conflict is play; play is conflict. Such play creates an exhilarating tension. It is the game of survival—with the element of life or death removed. The game of survival appears in a hundred relatively civilized forms—political, economic, or social. It is the social form that interests us here, specifically, television and motion pictures.

The struggle for survival manifests itself most frequently in the form of *conflict*. So attractive is conflict that it pervades all forms of drama, sports activities, quiz and game shows, many documentaries, even radio and television commercials. The need for conflict lies buried so deeply within us that we seek it out, even when unaware of the identities of the participants. Note the crowds that instantly gather when a fight erupts at a ball game or on a street corner. We watch commercials we would otherwise ignore simply because a husband fights with his wife over use of a deodorant or because two jocks argue over the attributes of a light beer. We watch bad movies, tepid commercials, sports broadcasts in which we have no rooting position largely because their gladiators are at war; our primordial instincts smell blood.

In drama more than in sports, we participate vicariously by identifying with protagonists—*we* fight to survive, struggling with modern saber-toothed tigers. In quiz and game shows, we sometimes identify with participants, but more often we actively enter the contest, competing with them, trying to win (vicariously) the Cadillac Eldorado or the trip to Tahiti.

Aware directors recognize the enormous appeal of conflict for audiences. With that insight, they must then assess the show's format, deciding whether or not the element of conflict logically pertains. (That choice begins with the producer and administrative staff.) The role of conflict in drama, sports, and quiz or game shows is universally accepted; its function in other entertainment forms often goes unrecognized. Finally, the director must search for methods of incorporating conflict within the show's legitimate parameters.

How, for example, could the director of a cooking show use conflict as a way of attracting viewers? Most traditionally, homemakers could be invited to mail in favorite recipes as part of a competition (conflict among viewers), with winners receiving a prize or appearing on the show as a guest. Some years ago I directed a cooking show with two hosts, husband and wife. We maintained a continuing sense of competition between them as each struggled to outcook, outperform, or outmaneuver the other, each teasing the other for small failures. Their behavior was good humored and loving. But it *was* conflict.

In newscasts, friendly rivalry between the anchorperson and the station's sports personality also provides conflict. Differing points of view about upcoming football games or the effectiveness of the home teams's players stimulate viewers and motivate them to take positions of their own.

Conflict doesn't have to be angry, filled with threats or violence. As we will discover in Chapter 4, it can be *internal*. Thus, the hostess of a cooking show might wage fights with herself, trying to break herself of certain habits (such as sneaking tastes before a dish is completed). Internal struggles in a program can be delightful, especially those involving foibles that viewers recognize in themselves.

When elements of conflict already exist in drama or a program format, it becomes the director's responsibility to make certain that they are not lost or minimized. In some cases, it is possible to add conflict to a dramatic scene where none exists, not by rewriting dialog but simply through performance. I recall a scene in a foreign feature film in which a government official questioned a young lady. Such an expository scene, clarifying final story details, usually is dull. The director made this one entertaining by giving the official a predatory manner. With glance, gesture, and innuendo, he insinuated a message that was never part of the dialog. The official was clearly interested in seduction; the lady would have none of it. The superimposed pattern was not dishonest. It provided subtle conflict in a scene that otherwise would have lacked it.

In sports programs and quiz or game shows, the participants are already clearly at war. Through the announcers or emcees, the competitive elements may be reinforced, making spectators aware that beneath the fun and good humor lies a battle between gladiators.

Male-Female Relationship

Because nature has implanted within us the curious and compelling magnetism represented by sexuality, many of the so-called rules of showmanship may be bent, broken, or knocked slightly askew. Graceless dancers, voiceless singers, and emotionless actresses have attracted and held (primarily male) audiences because of sexual appeal. For several decades, women who weren't dancers, singers, or actresses pranced about a stage in time (more or less) to musical accompaniment, displaying nothing more artistic than their bare physical dimensions. Stripteasers or exotic dancers demonstrated that personal involvement takes many forms. The spectators' voyeuristic pleasure belongs to the category of spectacle or sensual appeal, that is, satisfaction in viewing, sexual stimulation and arousal, and projection of self into a fantasy dream.

In the American theater, the only plays able to survive extended engagements with just two characters have focused on a pairing of the sexes. These include *Two for the Seesaw, Same Time Next Year*, and *Voice of the Turtle*. Viewers usually identify with the character of their own sex, experiencing the mating ritual vicariously. Identification in the case of male-female chemistry assumes such power that otherwise rigid rules of dramatic storytelling frequently go out the window. For instance, professional writers accept the validity of establishing a story problem early in order to "hook" an audience. Yet when an attractive male pursues an attractive female (or vice versa), audiences delight in the chase to such a degree that they are willing to suspend their usual expectations. The relatively routine situation of a boy and a girl flirting at an amusement park can be extended on videotape or film, if not indefinitely then certainly for twice the time a nonsexual encounter would take.

Television Commercials Several seasons ago, advertisements for blue jeans gained prominence as TV's most overt sexual expression. Young women thrust their denim-clad derrières into the camera, thus proclaiming the snugness and sensuality of the product. Billboard, newspaper, and magazine advertisements echoed the motif: The male-female relationship, brashly displayed on beaches, horseback, and mountaintops, would reach new heights in sensuality if only the participants wore designer jeans. Jeans manufacturers had discovered what cosmetics, deodorant, mouthwash, and a dozen other companies knew all along: The way to sell a product is through the promise of a richer, more sexually fulfilling life.

Sports Directors of sportscasts, aware of the nature of their predominantly male audience, sprinkle **honey shots** throughout their coverage: glimpses of attractive female cheerleaders doing their sensual routines for the home team. Sports director Andy Sidaris in the PBS program "Seconds to Go" was asked how he got the idea for honey shots. "I remember when I was seventeen, every time I looked at a girl I'd just tremble. And I thought, 'If I'm like that, maybe other people are, too.' And you know what? They are. They sure as hell are!"

Motion Pictures When television replaced the feature film as provider of mass entertainment, the movie companies faced the problem of enticing audiences out of their comfortable living rooms and into theaters. Research discovered that the bulk of the movie audience consisted of a surprisingly narrow age grouping: 14 to 24. Can you guess what males and females in that age group usually have on their minds? Certainly not Middle East politics! Their fascination with members of the opposite sex simultaneously provided motion picture producers with an answer to television competition as well as with a new focus for story material. In the sixties and seventies both domestic and foreign films for the first time featured frontal nudity as well as depictions of the sex act itself. There seems to be a diminution of such explicitness in the eighties, perhaps because audiences have become desensitized.

Miscellanea Songs, whether rock 'n' roll or old-time ballads, continue to deal with love and romance, the male-female relationship. Today's songs reflect today's culture in a more casual acceptance of sexual intercourse as a physical expression of love. The emergence of magazines such as *Playboy* and *Penthouse*, featuring graphic nudity and articles and stories dealing with sex-related issues, is another demonstration of modern cultural acceptance of more liberal standards of conduct.

Directors don't have to read textbooks to discover that sex sells tickets. Such awareness prompts some directors to call for abbreviated costumes in musical variety programs, to feature legs in dance numbers, to hire actors or actresses who will inject sexual overtones into movies or TV drama. Such awareness sometimes produces effective staging. And sometimes, usually through overstatement, it merely reflects bad taste. The need for taste and a proper balance of values must always remain paramount in a director's mind. Bare cleavage is expected in an advertisement specifically designed for *Playboy* magazine yet is totally out of place for a TV anchorwoman. Sensual kissing might fit the format of an adult film yet would be ludicrous in a Saturday morning kids' cartoon.

The key to a proper balance usually is *understatement*; the human imagination remains the richest dramatic source. If directors allow the mind to provide the unseen, then the unseen acquires sensuality and power. Concealed nudity usually conveys more sensuality than bare flesh does. By suggesting rather than showing, the director allows the spectator to participate.

Order/Symmetry

In the 1930s, America demonstrated its depression jitters through a jigsaw puzzle craze. Bleary-eyed citizens spent hours, days, sometimes weeks painstakingly assembling hundreds (or thousands) of tiny, intricately shaped puzzle pieces. When they were finished, they had a picture of rather ordinary scenic beauty in which they promptly lost interest. Why such tremendous effort for such meager reward? Because of a fundamental psychological human need to seek order, to put Humpty-Dumpty together again. Possibly the economic disarray of the thirties was responsible for the intensity of that need during that time. If Americans could not restore a world riven by an economic jigsaw, they would quietly restore order where they could.

Jigsaw puzzles are still sold in game stores. Although the craze died decades ago, the human needs responsible for it persist. Those needs today are gratified by dozens of different puzzle forms—and by literary and entertainment forms containing elements of the puzzle. Prominent are the whodunits: private eye or police shows in which the spectator discovers a world in disequilibrium. A murder has been committed. Order cannot be restored until the various pieces of the puzzle have been put together and the murderer has been unmasked. In horror films, a monster or a supernatural being commits a series of atrocities, creating a rupture in the dramatic equilibrium. Once the ghosts have been driven away or the monster destroyed, order is restored.

By definition, all drama concerns disequilibrium. In the traditional pattern, a ruthless antagonist fights to keep the hero from reaching a worthwhile goal; this creates anxiety. The problem worsens; the viewer's worry grows; the problem resolves; equilibrium is restored. Even in films that end unhappily, some measure of equilibrium usually is achieved: The hero regains his honor, a wife finds new respect for her husband.

Directors frequently gratify the audience's need for order or symmetry through rhythm. **Rhythm** in the visual media implies repetition. Artists use this principle in paintings, aware of the sensual pleasure in viewing the repetition of a line or form: trees, archways, a shadow pattern. When composing a shot, visually aware directors search for such rhythms in their staging of actors and in their choice of scenery and locales. A **raking** (side angle) **shot** of several faces, for example, provides a repeated profile, diminishing in size, that is visually pleasurable. Such shots are often used when photographing symphony orchestras, the camera angling down a row of cellists, the succession of virtually identical forms creating a visual pattern that is highly pleasing.

Repetition need not be solely visual. In music, a **leitmotif** is a recurring theme usually associated with a character or a concept (such as a love theme) that builds audience familiarity and emotion with each replaying. (For a fuller discussion see "Themes/Motifs" in Chapter 13.) The song "As Time Goes By" in *Casablanca* demonstrates the leitmotif principle dramatically. Recurring themes (motifs) may also occur in dialog, gaining impact with each new reading (for example, "Here's looking at you, kid," also from *Casablanca*). Similarly, locales may be revisited for additional impact as when in a love story the lovers return to their own private place (the small Chinese restaurant where they first met). In the motion picture *Greystoke*, Tarzan encountered death again and again among members of his primate family. He registered grief each time by placing the animal's "hand" atop his head. Through repetition, the gesture gained emotionality. When, late in the film, Tarzan's grandfather died and the apeman repeated the same grieving gesture, the effect was devastating. Such personal actions, called **business**, often reinforce the dramatic content of scenes. As a recurring motif, the action brings a sense of symmetry to a dramatic program.

Although symmetry tends to provide a comfortable and soothing state of mind, there are times when viewers enjoy being startled.

Surprise and Humor

How often have you watched a television program and said out loud the exact line a character was going to utter? How often have you predicted with accuracy the outcome of a movie or a twist in its plot? Such familiarity arises from repeated exposure to identical dialog and plot patterns.

For hikers, a new trail promises adventure and the excitement of discovery. Hiking the same path again and again produces only boredom. The ability to predict a show's outcome reveals an oft-traveled entertainment path that will inevitably lose viewers. In drama the plot's resolution must grow out of previously established elements, but skilled writers manage to accomplish this in an unexpected way. In entertainment as in art, innovation usually is welcome; freshness of concept or execution causes spectators to sit up and pay attention.

Clichés appear in many areas of entertainment. In drama, they begin with the script. When directors are offered clichéd material, they have two options: to try to freshen dialog, characters, or plot through a rewrite or to refuse the project altogether. Stereotypes also extend to casting. We have all seen an actor play the same character in a variety of shows. Such casting is lazy. The director, casting director, or producer is taking the easy, expected path by hiring the most obvious actor for a part without bothering to stretch his or her imagination.

When a television program becomes successful, its producers tend to cling to the story patterns that generated its success. While such patterns hold audiences for a while, eventually they become tiresome. The program's producers must either relinquish their security blanket, providing variety and surprise by deviating from expected patterns, or lose their audience.

Directors fall into stereotypical patterns of their own. For years, both on the stage and in film, whenever directors needed a piece of business they would instruct the actor to light a cigarette. Cigarette business has lately been replaced by drinking business. Watch three or four television soap operas. See how often the cast members pour drinks, sip thoughtfully at those drinks, add ice, or refill the drinks. Directors under pressure (like anyone under pressure) sometimes grab at quick, easy answers. Easy answers are usually clichés.

Patterns of staging and camera coverage also become stereotyped. Recently I spent several weeks on jury duty. Day after day, standby jurors watched television soaps. There was little else to do. In scene after scene, the staging was virtually identical: An actor entered (full shot), encountering another; as they approached each other, the director covered the action in over-shoulder shots; as they confronted each other, tension building, the director used closeups. When one of the characters exited, the director held close on the face of the actor remaining. A moment later, the director cut to another scene where other actors walked through an (almost) identical pattern. I was far more conscious of staging similarities than were the other standby jurors. They watched with casual interest, smiling occasionally; it was a way to pass the time.

Dramatic surprise is closely related to humor. When we expect a story to go in one direction and it unexpectedly veers in another, we are fooled; we laugh, perhaps in embarrassment at having guessed wrong. Charlie Chaplin used this principle repeatedly in his films. In *The Immigrant*, the audience discovers him on an ocean liner that tips dizzily back and forth in a storm. Poor Charlie is leaning far over the railing. After a moment he appears with a fish on his line. Instead of being sick as we had guessed, Charlie has been fishing! We laugh because we are surprised. In another film, we discover Charlie expertly spinning the combination dial on a huge safe. When the safe finally opens, he takes out a mop and bucket. Instead of being an executive as we had expected, Charlie is the janitor. He has fooled us again—and we laugh.

Humor constitutes another significant element of showmanship, another valuable directorial tool. One effective way of creating appeal in a character or performer is to provide him or her with moments in which to display humor. Audiences love to laugh. Anyone who gives them the opportunity becomes endearing. Of course, there is considerable difference between laughing *with* a performer and laughing *at* a performer. Performers with the capability of laughing at themselves become especially appealing.

Conscious of the power of humor to attract audiences, directors can provide performers with opportunities to display the brand of humor best suited to their personalities. In the fierce competition among television stations, newscast producers have encouraged members of their news teams to intersperse personal observations throughout the news stories and to find humor wherever possible. Their aim: to allow newscasters to become people rather than robots. By displaying a sense of humor, newscasters become real, dimensional, and infinitely more appealing.

INFORMATION

When we seek information voluntarily, we are motivated by an often demeaned human characteristic, curiosity. Present in all of us in varying degrees, curiosity takes two forms: morbid and healthy. The two cannot always be separated; one sometimes contains traces of the other; they may be inextricably interwoven.

Morbid Curiosity

An early memory: breaking my parents' edict not to cross streets so that I could sneak glimpses of the recluse who lived in the next block. She was an albino. In youngsters, such curiosity represents an understandable fascination with the different, the apparently aberrant; it is part of the learning process. Fascination with the different continues as we grow older. We visit freak shows at carnivals. We try not to stare at the beggar without legs or the old man with a hump on his back.

Morbid curiosity becomes almost obsessive in our attraction to the grisly and macabre. Watch spectators gather in excitement at the scene of an automobile accident and stare compulsively at the injured; if a bone protrudes through torn flesh, so much the better! Fascination with the grisly is why many are drawn to horror films such as *The Texas Chain Saw Massacre*.

Fascination with the macabre sets up a curiously conflicting reaction in most of us: We are simultaneously attracted and repelled. Squeamish spectators of horror films cover their eyes during particularly gruesome moments, sneaking glimpses at the screen through cracks between fingers. Part of our revulsion at the sight of a cadaver grows from this reminder of our own mortality. And our fascination perhaps signifies relief that the deceased is not us. Like freak shows, cadavers occupy a shadowed niche in our culture that is labeled "forbidden." Like Pandora, we strive to see the unseeable.

Fascination with the different also takes us to strange lands to watch Indian fakirs and Arabian camel drivers, to stare at the sphinx and pyramids and such awesome natural spectacles as volcanoes or waterfalls. Then curiosity loses its morbidity, reflecting the healthy desire to learn or discover.

Healthy Curiosity

The pioneering need to find out what's on the other side of the mountain has been responsible for humanity's greatest explorations. It took Columbus to the new world. It sends astronauts into space. It leads scientists into explorations of the atom and the nature of disease.

The need to learn prompts us to read the morning newspaper, to buy books or magazines, to visit foreign countries. It sends students to universities. It provides the motivation for watching educational and documentary films and some of TV's panel and game shows. Realistically, the desire to learn is often mixed with self-interest. Knowledge helps us climb the social or professional ladder, to gain respect from peers. When we watch "Wall Street Week" on TV, we may use some of the information to plan purchases or sales of securities. When we watch a documentary or panel show, we acquire knowledge that we may casually drop into conversations with friends, impressing them with our awareness of what's going on in the world. We watch instructional or industrial films usually to gain specific information that will help in our careers or hobbies.

Learning is not limited to educational programs. A significant fringe effect of drama is that audiences appraise the actions and appearances of role models, sometimes modifying their own actions and appearances to conform. Many find (or hope to find) solutions to personal problems in televised drama, adapting the solutions chosen by TV or movie role models to their own situations.[6] Such learning is not always positive or beneficial. Evidence gathered by H. J. Eysenck and D. K. B. Nias[7] and other researchers indicates that the mentally unstable in our society often imitate violent or aggressive actions taken by characters in TV and film. Directors must therefore be aware of a responsibility to their audience and to society, especially when role models undertake illegal or immoral actions. Life, apparently, does imitate art.

---

[6]Erik Barnouw traces such attempts at problem solving back to radio soap operas in
The Golden Web (New York: Oxford University Press, 1968), p. 97.

[7]*Sex, Violence and the Media* (New York: Harper & Row, 1978).

ESCAPE

During the great depression of the 1930s, attendance at motion picture theaters remained surprisingly strong. One-fourth of all factory workers were unemployed. Banks failed by the hundreds. Life savings went down the drain. Businessmen leaped from office buildings. Yet it was possible to escape from that world for a while and buy a dream. The more intense the pain, the more necessary it became to retreat into the theater's opiate darkness and become someone else. After 1932, weekly attendance figures rose to 80 million admissions per week and remained close to that figure for the rest of the decade. The more shattering life became, the more urgent became the need to live vicariously in another, more exciting world as another, more glamorous person.

During World War II, the same motivations recurred, this time grounded in fear and loneliness. Loved ones were away in the armed services; there was a chance they might not return. The friendly neighborhood movie theater offered respite, pictured the world not the way it was but the way viewers wanted it to be. Attendance during the war years climbed to 85 million per week, the highest in industry history, with the exception of 1946.

The Random House dictionary defines *entertainment* as diversion. To divert is to turn aside or distract from a serious occupation. Thus, when TV and film entertain, they take spectators' minds away from real cares and worries. Escape is the only entertainment element over which the director has no *direct* control. Remember, the categories are defined in accordance with audience needs. Directors have relatively little control over a spectator's ability to hide from life's pressures. The capability for escape is individual and personal, varying from viewer to viewer, depending on the gravity of life's pressures and the spectator's facility at submerging himself or herself in a fantasy world.

Indirectly, however, directors contribute enormously to audience escape. When we turn on our television sets or visit the movies, it is usually not for the purpose of viewing reality. Reality can be dull, dreary, boring. Go stand on a street corner and watch the passersby. Mildly diverting for a minute or two, they offer you none of the entertainment elements: spectacle, information, escape, or companionship. The parade of drab, disinterested people passes you by without involving you, either intellectually or emotionally.

Reality can also be painful. Witness the increasing number of battered wives, abused children, rapes, and homicides, and the rising tide of suicide.

No, it is not reality we seek in movies and TV. It is the *illusion* of reality—to use Alfred Hitchcock's phrase, life "from which we have wiped the stains of boredom." The stories that we watch are more structured than life, more economical in their dialog, scenes, and characters, more intense in their emotional stirrings. (See Chapter 4 for more on drama versus real life.) And yet the master illusionist (also known as a director) is usually able to convince us that we are watching life. And in a sense we are, for good drama always contains the *essence* of life. The master illusionist allows us to escape from boredom or desperation by fashioning intense emotional experiences and by performing his or her art with such sensitivity and skill that we taste the pain and joy and passion of those experiences.

COMPANIONSHIP

People often perceive or use radio and television programs as a permanent switched-on source of background noise or flickering images without really paying attention to what is going on.[8]

Companionship takes two forms: active and passive. The passive form described in the quotation occurs whenever entertainment provides a background to other activities. Thus, companionship may be applied to television but not to motion pictures seen in the theater. When entertainment provides passive companionship, the nature of the program is secondary to the mere presence of sound (music, news, or drama) relieving loneliness, helping to fill an empty home, or enlivening a dreary office.

The active form of companionship occurs when viewers make the TV set or motion picture screen their primary focus. Such viewing may be interspersed with conversation or snacks or (as in the case of television) trips to the telephone, kitchen, or bathroom, but the viewer's primary purpose is to relate to characters on the screen.

Personal Relationship

Involvement with a character or performer is the fundamental basis for drama as well as most other entertainment forms: *People relate to people*. Audiences applaud fireworks displays, delighted at the sound, motion, and color. Performing seals and trained elephants amuse them. They gasp and "oooh" and "aaah" at light shows and dancing waters displays. But seldom do audiences become genuinely emotionally involved unless the performers are human or exhibit markedly human characteristics.

Wherever we go, in the arts or in our daily lives, we seek people, driven (perhaps narcissistically) to find a mirror image of ourselves. When I go to a

---

[8]Fischer and Melnik, *Entertainment*, p. 16. Used with permission.

museum exhibiting nonrepresentational paintings, I invariably seek (and find!) human forms or faces in the tangle of colors.

So acute is this need for identification that when a motion picture or TV program includes no actors at all, as in some avant-garde films, the spectator strives to find some vestige of the human presence or, failing that, invests animals or machines with human characteristics. Walt Disney created a dynasty from anthropomorphism. For half a century he devised creatures that physically resembled animals, but who fell in love, ate spaghetti, primped in front of mirrors, smiled, got drunk, lusted, wept, and laughed. Disney's animals exhibited such clearly definable human characteristics that we could instantly identify with them.

Drama In drama, audiences identify with a protagonist or hero. If there is no identification—some genuine emotional involvement—then the writer, the director, or the actor has failed. The spectator–protagonist relationship often becomes profound, with audiences actually experiencing the protagonist's emotions: agonizing over his or her problems, rejoicing in the victories. Under ideal circumstances, the spectator loses orientation and sense of personal identity for the duration of the drama.

Psychologists and communications scholars have speculated on the reasons why audiences find such deep personal involvement in drama. Because "plays exist to create emotional response in an audience,"[9] the most accepted theory suggests that such emotional response affords a catharsis of guilts. Most of us during our lifetimes commit deeds of which we are less than proud. In order to continue to function, we push our guilty feelings deep into the subconscious where they fester, creating neuroses. When we watch drama, identifying strongly with a protagonist, agonizing as he or she suffers, we are able to exorcise some of the demons that torment us. This accounts for the pleasurable sense of release we experience when a story resolves.

> When we find ourselves deeply moved by a story, with an intensity we can hardly understand, it is because we have secretly accepted it as our own, *identifying* ourselves with the people in it. Identification—this is the mechanism that taps imprisoned emotions. If [audiences] weep, it is not for the heroine but for themselves. If they laugh, it is not because the hero's tensions are being relieved, but their own.[10]

Newscasts Although spectacle elements of sound, motion, and color will attract an audience's attention and hold it briefly, *human elements* generally will hold that

---

[9]From Yale's venerated teacher of dramatic writing George Pierce Baker in *Dramatic Technique* (Boston: Houghton Mifflin, 1919), p. 43.

[10]Erik Barnouw, *Mass Communication* (New York: Holt, Rinehart & Winston, 1956), p. 56.

attention far longer. Fires or floods provide temporary excitement on the six o'clock news, but a feature on a homeless family or a mother who has lost her child creates more substantial involvement. A mother who has lost her child represents drama in its purest form—a reminder that these entertainment categories necessarily overlap. Elements indigenous to one may apply just as significantly to another.

Because of the competition that has arisen between television stations, selection of a newscast host or anchorperson has become critically important. Some hosts attract viewers; others seem abrasive. While journalistic experience is essential, other characteristics appear equally significant: dramatic ability, a pleasing appearance, and personal charm. What is charm? Simply the capability of creating involvement in the spectator, a caring reaction, rapport. Of course, charm isn't limited to anchormen and women.

Performers Personal charm is a prime requisite in show business and an asset in every aspect of life. When we buy life insurance from salespeople whom we trust, they probably manifest that same elusive characteristic. Realistically, newscast hosts, comedians, singers, and dancers are also salespeople. The product they sell is themselves. If they are successful on television, we laugh or applaud. If not, we press the magic remote control button that sends them to oblivion.

Such personal appeal often seems related to energy. We are attracted to life and repelled by death. We are drawn to people who bubble with humor, high spirits, and dynamism. We edge nervously away from the "gray people," entombed within themselves. Musicians who play their instruments vibrantly, who radiate joy and enthusiasm, usually draw crowds. Musicians who are far superior in technical expertise yet who play stolidly, in impassive concentration, seldom create audience excitement.

The display of energy ("aliveness") is a part of showmanship; it stimulates a responsive aliveness in those who watch, sweeping them into the performer's mood, creating a sense of participation. Designers of the TV commercial, "Coke adds life," were certainly aware that life ("aliveness") constitutes a quality of enormous appeal when they equated the soft drink with that appeal.

A word of warning: A director's goal should not be the stimulation of artificial aliveness, encouraging performers to wear masks of gaiety, simulating emotions they do not feel. Most audiences are sensitive to false smiles and phoney mannerisms.

The director's goal should be threefold: (1) to encourage the performer to *feel*, actually to experience the exhilaration attendant to a public performance, (2) to *magnify* that exhilaration, and (3) to *externalize* it. Watch great singers, dancers, and instrumentalists. See how their faces and body movements reflect the varying musical moods. You empathize with them because of their sensitivity, their ability to feel the humor or sadness or exaltation of the music. You are caught up in their moods because they have not merely reacted to an external stimulus; they have become one with it, expressing the stimulus with all of the physical instruments at their command, externalizing moods and emotions through facial expression, gesture, and body movement. To project those feelings across a distance to a spectator, they must magnify them; the greater the distance, the greater the degree of magnification necessary.

A final factor, *authority*, relates to performers' ability to involve us as companions. Unfortunately, authority is a characteristic that directors can seldom instill. More usually, it grows with experience, relating closely to personal confidence. When some performers appear before a camera, the viewer worries, detecting a sense of nervousness or disquiet or perhaps only an inability to control the audience. Such lack of authority sometimes disappears with experience. Not always. A performer with authority "takes over." Johnny Carson and Joan Rivers demonstrate superb control of their audiences. They are comfortable, secure—or at least they convey that impression to the audience. When circumstances go awry, they take advantage of those mishaps, clowning, reassuring the audience through their own sense of security.

Directors can help performers gain authority by offering support and reassurance, bolstering their egos. Confidence is contagious. If a director exudes easy assurance, a sense of control, the performer will absorb it and demonstrate it. Similarly, if a director becomes upset or nervous, shouting at crew or cast members, that instability is communicated to the performer and often erases any composure he or she might have acquired.

An insecure performer can be protected from external factors that might exacerbate his or her insecurity. By double-checking each of the elements that will touch the performer (music, other performers, floor manager) and making sure that each has been meticulously instructed or rehearsed, the director can surround the performer with an atmosphere of certainty, a sense of control, a feeling that everyone on the production team knows exactly what will transpire. Such an atmosphere is vastly reassuring.

Companionship versus Escape

As indicated earlier, the separate elements of entertainment—pleasure, information, escape, companionship—overlap and support each other. When we form personal relationships with screen characters or performers, we enter their world, leaving our own pressurized world behind. Thus, the element of escape inevitably becomes a component of companionship; companionship becomes an element of escape.

ABOUT ENTERTAINMENT—A POSTSCRIPT

Pretelevision Hollywood films were occasionally criticized for their slickness: lustrous photography, unblemished dialog, rapturous music, and a predictable, structured, mass-appeal plot that sometimes left connoisseurs and critics dissatisfied. The films were glossy, appealing to worldwide audiences, yet they often lacked sensitivity and substance. They seemed to be all form and little content.

This chapter's discussion of the ingredients of showmanship does not imply that quality shows can be achieved merely by adding one part spectacle to three parts humor; mix well and add a pinch of conflict. Such an assumption would be absurd. Beneath the elements of audience appeal must lie the concrete foundation of *concept*. Without a substantial concept, most shows will eventually fail, no matter how glossy their production values.

By definition, shows also require someone to see them. When a Sender lacks a Receiver, communication fails. When directors or producers find that magic equation in which entertainment elements are appropriate to a concept, enriching the concept and simultaneously attracting an audience, they achieve genuine communication.

CHAPTER HIGHLIGHTS

— To function effectively, directors should understand the elements of audience appeal present in all entertainment forms. Researchers have divided the entertainment function into four categories: pleasure, information, escape, and companionship.

— A major ingredient of pleasure is *spectacle*, appealing primarily to our senses, essentially sound, color, and motion. We enjoy such elements in circuses, fireworks displays, and parades, but they also appear in sports (marching bands, costumes), in drama and commercials through use of colorful locations and lavish sets, and in newscasts through concentration on visual elements.

— A second component of pleasure is *survival* (conflict), which carries subconscious memories of our primitive ancestry and the need to kill or be killed. This primal force is reflected in athletic contests, in quiz and game shows, and in traditional drama in which a protagonist wages war against an adversary.

— *Symmetry* is another ingredient of pleasure. Some facet of the human mind seeks order. It is part of our fascination with puzzles. Most drama by definition creates a state of disequilibrium. Part of the spectator's pleasure lies in seeing equilibrium restored. Audiences enjoy the symmetry of certain rhythms in drama: in the repetition of musical or visual motifs, significant phrases, or actions by characters.

— *Male-female chemistry* appeals to another primal urge, the sex drive. Because of its raw power, it has become enormously persuasive in commercials, in attracting audiences to sensual performers or to stories accentuating the male-female relationship. Directors need taste in the staging of sexually implicit relationships. They should not forget the power of the human imagination to provide unseen detail.

— The final component of pleasure is *humor and surprise*. For an audience, knowing what to expect diminishes suspense. Stereotypical stories, dialog, casting, or directorial patterns quickly lose spectators. Unexpected twists and turns often produce humor by causing viewers to laugh (at themselves) for having guessed incorrectly.

— The second major function of entertainment is to provide information. This function gratifies the viewer's curiosity. Morbid curiosity drives us to seek out the deviant, the grisly, and the forbidden. It is double edged: We are simultaneously repelled and attracted. Curiosity also contains a healthier aspect that drives us to learn and discover. This aspect takes us to college and makes us watch educational programs. It is frequently based on self-interest, the need to improve our status in society.

— The third function of entertainment is escape, diversion from the pressures of everyday life. Short of striving to make characters or performers involving, directors can do relatively little to help audiences find escape. The need and the capability lie within the spectator.

— The fourth category of entertainment is companionship. Passive companionship is undirected; the TV simply provides a background for other activities. In the active form, the viewer directs attention at a TV or movie screen, seeking emotional involvement with characters, hoping to lose awareness of self and life's pressures.

— Audiences relate to people with whom they can identify. In newscasts they become emotionally involved in human interest stories far more than in displays of fire or flood. They become involved with performers (singers, comedians, dancers, musicians) who display such characteristics as personal charm, energy, and authority.

PROJECTS FOR ASPIRING DIRECTORS

1. Assume that you will be directing a feature film of the traditional Cinderella story. How can you use each of the following elements of pleasure to enrich your film?
   a. spectacle
   b. survival (conflict)
   c. male-female chemistry
   d. order/symmetry
   e. surprise and humor
   You may expand or change the Cinderella story, when necessary, to accommodate these elements, so long as your changes are faithful to the theme and spirit of the original story.

2. Use these same elements to enhance or enrich the story of Little Red Riding Hood.

3. Find three examples of the male-female relationship in TV commercials. How effective is this provocative element in selling each product?

4. Write an original movie premise (no more than a paragraph or two) that uses as its audience lure the element of morbid curiosity.

5. Of all of the motion pictures you have seen, what character involved you the most deeply and created the strongest emotional response? What factors were responsible for such a powerful emotional response?

6. Think of a successful motion picture or TV movie that totally lacked the element of conflict.

CHAPTER

3

THINKING VISUALLY

Students of drama trained in traditions of the stage often tend to approach the craft of film or television directing with disproportionate emphasis on the spoken word. In theater much of the narrative development occurs solely through dialog. Because of spatial limitations, major events happen offstage rather than on; they are described rather than shown.

The transition from the theatrical stage to the more visually oriented media of TV and film requires a change in emphasis. Such a change involves (1) learning principles of visual composition, (2) learning to "think visually," and (3) learning to use these principles to communicate a message. The first is relatively easy, but the last two elements require practice; they are developed skills.

Many art students (and some teachers) are reluctant to reduce composition to a set of mathematical principles. Because beauty lies in the eye of the beholder, they claim, composition must remain personal, varying from subject to subject and from artist to artist. It cannot be absolute. Those who chaff at the burden of discipline welcome such a view. While it contains an indisputable element of truth, the viewpoint ignores the fact that certain arrangements of line, form, and color have pleased the eye since the beginning of time. Harmony, balance, and rhythm have fundamental and continuing emotional appeal. The laws that govern them may be ignored—but at some risk.

Two and two make four. But a picture to be worth remembering must add up to more than the sum of its parts.

Albin Henning[1]

This chapter will describe some of the principles of visual composition. It cannot examine all such principles or discuss them in great depth. Such a discussion would require a textbook of its own. (See "Books That May Be Helpful" at the end of this book.) Instead, we will examine those few principles of visual composition most likely to help aspiring directors achieve the emotional impact they desire.

— VISUAL COMPOSITION: defining composition and stressing the need for a single, dominant center of interest

— BALANCE: an examination of the forces at work within a frame and how those forces must occasionally be adjusted for TV and film

— DOMINANCE: how characters or objects achieve dominance through contrast, placement, and line

---

[1]As quoted by Fritz Henning in *Concept and Composition* (Cincinnati: North Light, 1983), p. 4.

▬ RHYTHM: the repetition of line, form, or color for sensual effect

▬ GOLDEN MEAN: sites within a frame believed by some to represent the aesthetically perfect placement of characters or objects

▬ READING A PICTURE: how our eyes move within a frame to absorb its content

▬ ACQUIRING A VISUAL SENSE: suggestions for achieving visual awareness, including practical exercises

VISUAL COMPOSITION

A composition is a harmonious arrangement of two or more elements, one of which dominates all others in interest. Because of the greater attraction this element has for the eye, it becomes a focal point which we call the center of interest or climax.[2]

We start with a blank screen. In television its proportions are exactly three units high and four units wide, the same proportions that once shaped the motion picture screen. Upon the surface of this blank screen we will place shapes, lines, textures, and colors. By their arrangement, we will create balance or imbalance, harmony or disharmony, depending entirely on the psychological mood we intend to generate.

Because of its smaller screen size, television usually requires greater simplicity in its compositions than do motion pictures. The need for a single, dominant focus of interest becomes more critical. Secondary interests cannot be allowed to detract from the primary subject matter. Also, because the television screen is divided into 525 lines, a picture containing dozens of small shapes begins to lose definition. Such a picture would emerge crisp and clear on a gigantic movie screen. Recall some of the biblical epics you have seen replayed on television with thousands of extras, horses, and elaborate costumes. Remember how difficult it became in many scenes to distinguish detail? Think of the many times you have seen dresses or neckties with small stripes or intricate patterns that seemed to swim dizzily back and forth, moving stroboscopically between the lines of your TV screen.

A single, dominant focus of interest. But where should we place it?

---

[2]A. Thornton Bishop, *Composition and Rendering* (New York: John Wiley, 1933), p. 13.

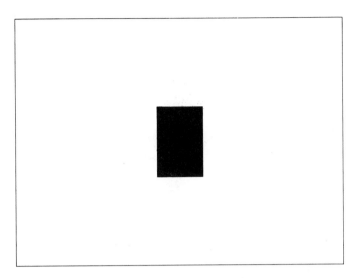

FIGURE 3.1   Placing an object in the exact center of a frame does not represent a composition. Compositions require two or more elements.

BALANCE

In Figure 3.1 the focus of interest has been placed exactly in the frame's center. Strictly speaking, such a placement does not constitute a composition because it involves only a single mass. But because television and film so often use a single figure within a frame (for example, closeups), let's temporarily expand our definition of composition.

The mass in Figure 3.1 appears secure, stable, comfortable. But such stability comes not from lack of tension. On the contrary, every composition, even one as simple as this, contains forces that exert pressure upon its elements.

> "Dead center" is not dead. No pull in any direction is felt when pulls from all directions balance one another. To the sensitive eye, the balance of such a point is alive with tension. Think of a rope that is motionless while two men of equal strength are pulling it in opposite directions. It is still, but loaded with energy.[3]

---

[3]Rudolf Arnheim, *Art and Visual Perception* (Berkeley: University of California Press, 1974), p. 16.

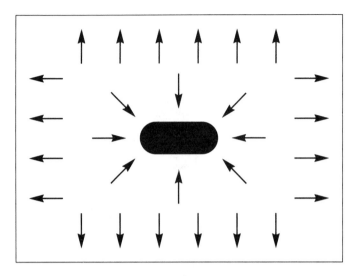

FIGURE 3.2   Magnetic forces (much like gravity) within a frame seem to pull objects toward them.

Those who research visual perception believe that certain forces exert a kind of magnetic pull on elements within the frame. The edges of a frame, for example, tend to pull objects toward them. So does the exact center. Thus the mass in Figure 3.1 is held in position by magnetic forces exerted by frame edges and by magnetism from the center. The corners don't seem to generate as much pull as either the frame edges or the center (see Figure 3.2). And no point within the frame is free from such influences. Note that these are psychological forces; they exist only in the mind of the observer.

Although center placement presents a figure held in perfect equilibrium, many artists arbitrarily rule it out, regarding such a composition as static. Like many rules, this one has been broken frequently and with great success by artists ranging from Leonardo da Vinci to Picasso.

In Figure 3.3 the mass has been placed above center. Now the space above and below it is divided unequally and is therefore more interesting than in Figure 3.1. The mass is suspended halfway between the frame edges and the center; the pull exerted by each of those magnetic sites again holds the mass in perfect equilibrium. The arrangement in Figure 3.3 corresponds to the head position in film and TV closeups. In most closeups, however, the head is supported by the shoulders, presenting a stable, solid triangular form.

Above-center placement of a face in closeups is acceptable and appropriate whenever the performer relates directly to the audience, looking into the camera

FIGURE 3.3   Placing the mass above center divides the space unequally, making the arrangement less static, more interesting.

lens. Such positioning is customary in commercial pitches, instructional shows, newscasts, and game or quiz shows. In drama such eye-to-eye contact with the audience has generally been regarded as taboo because it implies awareness of the viewer's (camera's) presence and thereby breaches the illusion of a fourth wall. Eye-to-eye contact makes the viewer a participant instead of an unseen observer. In a dramatic context, glancing into the camera lens would be appropriate only if an actor were to step out of character and speak to the audience in asides.

Adjusting the Balance

When photographing a closeup in drama, directors customarily "lead" actors, allowing slightly more space in front of them than behind—or more space in the direction actors face. This is sometimes called **nose room**. When characters stare off to the left or right, they create additional weight on that side of the frame and the director must allow extra space to accommodate it. Then the head is no longer equidistant from the sides of the frame, as shown by the position in Figure 3.4.

A character's **line of sight** is so powerful that it tends to direct audience attention in the direction of his or her glance. (See Chapter 6 for more on line of sight.)

Although allowing additional space in front of a figure has become more or less standard in cinematic closeups, the practice has been ignored as often as

FIGURE 3.4 Characters create additional weight in the frame when they glance (or face) in a specific direction. It is therefore customary to "lead" actors, allowing extra space in front of them to maintain pictorial balance.

observed by the world's great painters, confirming again that compositional rules can only be regarded as guidelines. Notice in Figure 3.5 how the artist, El Greco, has actually minimized the space in front of Saint Andrew, deliberately creating an uncomfortable tension in the viewer, echoing the despair on the apostle's face. Directors can use the same technique, intentionally creating a cramped or crowded frame. For example, characters may feel that dramatic circumstances have placed them in a vise, confining or restricting them. By crowding characters within a frame, the director conveys to the audience a sense of claustrophobia.

The Two-Shot

In Figure 3.6 two masses share a frame; they are equidistant from the center and of equal size and weight. Now we have a true composition, but a static one. Thornton Bishop's definition earlier states that in every composition one element must dominate. Because neither of the masses dominates, the composition here

FIGURE 3.5   In El Greco's painting of *The Apostle Saint Andrew*, the artist deliberately reduced the space in front of the figure, creating tension. (The Los Angeles County Museum of Art: Los Angeles County Funds. Used with permission.)

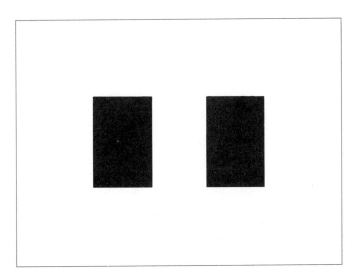

FIGURE 3.6   When two masses share a frame equally, the composition is static.

FIGURE 3.7   A medium two-shot in which audience interest is divided equally between the two figures. Such a shot tends to be static, lacking in visual dynamics.

becomes dull and flat, its focus of interest divided. From the point of view of pictorial composition, we have a balance. The masses are held in equilibrium by magnetic pulls from the frame edges, the center, and each other. Note that we now have a third magnetic force in action, that generated by objects within the frame: the larger the object, the greater the magnetic force generated. In Figure 3.6 each mass exerts its own magnetic attraction. Such a pull might be compared to the force of gravity.

Figure 3.7 (corresponding to Figure 3.6) reflects drama's standard flat two-shot in which two characters face each other, sharing the frame equally as the camera photographs them in profile. While such a shot is fairly common in TV and film, it tends to lack depth and dynamics; interest is divided; the picture is static. Watch what happens when the heads approach each other and the camera frame tightens (Figure 3.8). Now the shot becomes far more dynamic. The increased size of the subject matter, together with the loss of dead space, creates impact.

FIGURE 3.8   While interest is still divided equally between the two characters, this
closer angle tends to be far more dynamic than Figure 3.7. The greater size of the
faces contributes to the increase in impact.

Watch what happens again when we move the figures further apart (Figure
3.9). Now, with excessive space between them, they are pulled away from each
other by the powerful magnetism of the frame edges. The audience becomes
uncomfortable, feeling a need to "dolly back," to distance themselves, to balance
the space between the figures with space behind them to equalize the magnetic
forces.

Watch what happens when we move the figures close together, putting excess
space behind them (Figure 3.10). Again the audience is uncomfortable. Again
there is no balance. Now the magnetism exerted by the frame edges is almost
inoperative and the two figures are pulled toward each other. We feel the need to
"dolly forward," to move closer, to lose the space behind the characters to equalize the pressures.

In most circumstances, directors try to place characters within a camera
frame in perfect balance, so the magnetic forces pulling them toward each other
exactly match the strength of forces that pull them toward the frame edges. There

FIGURE 3.9   Here the figures seem to be pulled away from each other by the powerful "magnetism" of the frame edges. Such an awkward framing creates an uncomfortable feeling in audiences.

are no electronic meters that measure such a balance; it can only be gauged by the director's eye and sensitivity. In television and film drama, directors sometimes deliberately create imbalance, when a scene's emotional content suggests that the audience *should* feel uncomfortable. For example, the director may want to imply that characters are being pulled away from each other by forces beyond their control. Or if they are being confined, pressured, hemmed in by circumstance, a director might position them claustrophobically close to the frame edge. On such occasions, pictorial balance necessarily takes second position.

Balancing Images: Leverage

When the principal motif is placed farther from the center of the picture, additional secondary interests must be invented on the opposite side to establish a proper balance. Such a problem can be likened to one of

FIGURE 3.10   The two figures are pulled toward each other here. We feel the need to move closer, balancing the pulls between center frame and the frame edges.

leverage, where an object can be made to balance a heavier one by moving the lighter one farther from the fulcrum. The existence of a fulcrum is felt in all compositions where the proper balance of unequal interests is maintained.[4]

In Figure 3.11 the large mass balances the smaller one because the latter has been placed farther from the "fulcrum." In your mind, move the smaller mass closer to the center of the frame and see how the large one gradually assumes more weight. Imagine a balanced seesaw in which an adult sits on one side of the plank near the center and a small child sits on the other side at the very end. They are in perfect balance. As the child moves toward the center—or as the adult moves backward to the end of the plank—the seesaw loses its balance.

---

[4]Bishop, *Composition and Rendering*, p. 14.

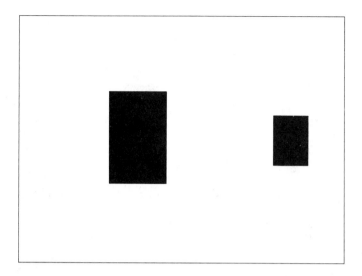

FIGURE 3.11   Like people on a seesaw, the masses here balance each other.

Drama's traditional over-shoulder shot may seem to contradict this principle. According to what we have just learned, the large head at the edge of the frame in Figure 3.12 should outweigh the smaller head facing us. In our study of dramatic emphasis (Chapters 6 and 7) we will learn that characters facing camera create far greater emphasis than figures facing away. Such emphasis more than compensates for any compositional imbalance. Thus, in Figure 3.12, our eyes instantly go to the young man facing us and almost ignore the back of the foreground head. What if the young woman turned around and faced us? Now we would see two faces, one very large in the foreground, the second smaller in the background. Note how the balance of weight changes dramatically. Because we have erased the difference in dramatic emphasis, the large head now creates far greater weight, or dominance, than the other.

DOMINANCE

In drama, perfect balance is not always desirable. Just as a claustrophobic frame creates predictable psychological reactions in the spectator, so does imbalance create impressions that may help the director achieve certain dramatic effects. In most scenes one character is dominant. By deliberately creating a compositional

FIGURE 13.12   A standard over-shoulder shot in which the figure facing camera assumes the greater impact. Were the young woman to turn and face us, audience interest would shift to her.

imbalance favoring the dominant character, the director may visually strengthen that character.

Just as one theme dominates all others in a script, so should one image dominate all others in a camera frame. When secondary interests detract from the primary focus of attention, the result is clutter. When several characters vie for attention, the director must determine which is of primary importance and then create dramatic or photographic emphasis. Photographic emphasis may be achieved through *contrast*, *placement*, or *line*.

Contrast

With objects of approximately equal interest, the one that contrasts with the others generally attracts attention. Contrast may be in size, color, tone, density, texture, or form or in the physical nature of an object. Viewing a picture of two children regarding a kitten, for instance, we probably fasten our gaze upon the kitten because it is different in both size and nature, other compositional aspects

FIGURE 3.13   In John Singer Sargent's *Portrait of Mrs. Edward L. Davis and Her Son*, our attention goes first to the boy because his sailor suit contrasts so markedly with the background. (The Los Angeles County Museum of Art: Frances and Armand Hammer Purchase Fund. Used with permission.)

aside. In Figure 3.13, the portrait by John Singer Sargent, notice that even though the mother is larger and occupies a more central position in the frame, the boy attracts our attention because his costume contrasts so markedly with the background. While this and other oil paintings are not in TV or film screen proportion, they still illustrate the principles described here.

With three or more objects, the one that contrasts with the others will gain primary attention. In Figure 3.14, note that the contrasting object may be either larger or smaller than the others.

Placement

Other factors being equal, the object nearest the center of a picture usually attracts the greatest attention. In the advertisement for Crown Royal (Figure 3.15) the advertiser has gained maximum impact by placing the bottle at center. Be-

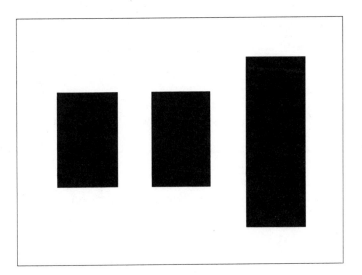

FIGURE 3.14   With three or more objects, the one that contrasts with the others will gain attention. The contrasting object may be larger or smaller.

cause it is also larger than the two glasses and different in form and character, it demands our attention.

Often, because of specific compositional needs or commercial requirements, an object or character cannot be placed at the center of a picture. The object also

FIGURE 3.15   Center placement usually provides maximum impact. The Crown Royal
   bottle is also larger than the glasses and different in form. (Warwick Advertising,
   Inc. Used with permission.)

may not provide a contrast with other elements in the picture. Then the viewer's
attention may be focused on the object through line.

Line

In nature, lines usually appear as the division between forms, such as the horizon
line separating ocean and sky or the slope of a mountain, its edge dividing land
and sky. Civilization has surrounded us with lines: streets (curbs, painted dividing lines), telephone poles, electric wires, railroad tracks, church spires, sky-
scrapers, and a thousand other structures.

The nature and position of a line make strong emotional statements about the
subject matter. Horizontal lines or planes generally tend to create comfortable
feelings of serenity, repose, and equilibrium—all's right with the world. Vertical
lines or planes provide greater strength and energy; they also suggest the spiritual,
a reaching upward to the heavens, as manifested in church steeples.

FIGURE 3.16   Frames created by intersecting lines (such as doors and windows) emphasize the subject matter they contain. In this scene from *Doctor Zhivago*, Rod Steiger and Julie Christie hold our primary attention. The observer's eye line creates additional emphasis. (An MGM release © 1965 Metro-Goldwyn-Mayer Inc. Used with permission.)

The Intersection of Lines   In photographic composition, lines represent one of the most powerful means of directing viewer attention. Their arrangement within the photographic frame determines the degree of their power. The simplest, yet most effective way of directing attention to a particular area is through the *intersection* of lines. Our eyes follow a line until it crosses another. We fix our attention there, as the director or photographer intended we should.

In drama, characters placed in doorways or windows inevitably attract attention because they are framed by intersecting vertical and horizontal lines. Figure 3.16 shows a scene from *Doctor Zhivago* with Rod Steiger and Julie Christie enclosed within a window frame. The rectangle focuses our attention on them. The rapt attention of the foreground figure further intensifies our interest. Similarly, in *Portrait of the Artist's Son* (Figure 3.17) the window frame holds our attention inside the painting even though the boy's stare tends to direct our

FIGURE 3.17   In Martin Drolling's *Portrait of the Artist's Son*, the window frame directs
our attention inward. Similarly, the frame of the canvas in background provides
emphasis for the young girl. Her glance to the left helps to balance the strong line
created by the boy's glance. Note that the violin's shape echoes the shape of the
birdcage. (The Los Angeles County Museum of Art: Museum Purchase with Balch
Funds by Mr. and Mrs. Allan C. Balch Collection, Mrs. Eloise Mabury Knapp, Mr.
and Mrs. Will Richeson, Miss Carlotta Mabury, John Jewett Garland, Mrs. Celia
Rosenberg, Mr. and Mrs. R. B. Honeyman, Bella Mabury, and Mrs. George William Davenport. Used with permission.)

attention outward. Notice that a second rectangle in the background, a girl enclosed within the frame of an artist's canvas, parallels the foreground window
frame, echoing it and providing three rectangles of approximately equal proportions: the outline of the painting itself, the window frame, and the canvas.

In Figure 3.18, a drawing by California artist Ellen More, the powerful lines
of the model's two arms form a pattern of strength and solidity that drives the
viewer's eyes to their intersection at the stool top. The lines of the model's leg as
well as those of the stool legs also direct our glance to the intersection.

FIGURE 3.18   The impact of intersecting lines is demonstrated in Ellen More's powerful *Male Nude*. (Used with permission of Ellen More.)

Converging lines frequently do not physically intersect; the intersection may be implicit, actual closure taking place at infinity—in the mind of the viewer. In Figure 3.19, for example, the architectural lines of the church, as well as the converging lines of the floor, direct our view to the far archway. So powerful are these converging lines that the viewer is almost unable to resist their impetus.

A line need be neither graphic nor literal. It may be suggested. Whenever a painted figure—or a character in drama—glances in a specific direction, the look creates a line and the spectator's attention will tend to follow that line. The foreground character in Figure 3.16 effectively demonstrates this principle. The closer a character is positioned to the camera, the more powerful the line of sight becomes. Such a circumstance is similar to seeing a group of people on a street corner staring upward. For a passerby, the urge to look up becomes almost irresistible. (Of course, curiosity is also a factor!) Lines also are formed by a

FIGURE 3.19   Lines may converge only in the viewer's mind but they rivet our attention nonetheless, as demonstrated in this *Interior of a Church at Night* by Dutch artists Anthonie de Lorme and Ludolf de Jongh. (The Los Angeles County Museum of Art: Gift of Dr. Hans Schaeffer. Used with permission.)

succession of objects, placed so that the eye is led from one to another. The sailboats in Monet's *Beach at Honfleur* (Figure 3.25) form a line that directs our attention to the lighthouse.

Grouping figures so that the eye travels along them in a direction toward the center of interest is but one scheme. The attention of a human figure in a composition, directed toward the person of chief interest suggests the direction for the observer to look. The posing of arms to define a line is frequent. In Da Vinci's *Last Supper* this scheme was followed, the gestures of the disciples being used to direct the eye of the spectator to the center figure.[5]

In Figure 3.20 the group of figures surrounding Christ direct their glances at him, making it almost impossible for us to look elsewhere. *Almost* impossible. The artist, Rocco Marconi, has created subtleties in his composition that make it more complex than it first appears. While the surrounding figures compel our attention to focus on Christ, his glance leads our eyes left, past the beseeching woman to the two sinister, plotting figures at extreme left. The character directly above them, framed in a window, directs our attention back to the Christ figure, establishing a circular movement within the frame. Such is the power of the look to establish a line.

The Diagonal Line   Of all directional lines, the diagonal is by far the strongest, the most compelling. Take a few minutes to leaf through a favorite magazine. Study the advertisements. Most use the principles of visual composition discussed in this chapter; many feature powerful diagonal lines to direct the reader's eyes to a product.

Many of the painters in the Dutch and Flemish schools used the diagonal line as a basic force in their compositions, allowing light from a window high in the frame to slant downward, directing our gaze and illuminating an object lower in the picture. In Figure 3.19 the architectural and floor lines in the church are diagonals, which accounts for their power in propelling our attention forward. In drama, a woman standing at the bottom of a stair rail would tend to gain audience attention because the rail directs our attention to her. Notice the strength of the diagonals in Figure 3.18.

Thus far we have discussed elements that relate to the formal compositional aspects of a picture. A compositional element that contributes to the viewer's sensual pleasure is rhythm.

RHYTHM

We recognize rhythm in the primitive dances of many cultures as well as in the markings on ancient Greek and Roman temples, Byzantine mosaics, and archeological relics such as vases. It is probably the oldest form of artistic expression. *Rhythm* usually implies some form of *repetition*. It is an appeal to the senses. In music, rhythm implies repetition of a beat in a prescribed cadence or repetition of a musical figure. In rhetoric it implies repetition of words or phrases (such as "of the people, by the people, and for the people"). In poetry it usually denotes a

---

[5]Bishop, p. 19.

FIGURE 3.20   When figures look in a specific direction, their glances create lines that
direct our attention, as demonstrated in Rocco Marconi's *Christ and the Women of
Canaan*. (The Los Angeles County Museum of Art: William Randolph Hearst
Collection. Used with permission.)

repeated metric structure (for example, iambic pentameter) or a rhyming pattern
whose repetition of sound is pleasing to the ear.

In the visual media, rhythm implies a repetition of line, form, or color. In
Figure 3.19 the repetition of columns and archways creates much of the painting's
beauty. It is sensually pleasurable to the viewer. When composing a shot, visually
aware directors often search for repetition in their staging of actors and in their
choice of scenery and locales. One of the reasons we see mirror shots so often in
films is that the reflections offer viewers a pleasing repetition, one image echoing
the other. The shot from *Midnight Cowboy* shown in Figure 3.21 is a typical
example. Note also how the mirror provides a frame for Jon Voight, holding our
attention within its periphery.

The Nike advertisement in Figure 3.22, part of a campaign celebrated for its
visual effectiveness, uses repetition as its key motif. The repeated lines of bleacher
seats lead our eyes unswervingly to the athlete wearing the Nike shoes. The artists
responsible for such ads understand how most viewers will "read" such a picture,
the direction in which their eyes will travel.

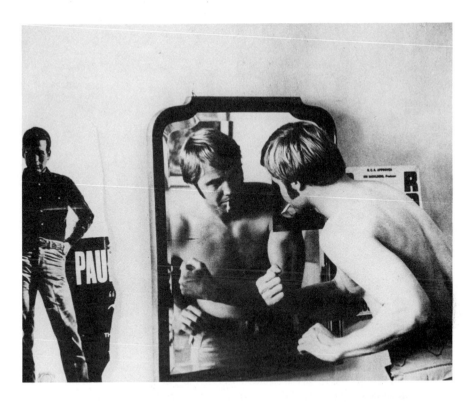

FIGURE 3.21   In this scene from *Midnight Cowboy*, the mirror provides a pleasing repetition of image. It also creates an inner frame, holding our attention on Jon Voight. (A UA release © 1969 Jerome Hellman Productions, Inc. Used with permission.)

The Crown Royal advertisement (Figure 3.15) provides repetition in the two glasses balancing each other on opposite sides of the frame and roughly echoing the shape of the bottle itself. The ad also provides an overlay of jigsaw puzzle shapes that create a pleasing repetition of form. Picture in your mind the same advertisement without the puzzle lines. It would lose much of its appeal.

In the Diego Rivera painting *Flower Day* (Figure 3.23), the repetition of form is most evident in the profusion of lilies. But notice that the lilies themselves echo the shape of the vendor's head and shoulders, and their color is echoed in the costumes of the seated women. Notice the repetition again in the leaves and stems in the vendor's hands and in the seated figure at the left almost exactly mirroring the figure on the right. Because repetition is far more interesting when there is variation, when one form does not *exactly* replicate the other, the seated figures are not identical. As you examine the painting further, you can find additional sensually pleasing repetitions of form.

FIGURE 3.22   This Nike ad, part of an award-winning campaign, uses the principle of repetition to provide sensual pleasure to the viewer. The bleacher lines also direct attention to the athlete. Note that she is positioned near the Golden Mean. (Chiat/Day Advertising. Used with permission.)

Flower Day reinforces other compositional principles as well. The seated figures direct their gaze to the vendor, creating lines of sight that compel us to do likewise. His head gains additional emphasis because it contrasts so markedly with the flowers behind him. In this triangular composition, the vendor almost becomes a holy figure, offering a benediction to the seated figures.

THE GOLDEN MEAN

Throughout history, artists have searched for a mathematical formula that would provide the basis for perfect design. One of these formulas, sometimes referred to as the **Golden Mean**, was first postulated by the Greek mathematician Euclid (although some historians have credited Pythagoras with the discovery).

FIGURE 3.23   Diego Rivera's *Flower Day* demonstrates the sensually gratifying element of repetition. Note that the two seated figures almost exactly mirror each other. The lilies endlessly repeat themselves and resemble the curves of the three heads and the foreground basket. (The Los Angeles County Museum of Art: Los Angeles County Funds. Used with permission.)

In the language of geometry: A straight line is said to have been cut in extreme and mean ratio when as the whole line is to the greater segment, so is the greater segment to the lesser. In Figure 3.24, line AC is as proportionate to CB as CB is to AB. Those who advocate the Golden Mean theory believe that the proportions provide an ideal basis for the placement of figures within a frame—at any of the four points where the lines intersect.

Many photography schools teach the principle far more simply. They draw lines dividing the frame into equal thirds, horizontally and vertically. The four points of intersection almost exactly coincide with the Golden Mean. In Figure 3.25, Monet's *Beach at Honfleur*, we focus our attention at the lighthouse, intersecting sea and land and very close to the Golden Mean. Notice that Monet has placed the figure of a man at another intersection of Golden Mean lines.

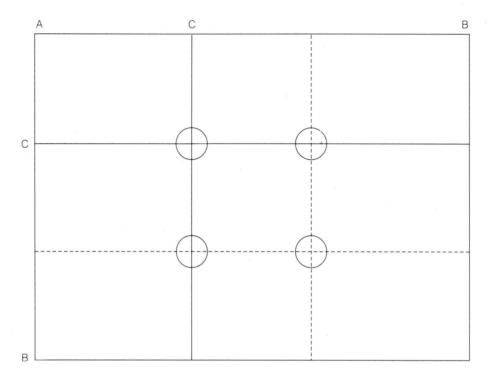

FIGURE 3.24 The intersection of lines designates the Golden Mean, one of four positions in the frame said to represent the aesthetically perfect focus of interest.

Remember that Golden Mean areas do not create emphasis; they perhaps represent the aesthetically perfect sites of interest; they create ideal proportions. In Figure 3.25 Monet has skillfully given the lighthouse importance through the intersection of several lines, the shore, the mountain, the line created by sailboats, the horizon, and the vertical line of the lighthouse itself. Importance derives from the intersection, not from the position of the lighthouse at the Golden Mean. In this painting the two coincide; the position of dominance is located at the point of the Golden Mean.

READING A PICTURE

In Western civilizations, viewers' eyes generally enter the frame at the left and move to the right and downward. Perhaps this pattern of viewing emerges from the way we read a printed page, starting at the top and moving from left to right.

FIGURE 3.25   In Claude Monet's *Beach at Honfleur*, the lighthouse is located near the Golden Mean. The male figure falls almost exactly at another location of the Golden Mean. (The Los Angeles County Museum of Art: Gift of Mrs. Reese Hale Taylor. Used with permission.)

In some Eastern cultures, reading patterns are reversed, moving from right to left or in vertical lines from top to bottom. In those cultures, television or film imagery may be "read" in a different directional pattern than in the West. Examine the slanting lines in Figure 3.26.[6] Which goes uphill? Which goes down? Because most of us read a frame from left to right, you probably interpreted the first slanted line as going uphill and the second as slanting down. If your eyes had entered the frame on the right, you would probably have interpreted the lines in an opposite fashion.

German art historian Heinrich Wolfflin claimed that because we read a picture from left to right, the right side thus becomes "heavier" than the left and must be balanced by additional weight on the left.[7] If we accept Wolfflin's premise

[6]Herbert Zettl, *Sight, Sound, Motion* (Belmont, Calif.: Wadsworth, 1973), p. 129.

[7]Described by Louis D. Giannetti in *Understanding Movies* (Englewood Cliffs, N.J.: Prentice-Hall, 1976), p. 62.

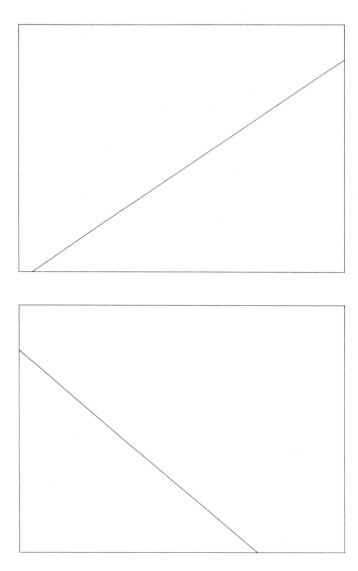

FIGURE 3.26   If you interpret these lines as going uphill in the first example and downhill in the second, you have demonstrated that your eyes enter the frame at the left,
the usual pattern for ''reading'' a picture in most Western cultures.

at face value, then we must conclude that in any two-shot the actor on the right
side of the frame receives more spectator attention than the actor on the left. Some
actors and directors disagree. They feel that a character at the left side of the
frame draws more audience attention simply because we see him or her first.

Actors preoccupied with their own importance frequently ask directors to stage them either at the left or the right side of the camera frame to focus attention on themselves and thus dominate a scene. In practice, many factors of dramatic emphasis come into play, modifying or subverting some of the principles of pictorial composition (see Chapters 6 and 7).

Some authors and critics claim that the upper portion of the frame is intrinsically heavier than the lower. When the horizon line exactly bisects a frame, the sky, being heavier, oppresses the earth, dominating figures in the lower half of the picture. John Ford in many of his Westerns used the powerful oppressive force of the sky to dramatize the frailty of his characters as they moved across the desert vastness. The sky's intimidating effect may be enhanced by placing the horizon lower in the frame, allowing the sky to occupy a major portion of pictorial space. When such a dominant sky is dark and angry, the effect becomes truly frightening.

ACQUIRING A VISUAL SENSE

Just as some musicians seem to possess an innate sense of melody, so do some directors seem to have a natural facility for composing a picture that expresses a concept or emotion in values that are strictly visual. Such talent possibly is innate, acquired through genes, but it may also be acquired through study and observation.

At the start of each semester, a few students in my writing classes despair of ever finding exciting story material. Weeks later, they are amazed to discover story possibilities everywhere—in their homes, among their friends, in newspapers, in their classes. The stories, of course, were always there. But in those few weeks the students had put on spectacles that suddenly enabled them to see: They had acquired story awareness.

Thinking visually concerns more than balance, dominance, line, and rhythm. In drama as well as in most other entertainment forms, visualization is intrinsic to the fundamental directorial task of communicating or interpreting a message. The director's primary goal is to entertain, inform, or persuade. To reach such a goal effectively, and with clarity, the director must consider the entire visual spectrum—everything that can be *seen* as opposed to everything that can be *heard*. Most significant within the visual spectrum are the performers' actions and their effectiveness in communicating an idea or emotion. Such actions communicate to the audience insights never possible with mere words. What performers *do* by far outweighs what they *say*.

Aspiring directors who despair of learning the principles of visual composition may take heart. Like story awareness, such knowledge can be acquired.

Education

The most obvious way to learn the principles of visual composition is to attend classes in art or in art history or both. Art classes offer acquaintance with the formal principles of color and composition, enlarging on the principles described in this chapter. Art history provides a look at the work of the great masters, instilling an awareness of visual aesthetics and compositional style and form. Either class (or both) enables one to put on magical spectacles and glimpse a world only dimly perceived before. Either awakens a new visual consciousness.

Trips to photographic exhibitions, galleries, or art museums will also enrich your knowledge. Those that exhibit the best in contemporary art will do more: They will help to update your taste. Words of warning: Be patient. Paintings that at first seem awkward, childish, or bizarre may take getting used to. Distrust those that you like immediately. Return to paintings that anger or confuse you and view them again. Read the brochures most galleries supply. In time you will discover appreciation where confusion once existed.

Observation

Watch the people who inhabit your world. See if you can tell what's going on inside their heads. Your clues are the actions they are frequently unaware of, little mannerisms that give them away. The gang member who has been arrested for car theft, desperately trying to play it cool, surreptitiously licks his lips because his mouth is dry. The girl who tries to conceal her partial deafness turns her head slightly in conversation, the better to hear. The finest directors are sensitive to such actions (more about this in Chapters 5 and 6). So are the finest actors and writers. Genuinely creative people are almost always keen observers of the human condition.

Games

Some games help build visual acuity. In my directing class I sometimes run short motion picture sequences for my students and then ask questions about them. Usually the sequences are action packed, crammed with exciting detail, so that students must struggle to absorb a variety of colorful images. Students who score poorly on early "tests" usually are delighted to discover that their visual acuity improves from week to week. Questions concern apparel of participants, their activities, details of the background, and use of camera. A familiar variation of this game is the picture puzzle, a still photograph crammed with a bizarre variety of visual elements. Students are given a few seconds to familiarize themselves with its contents and then are asked questions regarding the picture.

Another game consists of quizzing students about objects they see (or fail to see) in their everyday lives. For example, what letters are missing from the telephone dial? On traffic signals, is red at the top or bottom? What is the shape of a stop sign? Surprisingly, the most obvious questions often are the most difficult. The question missed most frequently in my classes: What is the color of the building we're sitting in? Such games teach us to be more visually aware of the world around us, an essential characteristic of the successful director.

Visual Story Telling

Television drama relies heavily on the spoken word to tell its story. Too often, instead of watching colorful dramatic incidents, we hear characters discussing them. One of the distinguishing characteristics of silent films was their ability to tell a story visually, without dialog. Although many of these early films used title cards to help define a story or to paraphrase dialog, the best directors (Chaplin, Griffith, Eisenstein) created films that were so visual in concept and form that they seldom required such crutches.

An excellent exercise in building visual awareness is to create a short script (five or six pages) totally without sound: no dialog, no narration, no music, no sound effects. You must tell your story completely with *visuals*, through actions and incidents. No title cards. No "talking heads." It should be written angle by angle, exactly as the final, edited version would appear. If you have difficulty in finding a concept, here's a hint: A chase is one dramatic pattern whose visual action tells its story.

Don't be concerned that such an exercise appears to be oriented to writers. A director's goal always is to find ways of visualizing. Such an exercise allies you with the great silent directors who invented actions and incidents that dramatized a concept totally without sound. Through gesture, facial expression, or business, they clearly conveyed to audiences a character's thoughts and emotions. That ability did not die with directors of silent films. Study current fine film or television drama. You will discover that today's best directors are highly skilled in the art of visualizing.

If the facilities are available to you and the costs not prohibitive, convert your short script into videotape or film. Often such projects become possible by enlisting the help of friends whose interests match yours. If you're a student, perhaps you can tackle such a project for a television or film production course. Directing a "visuals only" project will quickly reveal your ability to dramatize concepts or reveal hidden feelings through external actions. Not only will such a project be challenging and stimulating but it also will provide you with rich directorial experience.

CHAPTER HIGHLIGHTS

— Although the stage uses dialog preponderantly, television and film place greater emphasis on visual elements. Aspiring TV and film directors, therefore, should learn the classic principles of visual composition as well as the ability to think visually.

— Visual composition is a harmonious arrangement of two or more elements, one of which dominates all others in interest. Because of the difference in screen size, TV compositions generally require greater simplicity than do motion pictures.

— Placement of the primary focus of interest depends upon a number of compositional factors. Many feel that placing it at the exact center of frame creates a static picture. Moving the object higher makes a more interesting composition; the space beneath the object tends to support it. In closeups, when performers look to the right or left, the camera customarily leads them, that is, allows more space in the direction in which they are looking.

— Some artists and mathematicians consider the Golden Mean a perfect location aesthetically for the primary focus of interest. When lines are drawn through a frame dividing it vertically and horizontally into equal thirds, the intersections approximately mark the Golden Mean.

— When two objects share a frame equally, they do not create a composition; one object must dominate. Profile two-shots, for example, usually lack interest.

— The balance of objects within a frame is governed by principles of leverage, much like children playing on a seesaw. Smaller objects at the edge of a frame balance larger objects closer to the center. Principles of dramatic emphasis affect such a balance. Thus, performers facing the camera create more weight than performers facing away.

— When several factors vie for attention, one must be given dominance. Such dominance may be achieved through contrast, placement, or line.

— When an object contrasts in size, color, tone, density, texture, or form with other objects in a frame, viewers' attention goes to the contrasting object. With two large objects and one small, for example, we tend to look at the small object.

— Other factors being equal, the object nearest the center of the frame usually attracts attention.

PROJECTS FOR ASPIRING DIRECTORS

1. Study the people around you, then jot down your observations of actions and mannerisms that provide clues to their thoughts and emotions. You will probably find that tension or insecurity motivates many mannerisms such as nail biting and hair twirling. Society has trained us to conceal our weaknesses, but some small action or gesture usually gives us away.

2. Collect magazine or newspaper advertisements that demonstrate the following methods of achieving visual emphasis:
   a. contrast in size
   b. contrast in color
   c. contrast in shape
   d. diagonal line
   e. intersection of lines

3. Collect magazine or newspaper advertisements that demonstrate rhythm (repetition) of:
   a. line
   b. form
   c. color

4. Collect magazine or newspaper advertisements that demonstrate balance.

5. If you have written the visuals-only project described in this chapter, convert the most visual sequence into a **storyboard**. A storyboard is a series of drawings (frames) in screen proportion, three units by four, that depict the action angle by angle, exactly as the camera would view it. Use a felt-tipped pen.

6. If you have not written a visuals-only script, then write an original one-minute commercial giving special attention to its visual qualities. Now draw a storyboard of the commercial. See if the frames convey the message you intended.

DIRECTING
FICTION

NOTES ON THE SINGLE-CAMERA OPERATION

Part 2 deals primarily with the single-camera operation, sometimes called the "film method." The label has become a misnomer because the single-camera operation is now used extensively with video cameras and videotape. No matter what you call it, the method has two major advantages over a multiple-camera operation: quality control and communication with actors.

Quality control concerns performance, certain technical aspects, and an almost indefinable area I'll call "attention to detail." Most directors of photography (cinematographers) will tell you that it's almost impossible to light a scene effectively when photographed from three different angles. Yes, you can light comedy or situation comedy for multiple cameras because they require plenty of light for an upbeat, high key appearance (bright, with no shadows). But drama that requires shadows and modeling of faces is something else. When each setup is lighted individually, a director can achieve portrait studio quality.

The director stands beside the camera in the single-camera method, almost able to reach out and touch the actors. In this proximity, he or she maintains more than a physical closeness. Between takes the director can put an arm around the actor, whispering suggestions, offering support and guidance, challenging, stimulating, creating excitement for a scene and for a character. This is not possible when the director is half a block away, isolated in a sterile control booth, able to communicate only through a floor manager or studio loudspeaker.

Meticulous attention to detail is more possible in single than in multiple camera because the focus is on one actor in one angle rather than (possibly) three or four. Such considerations as hair, wardrobe, makeup, and hairline subtleties of performance can be examined and, when necessary, reshot.

But the single-camera method also has disadvantages. It is costly and time consuming. Multiple cameras can shoot three or four angles simultaneously. With a single camera, they must be photographed one at a time, with valuable hours consumed by new lighting and restaging. In addition, film and lab costs can be expensive.

Another disadvantage: The director must largely rely on the camera operator to assess the framing and camera coverage. Directors usually ask operators at the end of a scene, "How was it?" It seems ludicrous that a director should rely on someone else for such an assessment. In a multiple-camera operation, the director watches monitor screens and sees exactly what each camera is framing.

Today in film some directors use "slave" video cameras attached to film cameras so they can watch the framings for themselves. If material from the slave cameras is videotaped, the director can play it immediately after the scene is shot. If dissatisfied with any aspect, he or she can reshoot it.

Producers and directors have continued to favor film over videotape because film provides better definition. That seems to be changing. I have watched demonstrations of high definition television (HDTV) photographed and projected on videotape, using digital sound and a five to three aspect ratio rather than TV's traditional four to three. The picture quality, while not yet perfect, is excellent. Videotape may offer a less expensive use of the single-camera method in the future.

CHAPTER

4

WHERE DRAMA BEGINS: THE SCRIPT

The cliché that film is a visual medium, only partially true, sometimes leads devotees to believe that the camera is a deity and that lenses, lighting, and fancy angles are the major determinants of a film's success. They regard the director as a visionary, a kind of miracle worker who takes camera and actors to a favorite location where they make up the story in a fever of inspiration, improvising whatever dialog may be necessary. Out of this mystic experience comes an award-winning videotape or film, something of vast artistic merit, with visual images that whisper of Truth and Beauty.

Such a view, of course, is absurd. The most creative director in the world, given the most sophisticated hardware devised by the human mind, cannot create a superb motion picture or television show without a superb script. An imaginative director can perform sleight of hand, adding theatricality and tempo to a dull script, making it seem better than it is. But unless there is substance in the words, there cannot be substance in the final product.

Ernest Lehman, one of film's most creative (and most rational) writers, puts it this way:

I just happen to be one of those irrational persons who thinks a film cannot be any good if it isn't well written. It just *can't* be. And in all

If you are a young director, or wish to be, the number one thing you've got to have is a screenplay. You can't win your millions on the roulette table unless you've got a chip to put on the numbers—and your chip in the film business is a script.

Michael Winner[1]

likelihood, if it's bad, it was badly written. Most, but not all, bad movies can be traced to a bad script.[2]

A script (also called a **screenplay**) is the basic building block of any dramatic project, whether in theater, radio, videotape, or film. It is a carefully devised blueprint from which actors, director, and technicians will work. Through a succession of dramatic scenes, it presents the writer's vision of a fictitious world, a world in which characters fight battles, weep, fall in love, reach for the stars, or die . . . a world that will be shaped and interpreted by a director and communicated through actors to an audience.

Because the script is of such fundamental importance to the success of any dramatic work, the serious director must understand certain classic principles that are essential to its effectiveness as well as the dos and don'ts of contemporary

---

[1] As quoted by John Brady in *The Craft of the Screenwriter* (New York: Simon & Schuster, 1982), p. 179.

[2] In Eric Sherman, ed., *Directing the Film: Film Directors on Their Art* (Boston: Little, Brown, 1976), p. 52.

scriptwriting. The classic principles of drama date back to Aristotle. They will probably apply with equal force to scripts written a thousand years from now.

In this chapter, the principles of effective screenwriting will be examined in five discussions. The study of screenplay *characters*, because of its special concerns for directors, will be explored in Chapter 5.

— A PERSON WITH A PROBLEM: a phrase containing the structural underpinnings of a screenplay's narrative

— DRAMA VERSUS REAL LIFE: a comparison of two worlds

— SCRIPT PROGRESSION: the pattern of growth found in all drama—in acts, scenes, and speeches

— CONFLICT: an examination of the four categories of this most essential dramatic ingredient

— DIALOG: a study of characteristics of both good and bad dialog

— CHANGING THE SCRIPT: dangers inherent in "fixing" a script during production

A PERSON WITH A PROBLEM

An understanding of drama is basic to the writing of screenplays. Some texts have made the process appear formidable, dealing in structural paradigms and such abstruse intellectual concepts as unity of opposites and thesis versus antithesis. I bring good news: The basis for drama is contained in a simple five-word phrase.

Before discussing that five-word phrase, I would like you to meet two contrasting characters. The first is Wendy Curtis, a good-hearted, open-faced young woman of twenty, secretary to her junior class in college and big sister to half the women on campus. Wendy's not really pretty, but she has a grin that fools you into believing she is Miss America.

You should be aware of one additional fact: Wendy has missed her period two months in a row.

Once sure that she is pregnant, she hesitantly tells her boyfriend, Rick. He is delighted. Because of financial problems, the two have repeatedly postponed getting married. Now, Wendy's pregnancy will be the catalyst for the big event. How do her parents feel about the pregnancy? They too are delighted. They have been urging the marriage for over a year.

The wedding is held in the school chapel, followed by the throwing of rice, the splashing of happy tears, and a whirlwind forty-eight-hour honeymoon.

After graduation, Rick finds a high-paying job with a California computer firm, the baby is born healthy and strong, and the couple lives happily ever after. In time, their son grows up to become president of the United States.

Before commenting on Wendy's story, I want you to meet another, equally fascinating character. Sam Gutterman is a slob: about fifty, unshaven, smelling of stale perspiration, tangled hair falling to his shoulders. Regrettably, Sam sometimes lies, cheats, and steals. He lives two flights up in a disgustingly dirty flat, debris and garbage littering the floor. People tend to avoid Sam because he doesn't bathe often and his breath reeks of cheap muscatel.

Sam glances out the window of his flat to see two police cars pulling up at the curb. The cops have come to arrest him. Why? A week ago Sam beat up a frail old lady and robbed her. Last night she died.

Sam runs out of his apartment, down the back stairs, and into the alley behind the building. As he reaches a side street, he stops in shock. A grim figure strides toward him, gun in hand. It is the old woman's husband. He has come to kill Sam.

What a terrifying dilemma! If Sam runs back into the building, he will confront the police. If he remains in the alley, he faces certain death at the hands of the old lady's husband. What can poor Sam do?

Does anyone care?

These two stories may not win Emmy awards, but they provide some valuable clues to the nature of drama. Let's examine them individually.

An audience will watch Wendy Curtis's story for a while because viewers like her. They can care about her. They can root for her. But gradually the audience will grow restless, then impatient, and then will turn her off. Why? Because Wendy's story simply doesn't have a *problem*. Without a problem no story in the world will hold an audience for long. Audiences love to worry. They enjoy suffering over someone else's grief. That's what dramatic storytelling is all about. Delighted, we settled back in our easy chairs to watch her wage battles against boyfriend and parents. But instead of creating difficulties, Wendy's pregnancy eliminated them. Everyone lived happily ever after and the story became boring.

In Sam Gutterman's story we witnessed the reverse situation. Sam had a life-or-death problem: police on one side, vengeful husband on the other. Did we worry about Sam? Did we suffer over his anguish? Not for a moment. Because Sam was a noninvolving, repellent character, it became impossible to root for him. We never experienced his terror; we didn't care what happened.

By studying the contrasting inadequacies of these two stories, we may extrapolate a pattern that will define drama's most basic structure. That pattern is contained in a simple five-word phrase, *a person with a problem*, which may be used to measure the essential dramatic characteristics of most screenplays.

The Person

What kind of person does a protagonist (hero) have to be? A person with whom we can become *emotionally* involved. (I emphasize the word *emotionally* because emotion is what drama is all about.) Such a person does not have to be a Wendy

Wonderful; most human beings are far from perfect. But protagonists must exhibit qualities that either make us want to identify with them or generate other involving emotions within us.

Wendy arouses *sympathy* and therefore viewers identify, actually becoming Wendy, suffering her problems, experiencing her triumphs. Sympathy is the emotion that most often creates audience involvement. A second attitude that allows identification is *empathy*, an understanding of the emotions that drive characters, awareness of the factors that cause them to act the way they do. Characters who inspire empathy may be less than sympathetic, but if we "know where they're coming from," we can relate to them.

A third attitude also evokes audience involvement—*antipathy*. Throughout history, drama has created audience involvement with despicable villains. Shakespeare's tragedies are filled with such figures (Lady Macbeth, Richard III, Shylock). We may not identify with such characters, but they certainly generate strong emotions within us. We dislike them; we hate them; we may even despise them. Such strong emotions are surely as involving as sympathy. When we hate screen characters, we yearn to see them punished, getting what they deserve. Certain stars have built careers from such audience involvement: Erich von Stroheim ("the man you love to hate"), Bela Lugosi, Vincent Price, Peter Lorre, Sidney Greenstreet and others.

Thus, the "person" in the five-word definition of drama must generate audience involvement through *sympathy*, *empathy*, or *antipathy*. The greater the degree of emotional involvement, the more the spectators will forgive any script (or other) shortcomings. Witness the many foreign art films with a minimum of plot but with captivating characters that compel us to watch.

As a director, you will discover that the audience's need for personal involvement has many practical applications. For example, if a scripted protagonist performs actions or makes speeches that in any way diminish spectator involvement, those speeches or actions should be carefully examined, rewritten, or perhaps excised. Occasionally directors begin a project with a "gray" character, one weak in terms of arousing audience emotions. If such a character cannot be enriched in the script, the director has a final recourse: to cast an actor whose appearance, personal magnetism, or apparent vulnerability will capture an audience's sympathy.

The Problem

Problems arise in scripts in two different ways. In the first, a protagonist blithely attends the chores of everyday life when BLAM!—a problem descends unexpectedly from the blue. Suddenly the protagonist's life disintegrates. He or she must struggle to stay afloat, attempting to escape or to destroy the problem. A kidnapping story demonstrates this pattern perfectly.

Mary Lou Hennessy walks home from school with a friend. A black limousine screeches to a stop at the curb, two heavyset men leap out, shove the friend aside, grab Mary Lou, throw her into the vehicle, and drive off. The problem has descended from nowhere. Inevitably it will get worse before it resolves. The movies *E.T., the Extraterrestrial*, *Poltergeist*, and *Jaws* are examples of this kind of problem, as are most detective stories.

When Steven Spielberg was asked what constituted good film drama, he answered:

> For me, it's someone—a protagonist—who is no longer in control of his life, who loses control and then has to somehow regain it. That's good drama. All of my pictures have had external forces working on the protagonist. In almost every Hitchcock film, the protagonist loses control early in the first act. Then he not only has to get it back, he has to address the situation. That theme has followed me through my films, too.[3]

The second kind of problem, far more common, is created when someone in opposition (an antagonist) tries to keep the hero from attaining an important personal goal. Cinderella wants to go to the ball to meet the handsome prince. Her stepmother and ugly stepsisters forbid her to go. They become Cinderella's problem, the obstacle she must overcome. In many of the old fairy tales the prince asks for the hand of the beautiful princess. If you recall, her father (the king) always requires him to perform some herculean task first, such as slaying the three-headed dragon: a dramatic obstacle to prevent the prince from reaching his goal.

Note that in both patterns the protagonist has a *goal*. In both patterns the protagonist must overcome inimical forces to reach that goal. A series of struggles ensues, a tug of war; the protagonist either achieves the goal or fails. Or the goal changes.

The nature of dramatic problems seems simple enough, yet experienced writers sometimes lose sight of the need for that simplicity. They become entangled in character relationships, complicated subplots, and parallel story threads. Somewhere during the writing process the primary thrust of a story is lost. Frequently, early drafts of screenplays become more literary than dramatic in form, wandering from episode to episode. As a result, the script loses focus; it loses clarity; it loses energy. Most importantly, it loses or diminishes the requirements of "a person with a problem."

One additional point concerning the problem: *Establish it early*. Producer–director Nat Perrin describes a funny little man with a black beard and black hat who carries in his hands one of those round, black bombs you have seen in comic strips. At some point in every show, says Perrin, the little man walks on stage carrying his bomb. He deposits it on the stage floor, lights a match, and sets the fuse spluttering. Until the fuse—your story problem—has been lighted, the

---

[3] *The Newsletter of Directors Guild of America* 13, no. 8 (July–August, 1988): 6.

audience is free to leave the theater. Once it is spluttering, the audience will remain seated, waiting apprehensively for the explosion they know will come.

Generally, motion picture audiences are more patient than those seated before a TV set. They have parked their cars and paid their admission; they will give you time to introduce your problem in a more leisurely fashion. But in television, when viewers sit with remote controls in their hands, you'd better introduce that problem quickly, usually in the first three or four minutes, before they press the dreaded button that turns you off.

When reading a script for a directorial project, consider its *person with a problem*. Can you become emotionally involved with a major character? Deeply involved? Does that character have a clearly defined goal? If not, the writer will have to crystallize that goal. And is some person or force trying to keep your protagonist from reaching his or her goal, thereby creating a problem?

DRAMA VERSUS REAL LIFE

Some theorists believe that the closer a dramatic work approaches reality, the more effective it becomes. The director should understand that drama and real life are significantly different. Screenwriters try to make audiences believe that they are watching reality. In truth, spectators watch a carefully devised illusion. Dramatic re-creation is far from literal; characters and events only appear to mirror life. If writers are genuinely creative, however, their scenes will contain the essence of life.

Several characteristics separate drama from real life. Among them: (1) *economy*, selecting only those incidents, characters, and speeches that are significant, (2) *logic*, which implies a certain dramatic inevitability, and (3) *progression*, the heightening of tension or climbing action that is intrinsic to drama.

Economy

Some years ago, director Craig Gilbert took a camera and sound crew to the Santa Barbara, California, home of Pat and William Loud. His concept: to do a cinema verité study of an actual American family, filmed in their home day by day. For seven months Gilbert's camera recorded each day's most significant events, beginning at 8:30 or 9:00 in the morning and continuing sometimes until 10:00 at night.

Gilbert was lucky. While his camera was present, two extraordinary events occurred. Pat and William Loud decided to end their marriage of many years, and their son Lance, who had been living in New York, returned home to shock his family with the revelation that he was gay.

In all, Gilbert filmed 300 hours. His crew ignored the humdrum routine of everyday life, the banal details that would have little audience appeal. Instead, they turned on their camera and microphones only for those incidents that promised to be interesting, truly representative or significant. The cream.

When the program was broadcast on PBS as "An American Family," the 300 hours had been edited down to 12. The director thus had a rare opportunity to air only his most dramatic material, eliminating 288 hours, or 96 percent. But the cream somehow had turned sour.

While Gilbert's exercise was hailed by some as a sociological milestone, it was generally flayed by critics. Ratings were minuscule. I personally saw only two episodes. They were so dull that I could barely sit through them. It is true that the show was never intended to be viewed as drama, but it still failed to generate any substantial degree of spectator interest, much less excitement or suspense, despite two major dramatic happenings.

Why was the material dull? Alas, real life frequently is. Because real-life incidents are sometimes *dramatic* does not make them *drama*.

Listen to conversations at your dinner table, on the bus, or in restaurants. You'll find that people repeat themselves endlessly, wander all around the subject, and bring in meaningless and unnecessary details, even when their stories are exciting ones.

A skilled scriptwriter practices dramatic economy in three areas: *dialog*, *characters*, and *scenes*. The writer pares away unnecessary or cluttering words of dialog, boiling speeches down to their essence. Because writers strive to create the illusion of reality, speeches retain the sound of everyday life but the excesses have carefully been deleted. The writer economizes with characters, eliminating those not essential. Examine almost any novel that has been adapted into a motion picture. The film inevitably uses only a fraction of the book's characters. The award-winning film *Kramer Versus Kramer* retained only one-fourth of the characters in the novel. The writer also economizes on the number of scenes. If a scene does not contribute a new plot development or reveal something new about a principal character, it probably should be deleted.

If the scriptwriter has not practiced these economies, you, as director, must. The time to eliminate excesses is before a show is photographed. Shooting a scene that will eventually be edited out of a picture wastes everyone's time and money. Unnecessary characters clutter up the show, taking attention away from central characters. Unnecessary dialog destroys the effectiveness of speeches, dissipating their energy and muddying their ideas (see examples later in the chapter). Wise directors get rid of the "fat" and use only the "lean."

Logic

The second characteristic that separates drama from real life is logic. Drama is structured logically; real life is not.

Ben Chapman urgently needs $20,000. Two weeks ago his daughter was injured in a hit-and-run accident, placing her eyesight in jeopardy. The $20,000 will fly her to Boston for needed surgery. Ben doesn't have even $1,000. The medical bills have depleted the family savings, and Ben has been out of work for months. Now, on his way home from the employment office, Ben sighs at the gross impossibility of obtaining such a large sum of money. As he walks, he spots a bulging manila envelope on the pavement. He stoops and picks it up, opens the clasp, and glances inside. What do you suppose the envelope contains? How did you ever guess? It contains *money*! Almost $21,000!

Because Ben is an honest man, he reports his find to the police and advertises for the owner in newspapers. No one claims the money and so Ben is allowed to keep it. His daughter has the operation and regains her eyesight. What an incredible stroke of good fortune!

Such a human interest story might appear in a newspaper but never in a carefully crafted dramatic script. Why not? Because the solution to Ben's problem did not emerge logically from the fabric of his story. The writer conveniently handed it to him. In ancient Greek tragedy, problems sometimes were solved in the last act of a play when a golden chariot descended from the flies (area above stage). Out stepped a god who produced a miracle, instantly solving everyone's problems. Such convenient solutions, deus ex machina (god out of the machine), generally are rejected by audiences, who regard them as cheating; the writer apparently could find no legitimate way of ending the story.

Because drama is logical, events that occur in the third act should grow out of elements established in the first two acts. Events occurring in the second act should grow out of material established in the first. The solution to a problem cannot descend conveniently from the blue sky. As with Ben's discovery of the manila envelope, such easy solutions smack of coincidence. Because coincidence defies the laws of logic, it plays a limited role in drama.

Is coincidence ever allowed? Yes, a writer may use coincidence on only two occasions in a script. First, coincidence may appear at the beginning of a story. Many of Alfred Hitchcock's most successful films began with outrageous coincidences. In *North by Northwest*, for example, the advertising executive played by Cary Grant called for a telephone in New York's Plaza Bar, mistakenly convincing foreign spies that he was a CIA agent, thus plunging him into a bizarre series of misadventures. Audiences accept such initial coincidences with the apparent rationale that had the coincidence not occurred, the story would never have taken place.

Second, coincidence is also accepted by audiences when it worsens the protagonist's problem. When a small boy fleeing through a labyrinth of caves accidentally bumps into one of his pursuers, the accident (coincidence) not only is acceptable but also is savored by audiences. The writer isn't taking shortcuts to solve a protagonist's problem. On the contrary, the boy is now in deeper trouble than before and the writer must work harder to extricate him. As I indicated earlier, audiences love to worry. Coincidences that give them more to worry about are gratefully accepted.

As a director, when you encounter a script in which the writer has used coincidence unacceptably to solve a problem, you must either *eliminate it* by solving the problem in a different fashion or *set it up* in such a way that the coincidence appears logical. You can set up a coincidence by laying the ground-work for it earlier in the script.

Let's take the example of Ben Chapman's convenient discovery of an envelope containing money. How can we make that discovery appear logical, an outgrowth of what has gone before? Perhaps by revealing how the money came to be there. If, for example, the writer has woven another thread into the fabric of the story, a thread in which bank robbers from another state had lost a portion of their booty, then the audience would be prepared for Ben's discovering it on a city street. In this circumstance, the robbers might read Ben's newspaper advertisement but would be afraid to come forward, fearing a police trap. Once a coincidence grows out of earlier plot circumstances, it can no longer be labeled coincidence.

SCRIPT PROGRESSION

One reason why a specific, definable goal is so important to an effective script is that a goal implies *movement*. Movement is an integral part of drama; it keeps spectators attentive as the story grows or builds in a series of struggles. With developments favoring first the protagonist, then the adversary, the problem con-tinually worsens.

When we try to reach the treasure at the top of a mountain, we first have to climb that mountain. Our movement is focused in a single direction: upward. When an adversary (the weather, our own fears, the abominable snowman) tries to stop us, the forward movement becomes a series of skirmishes, taking us closer to the treasure or forcing a retreat down the mountain. That dramatic movement has traditionally been visualized as a diagram resembling a jagged fork of light-ning, as shown in Figure 4.1. The ascending line (resembling our climb up the mountain) represents drama's rising action. A story builds in tension through a series of minor crises to a climax that resolves the protagonist's problem. The diagram indicates the development of three major crises: the classic three-act pattern. Most feature motion pictures are constructed in three-act patterns al-though there is usually no curtain or other definitive separation between crises.

CONFLICT

I have established that one of drama's primary ingredients is a *problem*, but how should that problem be dramatized? Through a series of struggles between pro-tagonist and antagonist, a continuing tug of war that seems to favor first one and

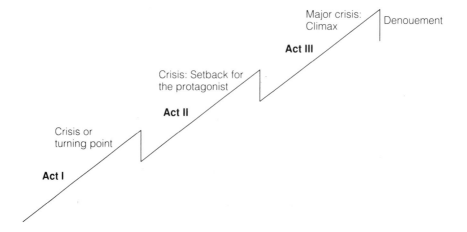

FIGURE 4.1   The ascending line of the lightning bolt represents drama's rising action
through a series of crises.

then the other. Such conflict is part of drama's essential nature. Although Chapter
2 discusses conflict as an element of entertainment and, accordingly, part of a
director's bag of tricks, we consider it here as a prime ingredient of drama. The
atavistic search for conflict is certainly one of the reasons for the theater's continuing appeal century after century, whether on stage, on a TV or movie screen, or
in a Punch and Judy show. Conflict lies at the core of drama: protagonist pitted
against antagonist, often in a life-or-death arena, with the spectator vicariously
experiencing the pains and thrills of battle.

Patterns of Conflict

Traditionally, conflict has fallen into three categories: conflict with others, conflict
with the environment, and conflict with self. With the release of so many fantasy
films in recent years I might add a fourth category that doesn't precisely fit the
others: conflict with the supernatural.

Of the four, conflict with others appears most often on TV and movie screens.
A sheriff in pursuit of a vicious desperado, a student bickering with her boyfriend, a boxer challenging a hated foe are all variations on the theme. Most well-
written drama includes more than one type of conflict.

Conflict with the environment usually is dramatized in terms of conflict with
nature or conflict with society. The sheriff pursuing the desperado changes categories when the scriptwriter sends him across a parched desert, fighting the
merciless sun, a dwindling water supply, and snakebite. The novel and feature
film *1984* ideally demonstrate the subcategory, conflict with society.

When the sheriff captures the desperado and begins the long trek home, the desperado reveals his humanity by saving the sheriff's life. (We've all seen this one, haven't we?) Now the sheriff faces a torturing dilemma. Can he turn the desperado over to a town that will certainly lynch him? Or must he betray the town's trust by allowing the desperado to break free? The story has developed into the richest kind of conflict, in which a protagonist must wrestle with his or her conscience—conflict with self.

Conflict: A Useful Tool

When scenes don't work, when they seem pale and vapid, they frequently lack the seeds of conflict. Some of you may find it difficult to believe that every scene is richer if it contains conflict. *Every* scene? Surely not *every* scene! I promise you: Conflict can only improve a scene, provided it has honest motivation. A scene between two people who completely agree with each other about everything probably should not contain conflict. But neither should a writer devise such an empty scene.

Conflict takes many forms. It may be overt, expressing itself in shouted words and violent actions. It may lie buried beneath the surface, simmering, steaming, adding pressure—coloring the words of the players or revealed only through a concealed gesture, fingers twisting paper or fists clenching. It may emerge as gentle teasing, as banter between a boy and a girl, mocking words that still express the language of love.

A detective format often creates problems for both writer and director, problems that may be resolved by adding a judicious bit of conflict. In the early stages of most investigative shows, the protagonist must gather clues. The process usually involves confronting people who may or may not shed light on the circumstances surrounding the crime. If the investigator secures answers quickly or easily, each scene is dull. A succession of such scenes would turn any show into a bowl of lukewarm jelly.

The investigator visits the murder victim's landlady. "Good afternoon, ma'am. I'm investigating the killing of Natalie Plum. Was she living here on July 7?"

"Yes, she was."

"Did you see her after that date?"

"No."

"Did she receive any visitors?"

"No. Never."

"Thank you very much."

End of scene. The investigator gets in his European sports car and drives off. An exciting scene? Suspenseful? About as suspenseful as watching whiskers grow.

If we make the investigator's job more difficult, the scene suddenly becomes interesting. Let's assume that the youthful landlady is attracted to the investigator and wants him to come inside and have a drink with her. But he's in a hurry. He doesn't have time for a drink. When he tries to obtain information, the landlady has trouble remembering. She invites him again to have a beer. The businesslike investigator refuses, hammering away with his questions. When the landlady continues to hem and haw, he finally realizes (he's a little slow!) that the only way he will get straight answers is to indulge this woman. Inside, over a beer and popcorn, he secures the information he needs; the landlady gets the companionship she needs, and they make a date to see each other again.

The detective then goes to a seventy-year-old inventor, seeking information. But the inventor doesn't have time for foolish questions. He must finish his invention today. He tries to shoo the investigator away. Our hero keeps pressing, humoring the flaky old man, and admiring his invention; eventually, he gets the information he needs.

He talks to a young child who has a vivid imagination. The boy has trouble separating reality from fantasy. Now the investigator must labor to find the truth, perhaps bribing the youngster with an ice cream cone.

Any of these scenes could be played in a dull question-and-answer pattern. Instead, they spark our interest through the addition of entertainment elements and colorful characters—and something more. Something far more significant than colorful characters. In each of the above situations I have devised a way for *conflict* to enter the scene. When the protagonist found his answers easily, the scenes were boring. When the going became *difficult*, the scenes became entertaining.

There's a major lesson here about drama and about life. Things easily obtained seem to have little value. Goals that we must struggle diligently to achieve, clawing our way past formidable obstacles, take on increased value. In any dramatic situation, make life difficult for your protagonist by keeping the goal elusive. Allow your audience the luxury of worrying. They'll love you for it!

Character Contrast

Characters with an identical point of view, the same ethnic, cultural, and political backgrounds, rarely argue. When they play scenes together, they agree with smiles on almost every issue. Such scenes inevitably are dull because they lack conflict. Both writers and directors should search for contrasting characters whenever possible because contrast tends to create richer scenes than similarity. Contrasting characters create texture through varying points of view. They create tension by making the scene's outcome problematic. They create entertainment through the sparks of conflict.

A marked contrast between leading characters brings such strength to any dramatic concept that it has formed the spine of countless stage plays and feature films as well as entire television series. Neil Simon's *The Odd Couple* is the prototype: the effete, meticulous, sensitive Felix Unger sharing an apartment with heavy-drinking, cigar-smoking Oscar Madison, a man of sloppy personal habits and animal appetites. No matter what situation a writer devised for the TV series, the odd couple faced it and struggled to solve it with conflicting attitudes. No matter what the plot progression, their opposing points of view placed them on a collision course, and stories inevitably erupted into open warfare.

Realizing that contrasting characters create dramatic tension, wise directors search for ways to build differing attitudes into characters in scripts and, whenever possible, to reinforce those contrasts in casting.

DIALOG

Because dialog is the most apparent of a script's elements, it generally receives more scrutiny than any of the others. Yet surprisingly, in spite of meticulous examination by writers, producers, directors, and actors, speeches in most motion pictures and TV programs are less than excellent. Such lack of quality undermines the illusion of reality that a production team works so hard to establish. It undermines spectator involvement with characters. If sufficiently off the mark, poor dialog can even invite embarrassed laughter from the audience.

Because dialog is such an overt expression of a character, aware directors give early attention to its effectiveness. If dissatisfied, they call for script rewrites prior to production, rather than wait to find out whether the words will fit comfortably into actors' mouths. Such directors understand the following critical dialog factors: economy, simplicity, characteristic speech, and invisibility.

Economy

In examining the differences between drama and real life, we noted earlier that economy is an essential dramatic principle. Nowhere is it more necessary than in the writing of dialog. When thoughts are restated, audiences become bored. When unnecessary words clutter a speech, ideas appear muddled. Experienced professionals examine dialog meticulously, stripping each speech to its essentials.

The fact that good dialog is spare, however, doesn't mean that you should remove words simply because they don't contribute to the plot. Often dialog helps create an understanding of a character or establish a mood. Such dialog should not be amputated in the name of economy.

Many television shows contain excessive dialog. Their writers probably have discovered that telling a story through dialog takes less effort than searching for visual ways to dramatize an idea. Such writers fall into the stage-play pattern in which characters tend to talk about offstage events. In the theater such descriptive dialog is necessary because limited stage space doesn't allow the proper dramatization of large-scale, physical events. In television and film, however, no such limitations exist. Good writers know that it's vastly more effective to show a piece of action than to talk about it. Dramatizing an action is vivid. For the viewer, seeing an event provides far greater impact than merely hearing someone describe it.

Occasionally, an event's effect upon a character becomes more significant than the event itself. In such circumstances, the character's description should certainly be dramatized. For example, when a newlywed's husband is killed in an automobile accident, she is devastated. Her broken description provides greater impact—certainly far more emotion—than a dramatization of the accident itself. Such a circumstance, however, is an exception. Generally, it makes good dramatic sense to show the event rather than talk about it.

Chapter 6 discusses the fact that dialog is often unreliable, that actions frequently are more revealing—and more trustworthy—than the words characters speak.

Simplicity

If you're like me, when you were writing essays in high school you tried to impress your English teachers by using as many big words as you knew—and perhaps a few you didn't know. (My particular favorites were *omnipotent* and *omniscient*. I managed to work them into every essay I wrote.) Although big words inevitably appear in high school essays and term papers, they don't occur often in everyday speech.

Listen to the conversations of your family and friends. How often do the words *omnipotent* or *omniscient* crop up? Not often. The words most of us use in everyday speech are short, simple, easy to say. They convey our thoughts easily, without getting in the way of meaning or calling attention to themselves.

Good writers are aware of this simplicity of speech. The characters they create generally speak in short, declarative sentences with one- and two-syllable words. If you have access to motion picture or television scripts read them carefully. (Many have been anthologized.) Note how simple the words are. Note that even the most profound thoughts can be expressed in everyday one-syllable words.

As a director, if you are unsure of speeches, if they seem false or manufactured, *read them out loud*. There is a great difference between dialog in novels and dialog in well-written scripts. The acid test is that the speeches must be capable of being spoken aloud. They must sound spontaneous, as if the thoughts they express had just occurred to the character. Stilted speech patterns, complex sentences, or difficult words will be immediately apparent when you read the words aloud. The solution to "unspeakable" dialog usually is to simplify it.

Most of us know people who delight in using long words. Some doctors, research scientists, and college professors use erudite or esoteric words as a badge of their professions. When these characters appear in a script, they may become valid exceptions to the rule of simplicity. They may. Among my acquaintances are about thirty college professors and half a dozen doctors. How do they speak? Mostly in one-syllable words. Pass a college professor on the street and he'll say,

"Hi! How you doin'?" just like everyone else. Face it: Most people, most of the time, speak in short, simple sentences of short, simple words. Even doctors, scientists, and college professors.

Characteristic Speech

Speech patterns vary, depending on where the speakers come from and the socio-economic world they live in. Most of us as we grow up tend to echo the speech patterns of our parents. If we live in the South or the East or the Midwest, our speech reflects the phrases and patterns and intonations of the area. When we move away to college, our speech is affected and partially transformed by our new friends and the speech patterns of the college area. When we enter the business community, our speech changes again, its vocabulary enlarging and absorbing the technical jargon of our new professions.

Each person has a way of speaking that is unique. One person watching a touchdown on television will exclaim, "Dynamite!" Another will say, "Awesome!" A third might mutter a four-letter word. Members of each subculture create their own speech fads, invent words, and change meanings. Many of these fads eventually drift into the mainstream of our language.

In a well-written script, each character has a distinctive speech pattern, consistent from scene to scene. Scripts in which speeches can be interchanged between characters usually reflect an inept or insensitive writer.

When you read a script dealing with racetrack touts or oil well drillers or bacteriologists and their dialog seems drab and colorless, lacking the pungent smell of reality, what can you as a director do? First, you can prevail upon the producer to hire another writer who either understands the characters or is willing to research them. Second, you can enter that world yourself, absorbing its color and flavor, talking at length with its inhabitants, studying their attitudes, listening to their speech patterns and vocabulary, familiarizing yourself with their lifestyle. Take along a tape recorder. Bring back conversations that will be helpful to your actors. Such research contributes far more than accurate dialog. It provides you with the ambience of your character's world, helping you to re-create that world with a vividness possible only from firsthand observation.

Lacking the time to conduct in-person research, you can visit the public library and read about the world your characters live in. Books and magazines often provide valuable insights. If you find particularly revealing material, make photocopies that you can give to cast members.

Invisibility

Directors, writers, and actors labor to make audiences believe that they are watching an actual event, to make them become emotionally involved in that event. The moment spectators become conscious of contrivance, either from intrusive music or a flashy directorial fillip or overacting, the illusion promptly disappears. Writing that calls attention to itself similarly shatters the illusion of reality. A writer showing off through use of "clever" dialog simply reminds the audience that the words are not the spontaneous, honest expressions of a character but speeches contrived by a writer.

What about the comedies of Mel Brooks and Woody Allen? What about such flashy comedy-adventure films as the James Bond series? The humorous, clever, or bizarre dialog in those films certainly calls attention to itself. Of course. Those shows seldom try to convince audiences that they are watching reality. Spectators usually remain aware that what they are watching is an exercise in theatricality, fun and games. Reality has been abandoned. In comedy, anything that gets a laugh is justified. In shows that seek genuine audience involvement, however, the production elements must remain invisible.

Bad Dialog: What to Do with It

The word *corny* is often used to describe bad dialog. It's an all-inclusive term that generally indicates artificial, melodramatic, or old-fashioned speeches. Usually there is a specific, definable reason why a line sounds corny. Directors working with writers have an obligation to define for them as specifically as possible why a scene doesn't work or what is wrong with its dialog.

The following short scene is guaranteed to win prizes at any film festival for "Most Disastrous Dialog of the Year." See if you can determine why each of the following lines seems so corny.

INT. APARTMENT—FULL SHOT—DAY

Greg watches a soap opera on television, sipping occasionally from a bottle of beer. Suddenly the door swings open and Lorna strides in angrily. Greg rises in surprise.

> GREG
> Lorna! What are you doing home from work two and a half hours early?

> LORNA
> I'm leaving you, Greg. I'm through, finished. I've had it up to here.

> GREG
> What do you mean, leaving?

> LORNA
> Did you think I didn't know about Julia Murphy? About the apartment over on North Elm Street that you two have been sharing for thirteen months?

She snaps off the television set angrily.

                              LORNA
                           (continuing)
How long did you think you could pull the wool over my eyes?
All those weekends you said you were visiting your mother!
Well, your mother has been in Europe for months, Greg, so
you couldn't have been visiting her!

                              GREG

Lorna, <u>listen</u> to me....

                              LORNA

What for? So you can lie to me again? Don't you understand,
Greg? I hate you. I can't stand the sight of you.

She takes a gun from her purse.

                              GREG

Where did you get that gun?

                              LORNA

I'm going to kill you, Greg. I'm going to shoot you. I'm going
to put a bullet through your heart. You're just so much excess
baggage now.

                              GREG

Lorna, in the name of God . . .

                              LORNA

Goodbye, Greg.

                              GREG

Lorna . . .

She fires the gun and Greg collapses to the floor.

An incredibly bad scene with incredibly bad dialog. Why is it so bad? Let's
examine the scene line by line and determine the reasons—if you think you can
stomach it again.

First line: "Lorna, what are you doing home from work two and a half hours
early?" Why does the line seem artificial? Because the writer has added unnecessary words just for the audience's benefit, to feed information. Although it may
be significant that Lorna is home early, the two and a half hours are totally
gratuitous. Would Greg (or anyone) honestly say that? Never.

In any script there is always certain information that an audience requires in
order to understand what the story is all about. Such information is called **exposition.** A skilled writer works it into dialog unobtrusively; spectators never become aware that they are being spoon fed. Indifferent writers dump exposition
on spectators heavy handedly. The worst offenses are scenes in which two characters state facts that both already know, facts that appear in dialog solely because

the author needs to explain them to the audience. That's why Greg asks, "What are you doing home from work two and a half hours early?" Wouldn't the line be more honest if he simply said, "Hi, you're home early."

Next line: "I'm leaving you, Greg. I'm through, finished. I've had it up to here." One of the major lapses in any kind of writing (of course) is the use of hackneyed figures of speech. This scene contains three or four of them. "I've had it up to here" is a particularly stomach-wrenching example. Others include "How long did you think you could pull the wool over my eyes?" and "You're just so much excess baggage." All of those lines are excess baggage. Clichés.

Lorna's speech contains other problems. As discussed earlier, effective dialog is usually *economical*. In this speech Lorna expresses the same thought in four different ways. The line would be vastly improved if she stated it only once: "I'm leaving you, Greg." Also, just as drama builds from scene to scene and act to act, so does healthy dialog build from the least important thought in a speech to the most important. (For example: A few of the soldiers had been wounded. A few had lost limbs. A few were becoming dust on some forgotten battlefield.) Lorna's speech does not build at all. It merely states and restates a single idea.

In another Lorna speech later in the scene, her lines actually diminish in importance. "I'm going to kill you, Greg. I'm going to shoot you. I'm going to put a bullet in your heart. You're just so much excess baggage now." Such a reverse progression—building downward—sometimes triggers audience laughter. Lajos Egri recalls a classic line that warns against murder because it may lead to drinking, which may lead to smoking, which may lead to nonobservance of the Sabbath![4] Reversing the normal dramatic growth makes excellent humor—but dreadful dialog!

Next line: "What do you mean, leaving?" Poor Greg doesn't have a single meaty speech in the entire scene; they are all **feed lines**, relatively meaingless words that have been inserted for the sole purpose of breaking up Lorna's speeches. We could erase almost all of Greg's lines and the scene would lose nothing. Occasionally every writer must invent words in order to keep a character "alive" in a scene, to prevent the dominant character's speeches from becoming a monolog. But "What do you mean" lines are a most tired and feeble invention.

Next: "Did you think I didn't know about Julia Murphy? About the apartment over on North Elm Street that you two have been sharing for thirteen months?" More dishonest exposition, words inserted solely for the audience's benefit. All Lorna has to say is, "I know about Julia."

Do you notice a pattern emerging? To improve each of these speeches the writer must prune, make the lines simpler and more honest, apply the first rule for writing dialog: economize.

As you read the scene, you may become conscious of a pattern that many inexperienced writers fall into—using character names too frequently. Do you

---

[4] *The Art of Dramatic Writing* (New York: Simon & Schuster, 1946), p. 242.

repeat the names of friends each time you address them? Of course not. It would sound ludicrous. Yet in the above scene, Lorna and Greg call each other by name in almost every speech.

"I'm going to kill you, Greg."

"Lorna, in the name of God . . ."

"Goodbye, Greg."

"Lorna!"

Names are essential in one place: when a character is first introduced. The audience needs help. Who is this new character? What is his or her relationship to characters we've already met? Viewers are grateful when someone pins a label on the new character by calling him or her by name. Clues to the character's identity—"Bob, I want you to meet my mother" or "Mom, this is my boyfriend, Tim"—will orient the audience.

Two final points may be helpful. The first concerns Greg's line, "Where did you get that gun?" Aside from being rather stupid (who cares where she got the gun?), the line is significant because it is a throwback to an earlier era. Today's TV viewers are generally too young to recall radio plays in which dialog had to describe what was happening because audiences couldn't see it. Radio drama was one dimensional: sound only. It forced audiences to use their imaginations; they created dramatic action, scenery, characters' faces, and costumes in their minds. But the writer had to give the audience clues. Therefore, the characters in radio plays used to utter lines such as "What's that on the floor? Good Lord, it's a body!" And "Darling, your face is all bruised. Have you been in a fight?" and "Where did you get that gun?"

Radio lines, of course, are unnecessary in a visual medium. Yet, surprisingly, they creep into scripts again and again. When they do, use your blue pencil.

Finally, much of the scene's phony melodrama arises from the fact that its dialog is "on the nose." **On the nose** is a phrase industry professionals use to indicate dialog (or other dramatic aspects) that is blunt and heavy handed. When a line is stated in the most obvious way, allowing the viewer no room for interpretation or participation, it's on the nose.

When a scene is as melodramatic as the Lorna–Greg confrontation, you can bring a greater degree of reality to it—and avoid being on the nose—by "playing against" the melodrama. The concept of **playing against** something simply means taking a different dramatic tack, going to the opposite extreme. For example, keeping the dialog simple and honest would undercut the potential melodrama in the above confrontation, making it more believable. So would the addition of wry or sardonic humor.

As a director, you can play against the thrust of a tragic scene merely by changing the attitudes of the players. If you allow your characters to wear mournful faces, feeling sorry for themselves, your scene will instantly turn to soap. If you play against the tragedy, if your characters pretend that the sadness doesn't exist, if they play it lightly, making jokes to cover their inner pain, then your scene will become extremely moving.

CHANGING THE SCRIPT

What do directors do if they're stuck with a bad script? Do they "fix" it on the soundstage?

No.

They do exactly what Sara did in Chapter 1. They talk to the producer. They communicate their concerns, seeking corrections before production begins. Solving problems on the soundstage can be enormously expensive because it takes time away from production. More importantly, hurried changes can be dangerous.

Most scripts that go into production have been examined repeatedly by the producer and writer (and often by many others). Before a director receives a script, it usually has gone through a minimum of five or six steps of appraisal and evaluation. Lines of dialog that appear to be of little value may actually be essential to explain a plot point or to set up a later action. Changes on the soundstage, approved by the director in the heat of battle, may overlook such long-range plot requirements. In television the script has probably been checked and rechecked by network censors and production executives. Changes on a soundstage risk violating one of their taboos.

When actors change uncomfortable dialog, it is often replaced by a cliché. Clichés are easy to say; they roll off the tongue. But they represent bad writing and weaken a film's effectiveness. Good directors must be alert to such changes by actors.

But let's face it, there are occasions when a scene doesn't work, when words or actions must be changed. If you're the director, what should you do? My advice: Get on the phone. Make the producer aware of the problem. If the writer is close at hand, bring him or her to the soundstage. Solve your problems *with* them, not in spite of them. Film and television production is an ensemble effort. Don't try to be a hero.

A DIRECTOR'S CHECKLIST

In examining a script for any directorial project, ask yourself these questions:

1. Is the concept original and fresh or is it clichéd? A genuinely fresh concept compensates for a dozen problems. An original concept justifies all of the work necessary to study, discuss, rewrite, and polish otherwise second-rate material. Stated another way, is the story worth telling?

2. Does the script contain a protagonist you can care about, root for, become emotionally involved with? Can you find a way to increase the depth of spectator involvement?

3. Does the protagonist face a single, major problem? Does the problem emerge from a powerful, perhaps frightening antagonist?

4. Is the element of conflict present in each scene and as an undercurrent (tension) throughout the script?

5. Does the story (and tension) continually develop? Does the script contain unnecessary scenes that simply restate plot or character points? If so, delete them.

6. Does the dialog meet the chapter's standards of economy, simplicity, vernacularism, and spontaneity? Does it fall into any of the traps described in the bad dialog scene?

CHAPTER HIGHLIGHTS

— Because the success of any dramatic project depends in large part on the effectiveness of its script, aware directors understand the classic principles of dramatic construction.

— *A person with a problem:* This convenient five-word phrase contains the essence of a screenplay. The person must involve audiences emotionally. The stronger the involvement, the more the viewers will care about a protagonist's goals or problems. Audience involvement may emerge from feelings of sympathy, empathy, or antipathy.

— The problem comes into a drama in two ways: (1) It descends from the blue unexpectedly, creating havoc in the hero's life, or (2) it results from someone's attempt to keep the hero from reaching a worthwhile goal. Note that in both patterns the problem is goal related. In the first instance, the protagonist's goal is simply to escape from the calamity.

— Drama traditionally builds in tension. The hero's problem worsens from act to act and audience apprehension grows. This fundamental climbing, building, growing action is central to drama.

— Drama is not reality; it is the *illusion* of reality. Skilled writers and directors convince audiences that the illusion is real. Two factors separate drama from real life: economy and logic. Real life is frequently filled with repetition, irrelevance, and meaningless detail. Drama avoids boredom by selecting the meat of a story and trimming away the unnecessary fat. Real life is frequently illogical and filled with coincidence. Drama, in contrast, builds into its structure only those elements necessary to its growth and resolution.

— Fundamental to drama, *conflict* is the natural manifestation of the protagonist's problem. When some antagonistic force tries to prevent him or her from achieving a goal, conflict is inevitable. Such struggles take one of four forms: conflict with others, conflict with the environment, conflict with the supernatural, and conflict with self. By building contrasts into characters—contrasts in race, background, or point of view—directors can ensure that conflict will inevitably emerge.

— Good dialog generally is *simple*. Regardless of their professions or life-styles, most people speak in short sentences of one- and two-syllable words. Good dialog is *economical*; it is not repetitive; it contains no unnecessary or irrelevant detail. Good dialog is in the *vernacular* of the character who speaks it. Good dialog does not call attention to itself.

— Dialog does not replace action. Skilled writers dramatize events rather than allow characters to talk about them. For audiences, witnessing an action usually has more impact than descriptive words.

— Bad or corny dialog results from clichés; awkward or inept exposition (giving information to audience); "on the nose" lines that are blunt, obvious, and heavy handed; "radio lines" that describe what viewers can see; and a failure to build speeches effectively.

— Good writers try to "play against" melodrama or tragedy to keep scenes from becoming corny or soapy. They take a different approach to dialog from the expected, perhaps playing light or humorous dialog in a tragic scene and simple, honest dialog in a melodramatic scene.

— Directors should try to correct script problems prior to production. Changes made during production can be both expensive and risky.

PROJECTS FOR ASPIRING DIRECTORS

1. Find an appropriate *person* for each of the following problems:
   a. The person must testify against someone he or she fears.
   b. Earlier in life, the person committed a serious indiscretion. Now the past incident suddenly emerges, threatening to destroy the person.

2. Find an appropriate *problem* for each of the following:
   a. An idealistic teenager blessed with loving parents
   b. A scientist who has discovered a cure for cancer
   c. A police officer nearing retirement

3. A real challenge: Write a short scene (two to three pages) that contains all four kinds of conflict:
   a. Conflict with others
   b. Conflict with environment
   c. Conflict with self
   d. Conflict with the supernatural

4. List six clichés you have heard recently and then find a fresher way of expressing each.

5. A vernacular exercise: Write two versions of a telephone conversation as it would be spoken (a) by a prostitute and (b) by a clergyman. In each, the character accuses a beloved brother of cheating her or him out of an inheritance. Note: we hear only one end of the conversation; we do not hear the brother.

CHAPTER

5

FINDING
THE CHARACTERS

The director's concept of a character begins with skeletal clues provided by the writer. In time, as that concept is discussed, evaluated, reshaped, and enriched by the director and actor, the skeleton acquires muscle and sinew, flesh and blood. On the soundstage, before film or video cameras, it acquires life.

Where do directors find the elements that transform skeletal clues into a dimensional character? Initially, in the same place that actors do: the reservoir of their personal experience. When a director reads a script, its characters stir vague memories, perhaps consciously, perhaps unconsciously, of everyone in the director's past life, real or imagined, who might somehow relate to the character in question: family, friends, casual acquaintances, characters in books or magazines or from the stage or screen. Sometimes only fragmentary characteristics spring to mind: a way of walking, a gesture, a facial expression. Imagination will synthesize both characters and characteristics, rearranging them kaleidoscopically into possible images of the scripted character.

This chapter will discuss ways in which directors build upon those imagined images, giving them dimension through personal notes, discussion with actors, and rehearsal.

You imagine yourself the person that the author has written, and you sink yourself into that. Then out of your mouth comes, perhaps, quite a different voice.

Lynn Fontanne[1]

Specifically, we will consider:

— A DIRECTOR'S NOTES: exploring significant aspects of a character's background

— THEME/SPINE: defining the character in terms of drives, needs, and goals, and relating character spines to story spines

— BEATS: determining that most scenes consist of smaller units, subscenes called *beats*. Understanding the content of beats helps both actors and directors in staging a scene

— HOW CHARACTER IS REVEALED: examining five ways in which audiences come to understand the true nature of a character

[1]*Actors Talk About Acting*, eds. Lewis Funke and John E. Booth (New York: Avon Books, 1963), p. 27.

▬ WEARING MASKS: discussing characteristics hidden from view that audiences must piece together from clues

▬ BUILDING A PERFORMANCE: eliminating tension; acting techniques that create believability and power

▬ TRANSFERRING CHARACTERS TO TAPE OR FILM: examining rehearsal patterns and problems that sometimes arise between actors and directors

A DIRECTOR'S NOTES

Before attempting to define characters, a director logically must study the script, determining the role of each character in relation to the screenplay's theme or central meaning. Serious directors often prepare pages of notes for themselves; the very act of putting words on paper helps to crystalize the goals and drives of primary characters. Such notes may include a wide spectrum of ideas and fragments: imagery that will help actors to find characterizations, props that reveal hidden character aspects, a background incident that may give a characterization richness.

To the best of my knowledge, directorial notes for a film or videotape production have never been published. But excellent notes for stage plays have been published occasionally and offer students a rare opportunity to gain insight into the thought patterns and procedures of topflight theatrical directors.[2]

Skilled writers usually prepare a few pages of **back story**, a history of each major character, before beginning a script. This exploratory material provides a comfortable familiarity with characters—an understanding of the basic emotions, attitudes, and goals that will help bring them to life. A director should do likewise for the same reason.

Here's how Academy award–winning director–writer Frank Pierson defines such a back story.

> What is needed is a biography of feelings: How does she feel about being rich or poor, having an old car or a new car?. . .
>
> It isn't important that she graduated from Radcliffe in 1968 summa cum laude or that her mother died in Hingham, Massachusetts, in 1978 and the whole family was there. What's vital is how she *felt* when her mother died and that she hated going to the funeral because her father would be there.

---

[2]See *Directors on Directing*, eds. Toby Cole and Helen Krich Chinoy (New York: Bobbs-Merrill, 1963).

It isn't important that she was fearful of her father because he abused her when she was young. What is important is that what she fears is the conflicted mixture of feelings that are aroused by her father's presence. This is a sick and sad and confused character, but you only know in what ways she is confused and sick and sad if you know how she feels.[3]

In his *Art of Dramatic Writing*, Lajos Egri discusses a dialectic approach to characterization in which characters change and grow because of contradictions in their nature or environment. Egri divides the study of a character's background into three categories: physiology, sociology, and psychology.[4] Directors might do well to use those same convenient categories in searching for events that shaped a character's feelings: *physiology*, a character's physical appearance; *sociology*, a character's place in society and the family; *psychology*, such internal values as temperament, moral code, personality, fears, drives, sense of humor, and imagination (or lack of it). For directors, this last category may prove the most valuable.

From such detailed notes you can develop a comfortable familiarity with your major characters. If the character is a young adult, you might ask such questions as: What kind of childhood did she have? What were her earliest fears? Was she obedient? Rebellious? How did she get along with brothers and sisters? Was she ever abused? Was she successful in school? How much schooling did she have? Did she go to church? Did she have many friends or was she a loner? When did she first have sex?

By answering such questions, you can develop dimension for each major character. A conscientious actress will also explore the character, discovering personal answers from her own reservoir of experience. Her answers will necessarily differ from yours, yet surprisingly often major background elements will coincide. It is not important that they do. The actress will test you (perhaps not consciously) to find out how thoroughly you understand the character. You will win her respect once she realizes you have done your homework.

Usually it is wise for a director to meet each star individually before the start of a project. Such a meeting over lunch or coffee gives each person a chance to learn what ideas the other has developed about the character. If their concepts are miles apart, there is still time to discuss and rethink. Such a discovery on the first day of shooting could be disastrous, with the resulting conflict consuming hours of valuable production time.

Experienced directors are seldom arbitrary about their concepts of characters. They respect good actors and welcome their input. Sometimes contributed concepts are so rich, so creative, so exciting that directors discard their own ideas,

---

[3]Frank Pierson in Alan A. Armer, *Writing the Screenplay—TV and Film* (Belmont, Calif.: Wadsworth, 1988), p. 138.

[4]Lajos Egri, *The Art of Dramatic Writing* (New York: Simon & Schuster, 1946), pp. 49–59.

changing scenes and dialog to accommodate the new direction. But if an actor's concept is misconceived, based on insufficient thought, then the director must make that clear. Without being abrasive, the director can suggest a rereading of the material and perhaps some alternate ways of developing the character.

Occasionally you will find actors who are unwilling to commit themselves too early to a characterization, requiring time to let a role simmer in their subconscious. Such hesitation becomes a problem only when the actor remains indecisive as the production date approaches. Then the director must use persuasion, authority, or gentle arm twisting to force a commitment.

THEME/SPINE

The central meaning or message of a screenplay is called its *theme* or **spine**, identical to what acting teacher Konstantin Stanislavski referred to as a *super-objective*:

> The central meaning of a play very often is stated in terms of the main action or goal of the principal character or characters. Before the director can crystallize his interpretation, he will need to analyze each character in the play, his motivation, and his relationship to all the other characters.[5]

Directors trained in the Stanislavski method will try to find verbs that express the basic action or struggle of their characters. They will discover that the spine of their screenplay relates closely to the basic wants, needs, drives, or goals of its characters. For example, in *Fatal Attraction* the goal of the protagonist (Michael Douglas) is to prevent the destruction of his family. The spine of the film may be stated as the struggle to avoid the consequences of sin. ("Whatsoever a man soweth, that shall he also reap.")[6] Note that the two spines are directly related. We should understand that spines may vary in definition, depending on individual interpretation. The director of a film, for example, may interpret the underlying message of a screenplay totally differently from the writer.

The spines of other major characters in *Fatal Attraction* bear a direct relationship to the protagonist's spine. This pattern is true in almost all screenplays. Thus, the protagonist's wife ultimately unites with him in fighting the antagonist

[5]David W. Sievers, Harry E. Stiver, and Stanley Kahan, *Directing for the Theatre*, 3d ed. (Dubuque, Iowa: W. C. Brown, 1974), p. 43.

[6]Gal. 6:7.

(played by Glenn Close) who threatens to destroy their family. The antagonist's spine is in direct opposition to the protagonist's: to destroy his family so that she can expunge personal pain—and perhaps gain the protagonist for herself.

In Bernardo Bertolucci's *The Last Emperor*, the protagonist seeks to regain his emperor status, foolishly ignoring reality. That's his character spine. The story spine might state: "Foolish pride causes its own destruction." In *Star Wars*, the story spine might state "Good ultimately overcomes evil." Luke Skywalker's character spine: "To rescue the princess from the forces of evil." Darth Vader's spine (in direct opposition): to prevent Skywalker from rescuing the princess. In every case, the story spine relates directly to the spine of the central characters.

Directors can understand screenplay characters if they know what they want, their drives, urges, needs, desires, motivations, and goals—not just in life but, more specifically, within the parameters of a screenplay. Writer–director Robert Towne (*Chinatown*) once stated that you can understand a character if you know what he or she fears most. Towne's premise may express the character spine concept in reverse. That is, characters fear most that their fundamental needs will be defeated.

Many of the best directors were once actors. Because of this practical, behind-the-footlights experience, these directors understand an actor's needs, the sudden panic when a line is forgotten, the hunger for approval, what it feels like to be directed. Such experience makes it easy for directors to reread a script from the perspective of the characters, as if they were playing each of the parts themselves. Such carefully focused reading can be enormously helpful in finding the spine for each character.

BEATS

Many actors, especially those trained in the Stanislavski method, break scenes into smaller building blocks called *beats*. They understand that most well-written scenes consist of a number of dramatic units, each of which usually builds to its own small climax. Beats are scenes within scenes; each new beat usually is marked by a change of direction in dialog or action. Sometimes beats end or begin with a character's entrance or exit.

What is the purpose of dividing scenes into beats? For actors, such analysis often helps to define changes in emotion or attitude. Beats help actors to define their emotional progression through a scene, from (say) humor to concern to anger.

For directors, defining beats helps them understand the real significance of the scene itself. Many directors actually *label* each of the beats. For example, (a) John is unaware of Debbie's anger (unawareness), (b) he tries to laugh it off (refusal to believe), (c) he finally fights back (counterattack), (d) she realizes she has hurt him (reconciliation). Each actor defines beats from the perspective of his or her character. Their definitions, therefore, may differ considerably from those of other actors. And the director, viewing the scene from a more objective perspective, may define beats differently from any of the actors. To demonstrate how subjective such a choice is, another director might define the beats differently from the first.

Defining beats often helps directors in the physical staging or blocking of scenes. A new beat suggests a different physical relationship between characters; it may, therefore, suggest movement by one or more of the characters or some physical adjustment to the change in relationship. Sometimes a new beat requires a new master camera angle. You will find additional discussion of beats in Chapter 6 as well as a practical demonstration of how beats help directors to determine blocking.

HOW CHARACTER IS REVEALED

Once director and actor have settled upon a character concept—or at least its rudimentary outline—they must consider how that character will reveal itself to an audience. Realistically, much of this character revelation happens almost automatically. It springs from an awareness in the actor's mind, a sense of *being* that character, experiencing that character's thoughts and emotions. Audiences form impressions of characters from the following overt information:

1. The character's entire physical aspect
2. The character's traits and mannerisms
3. What the character does or does not do
4. What the character says or does not say
5. How other characters react

Physical Aspect

In their scripts writers rarely provide more than a phrase or two describing characters. Frequently such descriptions are limited to name, age, and general appearance (for example, Elaine Garibaldi, 22, thin, painfully shy). Sometimes writers throw in clues to the essential nature of a character, but not often. Why not a more complete character analysis? Because writers correctly prefer to let a character's *behavior* provide that information. Moreover, a minimum of descriptive data allows considerable creative freedom for producer and director in casting and in delineating the character. The fact that writers haven't described every pore and wrinkle by no means implies that they haven't done their homework. The best ones prepare pages of detailed notes on each major character before beginning a script.

From the script's spare beginning point, the director adds coloration based on his or her examination of the character. Such coloration might concern the character's appearance, posture, dress, manner of walking, personal history, moral code, personality, hairstyle, thought processes, and speech patterns. These factors are affected by the physical appearance of the actor chosen to portray the character and by that actor's creative input. Eventually, actor and character will merge: The actor will take on the qualities and traits of the character; the character will subtly change to accommodate the style, manner, and appearance of the actor.

From repeated script readings and creative imaginings, the director forms mental pictures of all major characters, in considerable detail and covering wide-ranging aspects of their lives. While such minutiae may appear ludicrous here, it will prove enormously helpful in talks with the art director, the set dresser, and those in charge of makeup, hairstyling, wardrobe, and transportation.

Are the characters fat? Sloppily fat or just nicely padded? Are they thin? Hungry thin or elegantly slim? If a male character, is he bald? Is his hair neat or shaggy, long or short? Does he wear an afro? Does he have a beard or a mustache? If either, is it neat and trim or long and bushy? Does he wear spectacles? A hearing aid?

What kind of clothing does he wear? High fashion or old-fashioned? New or old, clean or dirty, freshly pressed or rumpled? Are his shoes scuffed or are they shined? Wing tips or moccasins? Are there holes in the soles? Does he wear a necktie? If so, is it wide or narrow, conservatively striped or covered with gaudy flowers?

What kind of car does the character drive? New or old? If old, has it been meticulously maintained or has it gone to seed? Is there a bumper sticker? What does it say? Is there a good luck charm hanging from the mirror?

What kind of commentary would the character's bedroom provide? Is the room tidy or are clothes thrown on the floor in disarray? Is the bed made? Are there stuffed animals on it? What kind of pictures hang on the wall: Picasso prints, *Playboy* centerfolds, or rock posters? Are there textbooks or magazines on the nightstand? What magazines: *National Geographic* or *Penthouse*? Are there pipes or cigarettes on the dresser? Ashtrays filled with cigarette butts? Is there stereo equipment? What kind of music: easy rock, big bands, or classical? What color is the room painted? Are there lace curtains?

Answers to such questions provide definitive clues to the nature of a character. Audiences cannot read the script. There is no godlike narrator to describe or define the qualities of a show's characters. Instead, audiences must play detective and interpret the characters from the information they are given. Piecing the clues together will form a composite picture, a mosaic that the director–actor team has carefully contrived.

Traits and Mannerisms

Traits and mannerisms acquire significance because they often reveal inner feelings or thought processes not always explained in dialog. In many cases, such feelings are actually *concealed* by dialog. (You will find an elaboration of such actions in Chapter 6 under the heading "Business.") Society has taught us that displaying our feelings is a sign of weakness. We therefore go to elaborate lengths to remain "cool," covering our emotions with a smile or a glib phrase. But inevitably we reveal ourselves, sometimes overtly, sometimes through small, virtually undetectable actions.

By studying the actions of people around you, you will become sensitive to the meanings of many of their mannerisms. What does it signify when people constantly polish their glasses? What do they reveal when they keep glancing at themselves in mirrors or combing and recombing their hair? Veteran film director Richard Fleischer recently pointed to *awareness* as a key characteristic of successful directors.

A director is basically an observer and has to be aware of everything that's going on around him, mostly in terms of human behavior. You have to be an observer of the current scene and watch people: how they behave in certain situations. You must be aware that there is a body language, that there are awkward pauses (in conversations).

I think you can develop this awareness if you don't become too involved with yourself. In a social situation you can watch the byplay between people in a room, see the little subtle changes that happen when people suddenly become hurt . . . or what makes them laugh . . . or who goes through a door first. Sooner or later, in some script, you're going to be reproducing that situation.[7]

As you become a habitual people watcher, you will discover the foot tappers, the fingernail biters, the teeth grinders, the frightened few who hide behind their hands or dark glasses, the lip chewers, the hair twirlers, the blinkers, and others. Although many of these mannerisms indicate some form of tension, others imply more significant mental states. Photographs taken during production of a TV movie in Utah show me repeatedly standing with my arms crossed protectively in front of my chest, hands resting on shoulders. The pose, totally uncharacteristic, revealed that I was unconsciously shielding myself from the world, from the harrowing pressures of an extremely difficult and problem-filled project.

Become aware of the way people walk. Tired people walk differently from excited people; old people usually walk differently from young people. Become

---

[7]From an interview with author.

aware of how people talk. Do they stammer? Do they speak quickly, words rushing out in a torrent? Do they speak carefully, examining the texture of each syllable? Do they speak loudly, as if to intimidate? Or do they speak quietly, timidly, afraid someone might take offense?

Action or Inaction

We learn much about characters from the actions they perform. As you will learn in Chapter 6, such actions are often a more accurate barometer of feelings than words are. Somewhat less obvious is that a *lack* of action can be equally revealing.

When a man desperately running for his life stops to help someone, at great risk to himself, we recognize that here is a character worthy of our respect. If he encounters children urgently needing help and walks on by, pretending not to see them, we regard him with contempt because he obviously thinks only of himself. If he encounters the children and hesitates, torn, aware of his own vulnerability, starts to help them and then suddenly hurries off, that indecision also creates a picture of the man and the pressures that weigh on him. Newspapers in recent years have reported countless stories of women being attacked, screaming for help from passersby who refuse to acknowledge them. Such lack of action eloquently describes the nature of such passersby.

Speech or Silence

Another criterion for evaluating characters is their dialog: what they say or what they don't say. Is a teenage boy quick to tell a girl that he loves her, or is he unable to share his emotions? Does a housewife report a murder to the police, or does she remain silent for fear of reprisal? Does a newlywed yell at her spouse for ignoring household chores, or does she remain quiet? And does she remain quiet out of timidity or out of love? Does an old woman confess that she was responsible for a fire in the retirement home, or does she clamp her mouth shut, allowing someone else to take the blame?

In dramatic terms, unspoken words can be louder than spoken ones.

Reactions of Others

A final clue enabling audiences to assess the nature of a dramatic character is how other characters in the play react to him or her. Their attitudes, however, may not be immediately definitive. If the women in a small town despise a sexy young widow, does that mean that the widow is a bad person? Perhaps. But not necessarily. It is also possible that the small-town women are jealous and mean minded. If such is the case, then the widow may actually merit our sympathy. We cannot gauge the validity of such a clue until we understand the nature of the peripheral characters.

In the early stages of a play, reactions of other cast members provide only tentative information, subject to change without notice. When a sympathetic protagonist violently dislikes a character, we automatically accept that hostility as our own—until a plot development redeems the character and proves our hostility unjustified.

WEARING MASKS

In 1925 Eugene O'Neill wrote *The Great God Brown*, a play in which characters wore masks. In most cases, the masks depicted youth or strength or cynical sophistication, concealing (and protecting) the characters' vulnerability. Characters fell in love, not with other characters but with the masks they wore.

O'Neill recognized that we all wear masks, presenting one face to our parents, another to our enemies, another to our mates. Which is the real us? In today's stressed society, most of us wear "cool" masks, hiding and protecting our anxieties. Sometimes, with loved ones whom we consider no threat, we take off our masks. But consider this: When you take off your mask, is there another underneath?

When an actor assumes a role, he or she fashions a multilayered, multidimensional person, wearing protective masks that conceal either vulnerability or some other emotion (or affliction) inappropriate to the occasion. Such layers (masks beneath masks) bring richness to a characterization; lack of them usually creates a character who is paper thin and whom audiences quickly tire of. Audience awareness of such hidden layers increases interest in a scene.

Sensitive directors concern themselves with the emotional stirrings beneath surface dialog or actions: the fear beneath the bravado, the shame beneath the smile, the guilt beneath the show of love. It is the face we cannot see that attracts and holds us, fascinating because it acquires richness in our imaginations.

The director's search for a dimensional character is really a search for the many characters who inhabit a single body, some in conflict with others, all adding texture and coloration.

BUILDING A PERFORMANCE

You have cast your show. With the blessing of your producer and the help of a casting director, you have selected a group of actors whose work you respect and who are suited to the roles they will play. You were careful not to select actors who were "on the nose," who had played similar roles countless times before. Such casting would be clichéd. You have deliberately cast a couple of the characters "against type," searching for freshness and defying audience expectancy. You believe that your actors will complement each other in the world of your screenplay, that they will create not only a believable chemistry together but one that is also imaginative and dynamic.

You have had preliminary meetings with your stars to explore the characters they will play. And now you are ready for the most exciting part of the directorial process: building performances.

First Reading

The theatrical practice of cast readings, also used occasionally in television and feature films, is an excellent way to allow characters to emerge naturally and unforced. It is a process of trial and error, tentative exploration, and happy discovery. Such readings by an entire cast—seated around a table with director, writer, and producer—also provide an early opportunity for improving the script. When scenes don't work, the writer can take them home and fix them, avoiding frantic last-minute changes on the set.

Inexperienced directors sometimes panic at first readings because the characterizations they expect seldom appear. Many actors deliberately resist committing themselves too early. They hide behind quiet, underplayed readings, giving themselves time to gain familiarity with dialog, their characters, relationships with other characters, and their director. As they become comfortable, they gain confidence and their characters gradually emerge.

First readings are also a time for discussion. If actors are confused about a character trait or an action or certain speeches, this is their opportunity to explore problems with the director and gain clarification. Occasionally, key scenes are also staged "dry" (without set or camera) in order to gain familiarity with characters' actions.

Rehearsal

In a single-camera operation, most directors rehearse actors immediately preceding their recording of a scene. During this rehearsal, the production crew watches and plans lighting, camera moves, and sound coverage. In a multiple-camera operation (sitcoms, for example), there is usually an extensive period of rehearsal prior to filming.

Some feature film directors, especially those trained in the theater or who have themselves acted, insist on a rehearsal period prior to production. They feel that such rehearsal inevitably results in richer performances. Award-winning director Sidney Lumet (who was a child actor) described his rehearsal experience with the film *Running on Empty* in a Directors Guild Newsletter.[8]

---

[8]Directors Guild of America Newsletter, October, 1988, pp. 2–3.

Newsletter: Rehearsal in all your films is important, but especially so in "Running on Empty," because the family had to be very close-knit.

Lumet: Yeah. And not only the interrelationships. River Phoenix had never rehearsed before. He's a movie actor. He was very nervous about (rehearsals). He thought it would kill spontaneity. I said to him, "River, I think you will find the opposite happens. Because if you know where you're going, you know where that character is supposed to be at each given moment, it releases you in a way. Instead of wondering, 'Where am I? What's this moment about? What was the last scene? (Because we never shoot in sequence.) What comes next?' Instead of all those worries, you know where you are. You're free to concentrate on the conversation. Then all the good, unexpected things happen. So it's actually more spontaneous."

Newsletter: So you do allow the actors an amount of input?

Lumet: Absolutely. It's a fairly rigid channel in the sense that: This is what this piece is about; this is where we're going with it. But then within those river banks, that river has a lot of freedom.

Performance

Unfortunately, a textbook on directing does not have room for any comprehensive examination of acting. The very best way for you to learn the fine points of acting is to *act*. Take acting classes. Join a little theater group. As many successful directors have stated, acting experience is essential for any director who plans to direct drama. Also, there are a number of excellent books on the subject. See "Books That May Be Helpful" at the end of this text. A few rules of thumb, however, may prove helpful.

Relaxation When actors walk onto a soundstage, they face an array of strangers: electricians, carpenters, camera crew, other (perhaps hostile) actors. At the outset, they have only one ally, the director. One of the primary tasks for a director is to make actors feel welcome, secure, comfortable—to allow them to relax.

Remember, it is far easier for actors to get in touch with their emotions when they are relaxed. Fear inhibits. So does tension. If actors are insecure, they will go through the motions, pretending to feel emotion, but their performances will usually be artificial, skin deep.

A surprisingly large number of actors tend to be insecure, including some of the very finest. It is they who must stand in the spotlight. It is they who must expose themselves to the scrutiny (perhaps ridicule) of critics and audiences. It is they who must cry, scream, or laugh on cue, convincingly, in character, sometimes over and over again.

As director, you can help to alleviate actors' anxieties in several ways. Initially, you can build security by discussing their characterizations with them, encouraging their input, reaching agreement on characters' background, style of dress, personality, moral code, traits and mannerisms, even the type of car they drive. Confusion about a characterization creates disquiet. Actors gain confidence when they feel that the director understands their characters as well as (or better than) they do.

In addition to understanding the characters that actors play, directors need to understand the actors themselves. Actors gain security when they realize that the director has seen them perform previously and is aware of their strengths and weaknesses. Because of this insight, actors can relax knowing that the director will never allow them to look bad. Once they realize that their director refuses to settle for mediocrity—demanding that they perform at the very peak of their abilities—anxiety frequently becomes anticipation, even excitement.

At the beginning of a project, facing a new crew, new cast members, and commitment to a new role, other apprehensions sometimes arise. A director can supply support on the set, protecting actors from intrusions into their privacy and allowing them the space to develop their characters and build a rapport with other cast members.

Some directors actually ask all crew members and guests to leave the stage when preparing to rehearse significant scenes. In the privacy of an empty stage, released from the scrutiny of outsiders, actors lose any feelings of self-consciousness. Now, guided by the director, they may explore offbeat stagings or adventuresome character developments without feeling they are making fools of themselves.

Energy Sometimes actors will display all of the appropriate emotions, but their scenes seem vapid, lacking in energy. What can you as the director do about it?

Simply telling an actor to use more energy works only to a degree. Moves may be faster, gestures more emphatic. Speeches may have more pace or volume. But that elusive quality known as "energy" simply doesn't appear.

> I believe that energy is the direct result of how much you care about what is happening. If the content of a scene—if what is happening in your performance life—is important enough, you will be listening with sufficient intensity, absorbing and responding with sufficient intensity, to create the necessary energy.[9]

Tony Barr's words provide the key. A former CBS vice president and owner of the successful Film Actors Workshop, Barr recognizes the critical importance

---

[9]Tony Barr, *Acting for the Camera* (Boston: Allyn & Bacon, 1982), p. 36.

of an actor caring about the outcome of a scene. We have all seen performers who automatically command our attention. We watch them even when they have little to do in a scene. They seem to radiate energy. And yet they are just actors, endowed with the same physical characteristics as others. They have no gland that secretes energy. What is the difference? Attitude. Even without words they participate because they have a stake in the scene; they care deeply about what happens. A 5-year-old watching a fight between her parents has no dialog, no actions. And yet our eyes are riveted upon her because her secure world is collapsing.

> Unless the material *demands* that you not care, *always choose to care about what is happening as much as you logically can within the context of the material.*[10]

Listening Fine actors listen. They listen with all of their senses. They hear more than the words spoken by other actors. They recognize the meanings underlying those words or concealed by them. They "hear" the words that are left unsaid. They hear the thoughts, moods, and emotions that motivate the words. Part of that listening is *perception*, awareness of a jaw tightening or the momentary narrowing of eyes. Listening in its larger sense represents a concentration on another person, a fastening of attention. Some directors have claimed that listening is probably an actor's most important ability.

When actors listen, they become involved. And when they become involved with one another, they react, they *feel*. They forget that they are actors on a soundstage waiting for someone to throw them a cue. They forget the fear of forgetting. They lose the self-consciousness that distracts inexperienced actors and gets in the way of honest reaction. When they listen, really listen, their performance immediately becomes richer.

Simplicity In recent years the phrase "Less is more" seems to have taken on increased meaning in many branches of the arts. In acting for television and film, the phrase is particularly significant.

Audiences tend to see on the screen what they expect to see. (See the discussion of "Closeness" in Chapter 7.) When a husband tells his wife that he is sick of their marriage, that he is leaving her to marry his secretary, the audience will empathize with her. In her close reaction shot, audiences will "see" grief and remorse and feel what the wife is feeling. If the actress exaggerates her feelings, contorts her face, creates the slightest hint of dishonesty, the moment will be ruptured.

On a theatrical stage, an actress would need to convey her emotions to the last seat in the balcony. She would have to magnify her reactions in order to communicate them effectively. Film and videotape present a totally different set of re-

---

[10]Barr, *Acting for the Camera*, p. 37.

quirements. Because a closeup is so intimate, so intensely revealing, the best acting for TV and motion pictures usually is simple. Understated. Less is more.

Clichés

Because of the tremendous volume of dramatic programs aired on television, character clichés abound. In searching for a characterization, actors sometimes fall into stereotyped patterns, unaware that the character they have selected is too familiar. The director must convince them that their feelings of security come from recognition. New directions are frightening; familiar ones (like old shoes) are comfortable. A good director encourages capable actors to be adventuresome, to try new directions, to take chances. Such a director protects them from failure, applauding their efforts, encouraging and steering them into a fresh and viable character pattern.

TRANSFERRING CHARACTERS TO TAPE OR FILM

The "film method" (also used for videotape) of shooting out of sequence poses a number of problems for actors and can be disruptive to characterizations. Shooting all of the scenes that occur in a single set (no matter when they appear in the script) provides tremendous financial advantages to a production company, but the pattern creates mental and emotional hazards for both actor and director.

In the theater an actor has time to "get into" a role, building a performance smoothly from scene to scene in continuous action. When shooting out of sequence, director and actor have no such luxury; they frequently face the problem of re-creating the emotional intensity, mood, or energy of a scene that was taped or filmed weeks ago. In the script, the previously photographed scene (scene A) occurs immediately before the one taped or filmed later (scene B). When the scenes are later spliced together, any lapse on the actor's (or the director's) part will be instantly discernible. The emotional level at the beginning of scene B must exactly coincide with the level at the end of scene A.

Such a recapturing of mood is even more difficult when scene B is photographed first. Now director and actor must calculate the emotional intensity to be reached when scene A will be photographed, perhaps weeks later. From this rough calculation, they must now begin scene B, hoping that when they later stage scene A, they will have perfect recall and that its climax will match the previously recorded scene B.

Such character and performance problems are more the director's concern than the actor's. Before any scene is photographed, usually while the crew is lighting the set, the director spends time with actors rehearsing dialog, discussing the scene's mood, character relationships, tempo, and potential staging problems—and ironing out any wrinkles. Usually at this time the director describes the emotional level of adjacent scenes, attempting to match those levels in the scene being rehearsed. If the matching is critical, the wise director will have screened the adjacent scene, perhaps a few hours earlier, familiarizing himself or herself with its emotional intensity. Sometimes directors ask that a moviola or videotape player be brought to the stage so that the earlier scene may actually be shown to cast members. In this way the director and cast may plant the previous scene's mood or emotional level firmly in their minds only moments before the new scene is photographed.

Despite the director's thoroughness in talking through a scene, some actors need additional time and space to prepare on their own. They sometimes retreat from the set, taking a few quiet moments to re-create the character. Like other creative artists, actors often have individual needs and very personal methods of preparation. Removing themselves from the real world, each actor enters his or her character's world, finding the character's mental and emotional state for the upcoming scene.

Just as individual actors prepare differently, so do they differ in the length of time necessary to achieve emotional depth. Directors must be vigilant to prevent them from reaching such depth too early. Occasionally actors try to play a scene with full intensity during rehearsal. Vigilant directors usually stop them, advising them just to walk it through, fearful that they may achieve a performance of such depth that they will be unable to achieve it again when cameras are rolling. In the days of live televised drama, I watched directors deliberately break up their dress rehearsals, pretending to criticize lighting or sound, to prevent the cast from expending their emotions, to keep them from going stale before airtime.

Director Delbert Mann tells of an experience with Sophia Loren in which she played the part of a mother who had lost a child. Because the role paralleled a tragic, real-life experience, the actress began getting misty-eyed during the scene's first walk-through. Mann understood the reasons for her sudden emotion and asked her just to walk through the scene casually without any attempt at acting. Even that was difficult for her, although she tried valiantly to do as the director had requested.

When it came time to film the master (wide) angle, Mann again cautioned her not to "act" the scene, advising her to play it coolly, with only minimal emotion. He understood the depths of which she was capable and wanted her to save her "performance" for closer angles. Moments later he rolled the cameras. The scene played beautifully until, about halfway through, the actress began sobbing. She had become the character, participating deeply in the scene's emotional content. It was a stunning, enormously moving moment—but it had played only in a very full shot, where the audience impact would be slight.

Later, when the director shot closeups, Loren summoned up as much emotion as she was capable of. But the moment had passed. While her closeup performance was technically flawless, it never quite reached the heart-tugging poignancy that she had achieved in the master angle.

Mann was confronted with a similar situation some years later. A sensitive young actress was so filled with emotion that the director sensed she would spill her tears in the master angle. He secretly directed his cinematographer to put a long lens on the camera so that he could photograph her in closeup throughout the scene. As expected, the actress gave a tear-filled, deeply moving performance. And this time Mann had photographed it in a closeup!

Some actors find a character's emotional depth easily, but others have tremendous difficulty experiencing inner pain. We have all heard stories of how directors "trick" certain actors into giving performances above their capabilities by abusing or shocking them, by relating untrue stories, or by driving them to emotional exhaustion. Although such gimmicks seem less reliable than emotion reached through an honest characterization, if they manage to create the necessary tears or anger or fear—and more conventional methods fail—then perhaps they are justified.

Although some directors obtain effective characterizations from harsh, authoritarian methods, most do not. Most contemporary directors understand the frailty of the human ego; they achieve their best results through support and encouragement. When directors need to make corrections after shooting an aborted scene, they usually offer some positive comment first. When such directors make corrections, they do so in privacy, taking the actor off the set.

Of course there are the so-called temperamental actors who quarrel with every directorial suggestion. Often when actors are "temperamental," it is because of some insecurity. Perhaps they are unhappy with their roles, the size of their dressing rooms, or their billing. The wise director tries not to deal with superficial symptoms but to discover the root causes of an actor's unhappiness. (See "The Parent-Psychiatrist" section of Chapter 1.)

Remember that a questioning attitude is healthy. Good actors like to dig into the underpinnings of a character so they can give a richer performance. But there's a substantial difference between questioning and quarrelsome. When time is short, quarrelsome actors must be dealt with firmly.

CHAPTER HIGHLIGHTS

— Directors find the nature of characters much as actors do: first from the script and then from their own background experience, synthesizing bits and pieces of people they have known. They share their concepts of each character with the actor chosen to play the role; together they fashion a final character concept.

— When directors study a script, they determine its *story spine* and its *character spines*. The word *spine* usually refers to a personal goal, need, or desire. A story spine defines the play's theme or root idea. Usually the spines of major characters are closely related to the story spine.

— Many directors and actors break scenes into smaller units, subscenes called *beats*. Each beat marks a change of direction in scene content and, therefore, a change in emotional attitude or staging.

— In creating character backgrounds, some directors use three convenient categories: physiology (describing physical characteristics), sociology (a character's place in society), and psychology (internal values such as moral code, fears, personality, sense of humor). By examining character backgrounds, directors may stimulate good actors to dig into their own imaginations. Directors usually meet with each major cast member prior to production to discuss character concepts.

— Once a character has been defined by the director and actor, that character is revealed to audiences by the character's entire physical aspect, the character's traits and mannerisms, what the character does or does not do, what the character says or does not say, and how other characters react to him or her. Good directors are keen observers; their observations help to enrich each of these categories.

— Most people wear "masks." That is, they conceal their vulnerabilities beneath protective masks that they present to the world. Sensitive directors and actors use the concept of masks to achieve multidimensional characterizations: the fear beneath the bravado, the shame beneath the smile.

— Many actors are insecure. Part of a director's function is to support and protect them, providing an atmosphere that will simultaneously challenge and stimulate.

— When building a performance, successful actors radiate energy by *caring*. When they care enough about the outcome of a scene, they attract audience attention. Successful actors also *listen*. By fastening their attention on another person, they lose self-consciousness and feel what the other is feeling. Finally, effective performances for TV and film usually are *simple*. Closeups magnify artifice or exaggeration.

— Sensitive directors understand when an actor reaches optimum performance and refuse to accept less. They prevent actors from reaching their optimum during rehearsal.

PROJECTS FOR ASPIRING DIRECTORS

1. Audition for a play, either at a university or a neighborhood little theater group. Whoa, hold on! Don't skip past this project. It's normal to be apprehensive about auditioning for strangers, especially if you've had little or no acting experience. But it's worth the risk, I promise. And who knows, you may surprise yourself and get the part.

2. Take an acting class.

3. At your university or a little theater group, get permission to watch a play in rehearsal. Stay with it for more than a single rehearsal. Study the director's relationship with cast members. Analyze how he or she blocks the physical action. Analyze how the director helps actors to build characterizations.

4. Interview either an actor or the director from the play you watched. Discuss how he or she works in terms of studying the material, finding appropriate characterizations, business, actions, and rehearsal patterns. If you're enrolled in production classes, see if you can get credit for a paper (no less than five pages) based on this interview.

5. Read the short dramatic scene in "Projects for Aspiring Directors" at the end of Chapter 6. Prepare notes detailing the backgrounds of both characters. Use the categories of physiology, sociology, and psychology, if they will help. Of paramount importance, what significant events shaped their lives and characters?

6. Dig even further into the characters of Amy and John. What are their favorite movies? TV shows? How are they dressed in this scene? What magazines do they read? What are their goals in life? Describe the living quarters of each in considerable detail.

CHAPTER

6

STAGING
THE ACTOR

Elia Kazan's words (at the top of the next page) establish the theme for this chapter dealing with the behavior of actors in front of a camera. A part of the process of turning psychology into behavior, translating thoughts and emotions into dramatic action, is called **blocking**. Directors block the positions and movements of both actors and cameras. I will discuss camera blocking later in this chapter and in Chapter 7.

In drama, actors are one of a director's key creative elements, the door through which spectators enter a story. Their physical movements, gestures, and positions in relation to other actors are governed by specific dramatic principles, many of them as old as the theater. This chapter will examine those principles in the following discussions:

— WHERE STAGING ORIGINATES: the director's role in planning and preparing scenes for rehearsal

— GUIDELINES IN STAGING: the importance of logic and showmanship

— EXTERNALIZING THOUGHTS OR EMOTIONS: the relative validity of dialog and action

A thought—directing finally consists
of turning Psychology into Behavior.
Elia Kazan[1]

— BUSINESS: specific character actions, traditionally called *business*; their dramatic purpose and function

— REVEALING INNER THOUGHTS: how directors provide insights into hidden thoughts or emotions

— INTERNAL MOTIVATIONS: three fundamental emotional states that impel character action

— EXTERNAL MOTIVATIONS: usually arising from the script or a need for emphasis; a comparison with stage techniques for achieving emphasis

— SYMBOLIC RELATIONSHIPS: physical positioning of characters is often a key to their relationships

— BEATS: smaller units within scenes that help directors to stage action

---

[1]Toby Cole and Helen Krich Chinoy, eds., *Directors on Directing*, 2d ed. (New York: Bobbs-Merrill, 1979), p. 364.

— BLOCKING CAMERA TO ACTORS: the importance of the viewer–actor relationship and the need for honesty in performance

— BLOCKING ACTORS TO CAMERA: the occasions when camera needs take precedence over performance needs

— PREPARATION: the steps in preproduction and production necessary to turn "psychology into behavior"

WHERE STAGING ORIGINATES

We're watching a scene on television. Todd, an earnest, balding man of forty, pleads with his wife to send her mother to a home for the aged. The wife, Debbie, angrily refuses, moving to the window. Todd pursues her, taking her arm. He's fed up. The old woman allows them no privacy. She's destroying their marriage.

Through the partially open doorway we can see Debbie's mother in the next room listening, hand trembling as she grips the back of a chair.

Debbie breaks away from her husband and sinks onto the couch, close to camera. She stares at her mother's crochet work on the couch. She picks it up almost tenderly. The camera tilts up from the crochet work to Debbie's face, her eyes glistening now. After a moment, she nods. All right, she will make the arrangements.

Camera pans to the doorway. The old woman's shoulders slump and she moves away.

If they stopped to think about it, most viewers would assume that actors' movements in such scenes are spontaneous, growing naturally out of the dramatic moment. Why does Debbie break away from Todd when he takes her arm? Spectators accept the action at face value: because Debbie's deep emotions at that instant impel her to move. Professionals familiar with dramatic staging understand that such action was adapted by a director from a script, rehearsed with actors, adjusted and refined and refined again. The finished product appears simple, natural, inevitable: art that conceals art.

Generally, the only time that audiences become aware of dramatic contrivance is when actors' movements seem false, in apparent disharmony with their characters' thoughts or emotions. Then audiences are disquieted only in a general sense, aware that "something's wrong," not that a director or actor has miscalculated. Such awareness creates a rupture in the audience's attention, causing spectators to disengage themselves from the magic world in which they are participating, to step back through the proscenium arch of their television sets or movie screens and again become dispassionate viewers.

GUIDELINES IN STAGING

When directors plan the staging of scenes, their blockings usually are based on two factors: *logic* and *drama*.

Logic is immediately understandable. Directors simply ask themselves, "What would this person logically be doing in this setting, at this time of day? What actions seem most honest and believable when this young woman confronts the father she hates or the man she loves?"

When characters move toward or away from each other, such movement arises from specific internal or external motivations. Such moves are also logical, triggered either by plot demands or emotions. Sometimes the script will indicate these actions. More often, the director will have to invent them from whatever clues are provided by stage directions and dialog.

The factor called "drama" is somewhat more complicated. We can lump some of its elements under the generic term *showmanship*, expressing story concepts in vivid, emotionally involving terms. Life as it is actually lived can be dull. Directors therefore seek ways of demonstrating (dramatizing) story values in terms that will stimulate audiences. Choice of locale, character actions, attitudes, even costumes can enrich the content of scenes. The term *drama* as used here also concerns the director's very practical need to reveal, clarify, or underscore a scene's emotional content.

These concepts will be elaborated in this chapter and the next.

EXTERNALIZING THOUGHTS OR EMOTIONS

How does an audience recognize honest emotion? By what characters say? Or by what they do?

The timeworn phrase, "Actions speak louder than words," applies forcibly in any visual dramatic medium. Although dialog usually provides reliable insights into a character's thoughts or emotions, there are occasions when words simply cannot be trusted. An obvious example: People lie about their weaknesses. A coward will stoutly deny his fear. A woman who is tense in a new job situation will pretend composure. Ask a dozen people how they feel. Some will exaggerate; others will understate; others, looking for sympathy, will baldly invent physical problems.

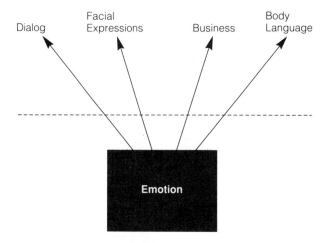

FIGURE 6.1 Emotion is the significant content of most scenes. But it lies beneath the surface, where audiences cannot see it. Therefore, it must be externalized through dialog, facial expression, business, or body language.

A shy boy and a timid girl in the moonlight may speak of a hundred casual matters that have nothing whatsoever to do with what they are thinking or feeling. But their thoughts and feelings are the significant content of such a scene. Therefore, the director's goal must be to *externalize* those feelings, somehow making the audience aware of their depth and nature.

The Validity of Dialog

Dialog often has little relationship to the essential thrust of a scene. Sometimes it is merely an accompaniment. I recall a scene from an episode of the TV series "The Fugitive" in which director Billy Graham panned his camera with a small, folded note as it was passed surreptitiously from one character to another, totally ignoring the faces of the characters. Yes, we heard their dialog. Yes, we got occasional glimpses of them. But the note represented the heart of the scene. Dialog merely provided a background.

Husband and wife at a party exchange casual dialog with friends. As they talk, the husband's eyes seek out a handsome dark-eyed woman across the room. The woman across the room becomes aware of his glance and stares back at him curiously. The husband smiles at her, unseen by his wife. The woman across the room lifts her champagne glass slightly, as if toasting him, then turns away. As in the previous example, the dialog provided nothing more than an accompaniment to the scene's significant content: the eye contact between the husband and the dark-eyed woman.

A more extreme example may be one in which characters say one thing but mean something entirely different. A mother attends her daughter's open house at junior high school. While visiting an art class, she's shocked to discover that the teacher is a man she had an affair with only a year before. They instantly recognize each other but because of the daughter's presence can only comment on the array of students' sketches and watercolors. Their dialog is meaningless. But beneath the banal words a second, more significant conversation takes place. The exchanged glances, the handshake held a moment too long, the veiled references, and innuendo are what the scene is all about. A spectator aware of the earlier love affair will focus attention on the subtext, dramatized not *by* the dialog but *in spite of it*, vastly more compelling because it is hidden beneath an apparently innocent teacher–parent encounter.

At the heart of most drama is emotion: what the characters feel and how they react to those feelings. It is a director's job somehow to externalize those feelings ("turning psychology into behavior"). An actor's facial expression is one way to reveal emotion. Business and physical actions are another. And dialog is still another. Note that dialog is just one of several ways to express the significant content of a scene. (See Figure 6.1.)

Former CBS vice president Tony Barr conducts a successful workshop in North Hollywood for film and television actors. One of the exercises he presents to students is a short scene whose dialog is banal, capable of various interpretations.[2]

                              HE
Good morning.

                              SHE
Good morning.

                              HE
How do you feel?

                              SHE
Great.

                              HE
I'm sure.

                              SHE
What do you want for breakfast?

                              HE
Whatever.

---

[2]As described by Tony Barr in *Acting for the Camera* (Boston: Allyn & Bacon, 1982), pp. 68–74.

> SHE
>
> I'll fix you some scrambled eggs.
>
>   HE
>
> Fine.
>
>   SHE
>
> You going to work this morning?
>
>   HE
>
> Have to.
>
>   SHE
>
> Oh.
>
>   HE
>
> You want me to stay home?
>
>   SHE
>
> It's up to you.
>
>   HE
>
> Can't.
>
>   SHE
>
> Like I said—it's up to you.

Tony Barr asks his students to play the scene several times for his cameras. For each enactment he gives them a totally different **preparation** (emotional attitude). The first time he suggests that they have just returned from their honeymoon and days of passionate lovemaking. For the second enactment, he suggests that the husband came home at 4:00 in the morning after an affair with his secretary. Husband and wife fought bitterly. For the final enactment, Barr suggests that the couple learned only yesterday that the husband has a terminal illness.

In all three versions, of course, the words are identical. The actors on their own, without direction, seek to express the emotion behind the words. Even with relatively inexperienced actors, the differences among the three versions are astounding. In the first, dialog and actions emerge as sensuous, steamy, filled with innuendo. The actors cling to each other, touching. The camera is close, suggesting their intimacy.

In the second version, words are clipped, brittle, concealing depths of resentment and bitterness. Husband and wife remain apart, separated by the breakfast table. Guilt makes it difficult for the husband to remain seated. He does not look at his wife.

In the third version, actors sometimes play tragedy overtly; sometimes they try to conceal it, allowing the heartbreak to appear only occasionally in a broken word or phrase. Sometimes wives choose to play gentleness. Sometimes they try to ignore the bitter news and discover they cannot, turning abruptly away from their husbands to conceal tears. But whichever attitude the actors select, the underlying emotion dictates not only their vocal colorations and facial expressions but also their physical movement in relation to each other.

Despite the fact that the dialog is identical in each, the three enactments emerge as totally different scenes, underscoring again that words are often unimportant; what counts is the emotion beneath the words.

The Validity of Actions

Just as dialog is not totally trustworthy in revealing thoughts or emotions, neither are actions. Timid characters may strut. Lovesick characters may pretend disinterest. Most of us act casual in order to conceal our anxieties or fears. (See the discussion on wearing masks in Chapter 5.)

Actions sometimes distort the truth, but they still provide a far more accurate barometer than words do. Thus, Mr. Jones plays it flip and unperturbed when his boss fires him. But we see his fingers tearing at the eraser on the end of his pencil. Lucy Ann pretends fondness for her date, yet she constantly moves away from him, glancing furtively at her watch.

In the short scene that began the chapter, we understood Debbie's agitation as much from her actions as from what she said. We understood her mother's anxiety from her trembling hands and realized her disappointment from the sudden slumping of her shoulders.

As kids we all played "Show and Tell." As directors, we should realize that showing usually proves superior to telling. It's more visual. It's more honest. Most importantly, it's stronger. Why stronger? Because showing requires some degree of spectator *participation*. Audiences must interpret actions. They must draw conclusions as to their significance.

If Debbie overtly states, "I love my mother," the information is dumped in our laps. There's little interaction between Debbie and the audience. But when Debbie picks up her mother's crochet work and holds it to her cheek, eyes glistening, we immediately participate. We have witnessed earlier scenes in which the mother continually crocheted. The delicate handwork thus becomes a symbol of the older woman. When Debbie holds the half-finished sweater to her cheek, she is holding her mother. Her tears complete the unspoken thought: an expression of grief that the old woman will be sent to a home.

Although reciting the steps of such a sequence seems obvious and unnecessary, it nevertheless enables us to appreciate the contrast between the action, simple as it is, and Debbie's overt statement, which allows no spectator participation at all. Actions require the audience to contribute from remembered experience. By contributing, viewers automatically become more deeply involved.

BUSINESS

Debbie's action with the crochet work is called *business*, a term inherited from the theater. Business really means "busy-ness" and refers to characters' actions within a scene, such as sewing, reading a newspaper, polishing fingernails, cleaning a gun, or washing a car. Ideally, business should make a statement either about the character who performs it or about the scene itself. A man who continually combs his hair, for example, offers insights to his character. A woman who sculpts probably has different instincts than one who practices witchcraft or one who mud wrestles in a neighborhood nightclub. What characters *do* suggests what they *are*!

Russian director and theorist V. I. Pudovkin used the term **plastic material** to indicate visually expressive actions that help audiences define a character. He described a sequence of shots from the silent film *Tol'able David* to illustrate his concept:

1. The tramp—a degenerate brute, his face overgrown with unshaven bristles—is about to enter a house, but stops, his attention caught by something.

2. Closeup of the face of the watching tramp.

3. Showing what he sees—a tiny, fluffy kitten asleep in the sun.

4. The tramp again. He raises a heavy stone with the transparent intention of using it to obliterate the sleeping little beast, and only the casual push of a fellow, just then carrying objects into the house, hinders him from carrying out his cruel intention.

Thus (using Pudovkin's words) "the heavy stone in the hands of the huge man immediately becomes the symbol of absurd and senseless cruelty," providing viewers with an obvious insight into the tramp's nature.[3]

Business may also contribute to the dynamics of a scene. The action of cleaning a gun adds an unspoken threat to a scene in which one man makes sinister demands on another. A woman's declaration of love for her husband is contradicted when she ignores him to apply fingernail polish. Rubbing suntain oil on a girlfriend's back adds sensuality and innuendo to a young man's suggestion that they live together.

---

[3]V. I. Pudovkin, *Film Technique and Film Acting* (London: Vision Press, 1958), p. 56.
(First published in English in 1929.)

Probably the most obvious reason for creating business is that it contributes to the illusion of reality. In real life people seldom stand in the center of a room and make speeches to one another. Instead, they busy themselves with the thousand necessary tasks that clutter their lives. A woman in her office may write notes, straighten papers on her desk, or make telephone calls while playing a scene with a business associate. If those actions heighten the dramatic thrust of the scene, so much the better. A man at home may trim hedges, mow, or water his lawn while playing a scene with a neighbor. Those actions contribute to the illusion of reality, but they also comment on the nature of the man. He obviously cares about his home and its appearance. A different character might say the same dialog while swilling beer and watching a porno film on cable TV.

Business is a major concern of directors because writers seldom describe it in their scripts. Business should always appear as a natural and legitimate part of any dramatic scene and should be appropriate to (and define) the character who performs it.

For years, both on the stage and in film, whenever directors needed a piece of business they would instruct a character to light a cigarette. Cigarette business became a directorial cliché. Today the business of preparing or sipping drinks has fallen into that same odious category; it has been overworked so much that the action becomes almost meaningless.

REVEALING INNER THOUGHTS

Debbie's picking up her mother's crochet work, beyond constituting a simple piece of business, also demonstrates another directorial technique. How can a director convey to an audience what a character is thinking about? In a novel or a short story, such communication is easy: The author simply *tells* the reader what's happening inside a character's mind. For a director, the problem becomes somewhat more difficult.

One effective solution is to select a symbol whose significance the spectator will understand. Our lives are filled with such symbols: a funny T-shirt given to us by someone we love, a high school baseball trophy, a bundle of yellowing letters, a worn teddy bear from our youth. The manner in which characters then relate to those symbols provides the audience with insights into the characters' thoughts and feelings. For example, when a former heavyweight boxer looks wistfully at his gloves hanging on the wall, viewers know what thoughts are filling his mind. When a woman cleans and polishes the bicycle that once belonged to her dead son, the audience can understand her emotions.

Audiences are brighter than show business cynics seem to believe. Viewers seize upon any clue the director gives them, eager to understand and contribute.

But they need a beginning point—a symbol of a place, an event, or a person—that the director must provide.

INTERNAL MOTIVATIONS

When I was handed my first dramatic directorial assignment, I experienced a few hours of panic. My knowledge of the art was minimal. Few books on the subject existed and I was reluctant to confess to colleagues that I needed advice. One experienced professional understood my anxiety and volunteered a rule of thumb: Every scene is a chase. A man pursues a woman around a kitchen, he catches up with her, and they face each other for a showdown. Then, perhaps, the man begins to retreat and the woman pursues. What a simple solution to my problems! I was home free!

Well, it took me fewer than twenty-four hours to realize that this so-called rule of thumb is far from infallible. It does work—but only on occasion. The Debbie-Todd scene is one of those occasions. Todd pursues Debbie, driven by anger and his need to save their marriage. Debbie retreats from Todd and from her own fears until, finally, backed into a corner, she makes a decision.

The fact that a chase works sometimes and not others set me to thinking. What dynamics are involved that make a character a pursuer in one scene and a retreater in another? Study of a number of different scenes in different dramatic modes revealed that movement is most often triggered *within* a character. Such internal motivation—thoughts, feelings, emotions—is generally one of three types: involvement, disinvolvement, or emotional turbulence.

Involvement

Movement *toward* another character, seeking involvement, may be motivated by a wide variety of emotions: rage, anger or annoyance, sadism, lust, desire, love or affection. Also, one character moves to another out of need: for understanding, information (or to impart information), forgiveness, love's favors, or a physical object. Generally speaking, the more urgent the emotion, the more intense will be the action it generates. When greed or rage become psychotic, for example, the forward movement may become an attack, inflicting bodily injury or death.

This list of aggressive or seeking motivations (as with those that follow) is far from complete, but it provides aspiring directors with a beginning point in understanding the motivations for character movement. By studying well-directed television or feature films, the student director will undoubtedly discover other emotions that impel one character toward another.

Disinvolvement

Moves of avoidance, rejection, or retreat—movement *away from* another character—generally are triggered by emotions of fear, shame, guilt, or timidity; hatred, dislike, disgust, or disinterest; rage, anger, annoyance, coquetry, or teasing (entrapment). Notice that the emotions of rage, anger, and annoyance appear as motivations for both advancing and retreating moves.

In any chase, the pursuer will frequently (but not always) be driven by an emotion of involvement and the retreater by an emotion of disinvolvement. In the earlier scene, Todd is driven by anger and frustration, Debbie by guilt. Such labels are simplistic, of course. In any successful dramatic work, motivations inevitably become a complex blending of emotions with individual character traits created by the specific actor for a specific role in a specific script. A timid character generally tends to retreat; an aggressive character tends to advance. Thus, built-in character values must be superimposed on values generated by scene content. When modified in such a way, the labels may change slightly or enormously or not at all. A tense, highly neurotic character, for example, may act atypically, attacking someone when afraid rather than retreating. In directing as in other creative crafts, rules cannot remain inflexible. All I can suggest is that most character moves are triggered by the listed emotions most of the time.

Frequently, characters retreat into *business*. When Ellen doesn't want to hear her husband's angry complaints, she escapes by paying close attention to her cooking. When Michael wants to escape from the clamoring of his kids, he immerses himself in the evening newspaper. Sometimes characters retreat by withdrawing mentally from a boring or uncomfortable situation, escaping into the snug harbor of private fantasy. Students in my classes (alas) occasionally lapse into a glassy, alpha wave state. Although they are physically present in the classroom, their minds have taken them back to last night's date or forward to Saturday's basketball game.

Emotional Turbulence

Moves that are *nondirected*, that have no relationship to other characters, frequently emerge from emotional turbulence or agitation. Such anxiety motivations as guilt, worry, apprehension, anger, and loss of love or other security may trigger the movement. But positive emotional states such as ecstasy, joy, excitement, eagerness, or exhilaration may also create nondirected movement. Disorientation or disequilibrium—as from alcohol, drugs, or psychosis—can also cause nondirected movement. Sometimes, as with aggressive and retreating behavior, nondirected movement is built into a character. We all know people who by their very nature are restless, constantly in motion.

Professional directors understand internal motivations—but not always on a conscious level. Frequently they suggest movement to an actor simply because it "feels right." Such intuitive feelings usually derive from long experience, from observation and study, from discipline, from well-honed dramatic instincts, from projects that failed and projects that soared. Most professional directors may not be able to articulate in so many words the nature of the motivations described here, but they will demonstrate those principles inevitably in every scene of every project they direct.

EXTERNAL MOTIVATIONS

Although some character movements are triggered by thoughts or emotions—by internal motivations—others are generated by considerations *outside* the character.

Script

The most common external motivations arise directly from the script. A major part of a director's job is simply to tell the story. The staging of scenes, the manner in which characters move, their business, gestures, and facial expressions all serve that primary goal: to act out the plot points as defined by the writer. Thus, characters kiss, kill, fight, drink poison, dress, undress, and enter and exit scenes all as part of the basic requirement of telling a story. When the script calls for a ringing telephone, a protagonist is obliged to answer it; the move to the phone is thus dictated by script, although the director has a number of options: where to place the phone and where to position the protagonist prior to the ring, for example.

Although the writer may suggest blockings for a scene, usually very roughly, sometimes in enormous detail, the director has the freedom either to use the suggested blockings or to modify them. Because staging is acknowledged to be primarily a directorial skill, the director is frequently able to improve on a writer's suggestions. The writer and the director work together as a collaborating team.[4] The director constructs a building from a blueprint provided by the writer. If both have performed their jobs well, the skyscraper will emerge not only structurally strong but also aesthetically satisfying.

---

[4]A concept articulated by Wolf Rilla in *The Writer and the Screen* (New York: William Morrow, 1974).

Emphasis

A major directorial function is to steer spectators' attention to subject matter that carries the greatest dramatic impact at each moment in the story progression. Primarily through blocking of actors and camera, directors maximize elements they consider significant and minimize those of lesser importance: accentuating the positive, eliminating the negative. Additionally, through a variety of techniques, they are able to control how much impact their material will have upon a spectator.

Stage Techniques To define how a television or film director accentuates significant elements, let's begin with an examination of traditional stage techniques for creating emphasis. There is a surprisingly large number of them.

— AREA OF STAGE: When actors stand downstage center, they attract the greatest degree of audience attention. As they move to the side or upstage, their impact gradually diminishes. (Note: Downstage is nearest the audience, upstage farthest away.)

— POSITION OF ACTOR'S BODY: When actors *face* the audience, they achieve the greatest emphasis. As they turn away, they gradually lose impact until, finally, with their backs to the audience, they attract the least audience attention. (Exception: If an actor is the only cast member facing upstage, he or she may attract some degree of attention.)

— DIRECTION OF ACTORS' LOOKS: The actor at whom other actors focus their attention receives major emphasis.

— SEPARATION FROM GROUP: An actor isolated from others usually gains attention from the audience.

— MOVEMENT: An actor in movement attracts more attention than do actors who are stationary.

— LEVELS: An actor on a different stage level than others gains emphasis. Similarly, a standing actor attracts more attention than seated actors. If all stand, then the seated actor may have the greatest emphasis.

— LIGHTING: An actor more brilliantly lighted or an unlighted actor in silhouette attracts more attention than do those lighted normally.

— COSTUME: An actor costumed differently from others inevitably attracts attention.

— PICTORIAL COMPOSITION: Elements of scenery may direct the spectator's eye to a key actor. The diagonal line of a staircase, for example, directs attention to an actor on a landing. Arches, doorways, and windows tend to frame (and therefore emphasize) actors.

Notice that in a number of these categories, emphasis is achieved through *difference*—that is, when one character somehow is separated or distinguished from the others by costume, lighting, and positioning.

Through use of these stage techniques, a director can emphasize certain actors at certain key moments, then divert that attention to others as the dramatic focus of a scene changes.

Stage Versus TV and Film

Most of the stage techniques just listed can be directly transferred to videotape or film. When three actors in a film scene direct their attention to a fourth, for example, viewers will focus on that fourth actor. An actor standing in profile has less emphasis than one facing the camera. The primary difference between stage and screen is that a camera replaces the live audience. Thus, in TV or film, actors' relationship with their audience is in direct proportion to their relationship with the camera. Generally speaking, the closer they are to the camera, the more impact the actors will have, the more powerful will be their actions, reactions, and words they utter. Note that the closest shot in *Citizen Kane* is of Kane's mouth as he says "Rosebud." Such extraordinary emphasis signals the importance of that word. (See Figure 6.2.)

If, for reasons of staging, it becomes awkward to bring an actor to the camera, the director may take the camera to the actor. A camera dollying in on an actor is similar (in the degree of emphasis only) to an actor in a theatrical presentation moving downstage (toward the audience). By moving closer, the actor achieves greater impact. A director may achieve emphasis instantaneously by cutting to a closer angle (for example, from a medium shot to a closeup).

Entrances and Exits

Entrances usually are more dynamic than exits. When a character enters a scene, he or she creates audience expectancy; new elements create new story developments. Entrances that are directed toward camera seem to have greater impact than those that move laterally across screen. To extend this principle, *most* actions that are directed toward camera have impact. When one character strikes another, for instance, the blow carries greater power if it is directed toward the audience/camera (perhaps in an over-shoulder shot) than if played in a reverse angle.

Entrances are effective ways of beginning a film or, in television, of beginning an act. Psychologically, as characters enter the narrative world, the audience enters with them. Similarly, exits are often an effective way of ending a story. As characters move off toward the horizon, the audience simultaneously is withdrawing its attention from the story. Think of the many Charlie Chaplin comedies that ended on shots of the little tramp trudging off down the railroad tracks until he became a speck in the distance.

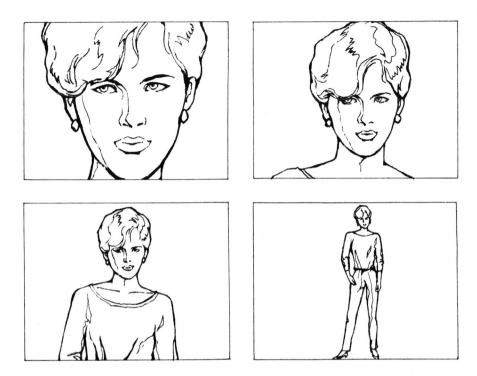

FIGURE 6.2   In TV and film, as on the stage, the closer performers get to their audience, the greater the impact they create.

Within a dramatic structure, however, exits can delay the action. Watch how
the director and editor treat exits in the next movie you see. Most editors eliminate
them because the scene's action has been completed and impatient audiences want
to move on. Occasionally, exits have special significance to the story. When they
do, of course, they should be played for their full value. When someone has been
defeated, for example, and the exit is a symbol of that defeat, then the exit has
meaning and should be given appropriate emphasis.

Emphasis Through Business

Emphasis can also be gained through the relationship between dialog and business or between dialog and movement. An old theatrical rule of thumb states:
Action before dialog points (emphasizes) the dialog; dialog before action points
the action. In the scene that began this chapter, if Debbie rises from the couch,
crosses slowly to Todd, sighs, and states, "I'll have her out of here by Friday,"

her *line* receives the impact. However, if she first tells him, "I'll have her out of here by Friday," and then crosses to the window, presses her forehead to the cold glass, and fights back tears, her *action* achieves the greater impact.

Consider:

> BLAISDELL
> Why am I going to kill you?

He stares at his knife, tests the sharpness of its blade.

> BLAISDELL
> (continuing)
> Because you bore me.

Compare that relationship between line and business with:

> BLAISDELL
> Why am I going to kill you? Because you bore me.

He stares at his knife, deliberately, tauntingly tests the sharpness of its blade.

In the first example, Blaisdell's line "Because you bore me" gained impact because he set it up by his knife business. The business whetted the viewers' appetite, built apprehension. In the second example, the knife business gained greater impact; the dialog had set it up.

Determining Significance

To achieve appropriate emphasis, directors must be aware of which moments in a play are significant and which are not—and the *degree* of significance generated by plot and characters. Although such awareness equates with an understanding of drama itself and is largely subjective, varying from director to director, a few clues here may be helpful.

Any drama cannot be continuously intense; audiences need periods of relaxation in which to recover from moments of dramatic stress. Without those contrasting moments, tensions would lose much of their impact. Shakespeare's use of farce in many of his tragedies demonstrates the effectiveness of such contrast.

The growth of tension usually begins in the script. As indicated in Chapter 4, each act generally starts on a quiet note. Then gradually, from scene to scene (and within each scene), emotions build to higher and higher levels of intensity, peaking in a moment of genuine crisis at the act's end. As emotions build, so does emphasis. In moments of emotional pressure, cameras usually are close. In moments of *extreme* pressure, cameras are *extremely* close. Thus, a director's first requirement: Emphasis should match the degree of dramatic tension generated by a scene. Let's examine a few other requirements for emphasis.

Pain   Pain is closely related to dramatic tension but is significant enough to warrant a category of its own. The degree of emphasis always should reflect the anguish generated within the characters. As the intensity of pain builds, the need for impact increases. Thus, in most scenes, the anguished character usually warrants greater emphasis than the others. This principle will be directly affected or modified by the degree of the spectators' emotional involvement with that character. If they care more, his or her anguish becomes significant. When they don't care, it becomes relatively meaningless.

Decisions   Many scenes build to a moment in which a character must make a key decision. The moment *before* that decision is made becomes the scene's climax and should therefore be emphasized, often by extending it. Major characters' decisions, however, do not always occur at the end of a scene. Whenever characters make decisions serious enough to alter their purpose, direction, or goals in a play, such moments merit emphasis.

As mentioned earlier, labels tend to become arbitrary. They oversimplify. We must realize that making a key decision is necessarily a pressured moment. Such a moment generates emotional tension and perhaps engenders mental anguish. Thus, all of the categories listed here for determining emphasis necessarily overlap, reinforcing and supporting one another.

Revelations   Directors generally emphasize major revelations in plot or character. Such revelations often become turning points for a plot (reversals) or change the audience's concept of a character. In the motion picture *The Third Man*, for example, Harry Lime (played by Orson Welles) is "killed" prior to the film's opening. Later, when the director reveals that Lime is alive and well, the moment warrants special emphasis.

Revelations sometimes give viewers special privilege, providing them with greater knowledge than the film's characters have. When a grieving son attends his mother's funeral, for example, and we see him try to conceal a smile, that moment requires emphasis because (1) it reveals a significant aspect of his character and (2) it gives us special, privileged information denied those inside the drama. D. W. Griffith's extreme closeup of Mae Marsh's hands in *Intolerance*, fingers twisting nervously, revealed the depth of her feelings at her husband's murder trial.

Directors often must walk the narrow line between emphasis and overemphasis, especially in situations that border on melodrama. When we discover that a loving wife is actually plotting her husband's demise, for example, or that a timid bank teller is really a psychotic assassin, the director must be careful to avoid the overstatement that cheapens and blunts such insights.

The next chapter will discuss emphasis at greater length, specifically in its relationship with the camera.

SYMBOLIC RELATIONSHIPS

Have you ever watched an airline movie without earphones? Or studied a TV soap opera with the sound off? Chances are, if the show was well directed, you were able to make some fairly shrewd inferences about the relationships between characters—not so much from their gestures or facial expressions as from their movements and physical positioning in relation to one another.

Let's return to the Todd-Debbie scene one last time. We'll turn off the volume and pretend to know nothing of their story. We see a balding man pursuing an attractive, middle-aged woman. She breaks away from him and sits on a couch. He towers over her, staring down at her. She says something and starts to cry. Maybe he's accusing her of unfaithfulness, maybe of squandering their money. We don't know, of course, without hearing the words. But the visual image provides clues to their *relationship*.

Clearly the balding man is the aggressor, the dominant person making angry demands, and she is troubled, retreating, unable to face him. The picture alone provides insights. Ideally, the behavior and physical positioning of characters in every dramatic scene should provide that same kind of dramatic context, should create a recognizable representation of the characters' emotional or symbolic relationship.

Let's examine a typical soap opera situation and see how the physical positioning of characters might symbolize or reflect their emotional relationship. Larry's employer-mistress pays a visit to his wife. Her purpose: to convince the wife to divorce Larry so that he will be free to marry her. Larry's wife is shocked; this is the first she has known of her husband's affair with his attractive boss. More than shocked, she is blazing angry. At that moment, Larry enters, home early from work.

As director, where do you position Larry in this scene? Initially, probably in the middle, employer-mistress on one side, wife on the other, pulled in two directions by the women who love him. How does Larry feel about these women? If he truly loves his wife, he will go to her, trying to explain his actions, seeking forgiveness. Will she remain beside him, permitting this closeness? Probably not. Motivated by anger and resentment, she will move away. She has distanced herself from her husband emotionally and now she will distance herself from him physically. The external mirrors the internal. What of the mistress? If she loves Larry, she will go to him. Perhaps he moves away, rejecting her. Perhaps the mistress will place herself between husband and wife, physically as well as symbolically the divisive force in their marriage.

Presented in such naked terms, with one-dimensional characters, such staging appears simplistic, old fashioned, melodramatic, and arbitrary. With the complexities offered by more mature characters and the introduction of plot subtleties, however, the scene achieves some degree of dimension and believability.

When blocking a scene, begin with questions. What are the emotions that drive your characters? Are they angry with each other, emotionally far apart? Then perhaps your staging should place them far apart. If the scene's progression allows them gradually to rediscover their love, to become emotionally closer and closer, then perhaps your blocking should parallel that pattern, bringing them physically closer and closer, creating in the scene's final moments an atmosphere of delicious intimacy.

Examine your scene as written. Try to visualize the movement of characters in your mind. Let their emotions create your mental imagery. Let their relationship determine their relative positioning. Draw a quick diagram of the scene, just as you saw it in your mind. Then apply this test: Would an audience be able to deduce the relationships of your characters just from your staging?

Watch dramatic programs on TV. Turn off the volume. If the programs have been skillfully directed, the silent pictures will speak loudly.

BEATS

In Chapter 5 you learned that most scenes are composed of smaller units called *beats*. These are scenes within scenes, each of which usually builds to its own small climax. Each beat marks a change of direction in a scene and often introduces new subject matter or a new approach.

Actors trained in the so-called Method school of acting usually prepare their work by breaking scenes into individual beats, sometimes labeling those beats— for example, "disgust" or "new understanding of father." Such preparation helps actors to understand the emotional attitudes they will play and to plan the transitions between beats.

Defining beats is subjective. That is, two actors may disagree on the definition of beats in their scene because each plays a different character and enters the scene with a different emotional attitude. Because directors take a more objective view, they may define beats differently from either of the actors.

What is the significance of beats? First, they help us to understand a scene's content. Secondly, they help us in staging. Because a new beat veers in a new direction, it suggests a new pattern of movement for actors. A new beat might provide the opportunity for an actor to rise and move to the window or put aside a magazine he or she has been leafing through.

Read the following scene. When you have finished, see if you can pinpoint the beats, the smaller dramatic units, each of which is marked by a change in direction, attitude, or subject.

FINAL CURTAIN[5]

THE SCENE: BACKSTAGE, LATE EVENING.
The play is over. Actress ANN FOLEY crosses to pick up her coat, turns to exit, discovers ERIC, the play's director, staring at her.

ERIC

Congratulations, Ann. You really outdid yourself tonight.

ANN

Thanks.

ERIC

It was probably the most superficial performance you've given.

ANN

Oh, come on, Eric. The audience loved me.

ERIC

Sure they did. But you can do better.

ANN

Look, I'm tired. Let's talk about it tomorrow, okay?

ERIC

This could be a fabulous play, Ann. But it takes work. Tonight we looked like a bunch of amateurs.

ANN

Thanks.

ERIC

I have the feeling you're acting when you're out there. Putting on a big show, trying to impress everyone with how talented you are.

ANN

Somebody was impressed. Did you hear the applause?

---

[5]Courtesy of Samuel French, Inc.

ERIC

It's not what the audience thinks.
>   (moves chairs)
All right, this is the couch. And this table is the bar
where....

ANN

Dammit, Eric, I'm tired. If you want to rehearse, we'll
do it tomorrow.

ERIC

Tomorrow you'll be cold. We'll do it tonight. All right?

ANN
>   (after a pause)
All right.

ERIC

Fine. This is the table where the knife is.

(Gets knife, puts it on table, arranges scarf around his neck)

All right, let's take it from my entrance—middle of
the last act. You're on the couch.

Ann sits on the "couch." Eric crosses to her.

ERIC

Cue.

ANN

Jeff!

ERIC

So this is where you live. Pretty fancy for a cocktail waitress.

ANN

This place belongs to a . . . a friend of mine.

ERIC
>   (grabs her roughly)
Who?

ANN

What difference does it make?

ERIC

I want to know who.

ANN

Jeff, please, I don't want to hurt you.
>   (then)
All right! It . . . it's David.

ERIC

My brother? You're lying.

> ANN
> No. Jeff, we love each other. We're gonna be married.

> ERIC
> Come here. Closer. That's nice. You're not going to marry David. I'm sorry but I can't allow it. He's always gotten everything that was meant for me. Ever since we were kids. Well, not this time.

> ANN
> Eric, you're hurting me . . .

ERIC clutches her tightly, fingers biting into her skin. She breaks away but ERIC pursues her, hand still on scarf, staring at her sadly.

> ANN
> Eric, <u>stop</u> it!

He comes out of his part slowly, hand dropping from the scarf.

> ANN
> You can't live the part.

> ERIC
> Why not? Maybe that's your trouble. You won't make it real. How can you call yourself an actress if you won't make it real?

> ANN
> Because . . . well, because it's a phoney scene. It couldn't possibly be real.

> ERIC
> No?

> ANN
> Just think about it. You're going to strangle me with your scarf. And lo and behold, there is a kitchen knife conveniently lying on the bar, which I pick up and stab you with. You call that real?

> ERIC
> You could make it real.

> ANN
> No way. It's contrived and it's phoney.

> ERIC
> Well, maybe you're right. Maybe it's the scene's fault. You know, I was just thinking. . . .

> ANN
> What?

ERIC

I was thinking how the play is a lot like....

ANN

Like what?

ERIC

Never mind.

ANN

No, go ahead. Like what?

ERIC

That the people in the play—Jeff and Marlene—are a lot like ... like you and me.

ANN

Come on, that ... that's crazy. They're nothing like you and me. She's a cocktail waitress and....

ERIC

No, I mean the way they feel about things. The way ... the way Jeff feels about Marlene.

ANN
(nervously)

I don't know what you mean.

ERIC

Yes, you do. Jeff is crazy about Marlene. And I ... well, you know how I feel about you, Ann. You've known it all along.

ANN

You can't be serious. You never paid that much attention....

ERIC

You knew how I felt. How I feel.

ANN

It's really getting late, Eric. We can talk about it tomorrow.

ERIC

And you're just like Marlene, aren't you? Smiling, inviting ... until I give you the lead in the play. Then it's "Sorry, Eric. We can talk about it tomorrow."

ANN
(hotly)

That's not fair. I never played you for the lead. I never "invited" you! Look, I appreciate all your help, but....

                        ERIC
But you don't love me.

                        ANN
Of course I don't. That's ridiculous. You've never given me a
chance to.

                        ERIC
You see? It <u>is</u> like the play.

                        ANN
No, it isn't. Not at all. Goodnight.
            (starts to exit)

                        ERIC
And now you're Marlene. Saying goodnight,
            (grabs her)
walking away. Pushing me out of your life. . . .

                        ANN
Eric, let go of me. I'll scream, I swear. . . .

                        ERIC
Go ahead, Marlene. Scream.

                        ANN
            (struggling)
This is crazy. What's happened to you?

                        ERIC
If you won't love me. . . .

He clutches her tightly with one arm. The other reaches to his
scarf. He pulls it from his neck.

                        ANN
Stay away from me! Stay away!

                        ERIC
            (advances, twisting the scarf)
. . . you won't love anybody.

ANN looks around wildly, spies the knife on the table, grabs it.

                        ANN
Don't come any closer!

He lunges forward and twists the knife from her hand.

                        ANN
Eric!

Now, half smiling, he flings the knife to the floor.

> ERIC
>
> So, the play isn't realistic. It doesn't ring true. And the knife is phony and contrived. You've never given a better performance in your life.
>
> ANN
>
> Performance?
>
> ERIC
>
> Your reactions were very honest. Let's see if you can do as well on stage tomorrow night.
>
> ANN
>
> But . . . you mean you? . . .
>
> ERIC
>
> I'm going home now. My wife will be waiting for me.
> (gentle smile)
> Goodnight, Ann.

He exits. ANN sinks into a chair, exhausted.

Before reading on, take a moment to figure out the beats in the scene between Ann and Eric. Draw lines between the beats to separate them.

Now let's see if your breakdown agrees with mine. We might label the first beat "preparation." It ends as the actors take their places to begin the rehearsal. The actor playing Eric might call the first beat "disgust," "anger," or "disappointment." The actress playing Ann might call the first beat "impatience."

The second beat, clearly, is the play within a play, ending with Ann's lines "Eric, you're hurting me" and "Eric, *stop* it!"

The third beat is tricky. I asked myself, "At what point does Eric conceive the idea of his charade?" For me, it happens between "Well, maybe you're right. Maybe it's the scene's fault," and "You know, I was just thinking." Draw your "beat" line through the middle of that speech.

The fourth beat (and most scenes don't have this many) is the charade itself in which Eric pretends to be in love with Ann, building to his knocking the knife to the floor.

The final beat is the aftermath, when Ann realizes that Eric's pretense of love was a hoax perpetrated to help her performance in the play.

Did your selection agree with mine? It doesn't have to. There can be total disagreement among directors concerning dramatic beats because each director has a separate and personal vision of how a scene should play.

Now try to visualize the staging for "Final Curtain." Read the scene again, visualizing the action beat by beat. You will find, I think, that each beat marks a change in movement and attitude by the actors. For example, when Eric conceives the idea of his charade, that idea will move him toward Ann (involving move), and as Ann becomes nervous and then apprehensive, she will back away from him (disinvolving move).

BLOCKING CAMERA TO ACTORS

Most experienced directors begin the taping or filming of a scene by blocking their actors. Once satisfied that movement and business are comfortable, honest, and dramatically effective, they then block their camera(s) to cover the action. Why not the other way around? Why not start with dynamic camera angles and then block actors to camera? Because, as indicated earlier, audiences become involved with *people*. So close and sensitive is this relationship that, for the duration of the program, the spectator *becomes* the protagonist. Dishonesty in staging or performance creates obstacles to that rapport.

Although the director and crew remain acutely aware of the camera's important, extremely creative functions, the average spectator notes them only as part of the production mosaic, aware of angles and camera movement on a subconscious rather than conscious level. Surprisingly, even production veterans tend to get lost in drama, temporarily oblivious to its technological aspects, submerged in a fantasy world until some extraordinary camera movement or editorial fillip snaps them out of it.

When a director begins blocking with the camera rather than with actors, staging can appear manufactured. An actor's movement, forced into circumscribed patterns, then must accommodate the needs of the glass-eyed monster instead of emerging honestly from thoughts or emotions.

The director's hand should remain invisible. Action that appears contrived or arbitrary makes it difficult for an audience to forget that those people on the screen are actors wearing costumes and makeup, speaking scripted words in front of painted sets.

Occasionally, inexperienced directors loudly proclaim their presence with flashy camera angles and attention-getting gimmicks. Every overstated shot shouts, "Look at me! I'm directing!" They are showing off; they are also destroying the show, making the audience aware of contrivance and artifice by calling attention to themselves.

Sensitive acting, well directed, can make a show succeed in spite of mediocre camera work. Mediocre acting and staging, however, usually condemn a show to failure despite the most brilliant camera work in the world.

Actors first, then camera.

BLOCKING ACTORS TO CAMERA

Is it ever appropriate to think of camera first? Yes, dozens of situations require that directors begin their planning with the camera's perspective.

At the start of a show or sequence, *orientation* frequently becomes a concern. Where is the action taking place? On another planet? In an apartment building? In a submarine? When orientation is more important than human values, select an angle that will clearly establish the locale for your audience. Such footage, often appearing under opening titles, also establishes the mood or atmosphere of the show, much as the overture in a Broadway play prepares the audience emotionally for the action to follow.

On occasion, the camera must make a *plot point*. Perhaps an angle past the leading lady must include a closet in which the villain hides. Perhaps you must begin a scene on a cigar butt in an ashtray to establish that the villain has been here. Perhaps, when the villain is present in a scene, his identity must be kept secret. Then the camera can shoot from a special angle that reveals only his shadow or his legs or his hand wearing a distinctive ring.

In shows where believability is not a factor, where honesty of performance is irrelevant, you may begin anywhere you choose. In farce, for instance, the director's only goal is to get laughs. If that means phony action, dishonest line readings, and distortion lenses, so be it. As comedy moves away from farce, however, and approaches reality (such as the situation comedy), honesty of performance becomes a more critical factor.

Sometimes camera location is determined by a scene's mood or atmosphere, as when a camera begins on a window or on a screen door banging restlessly in the wind to establish stormy weather.

Occasionally camera position is determined by optical effects (**EFX**) or animation. Perhaps a factory on fire will be *matted* (optically printed) into your scene after you have photographed it. To accommodate the matting, you must lock your camera down in one master angle. Then, as in the previous circumstances, the blocking of actors cannot be the primary consideration.

Once a scene has been blocked, rehearsed, and lighted, actors frequently must adjust the blocking so that camera angles will be clean and effective and will make the dramatic statement required by the director. Sometimes one actor will obstruct another while crossing downstage or will throw a disruptive shadow. Any professional actor can comfortably make simple adjustments that will not affect the integrity of his or her actions. When actors' positions are critical, be sure that they are clearly marked on the stage floor with chalk or tape. (See Figure 6.3.)

PREPARATION

Most directors plan their blocking ahead of time, the night before shooting or even earlier. As I have indicated and will certainly indicate again, meticulous preparation is a hallmark of the successful director.

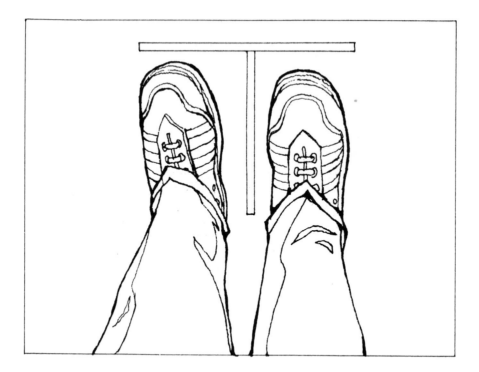

FIGURE 6.3   Actors' positions are often marked on the stage floor with chalk or tape. Two strips of tape in the shape of a T make it easy for actors to "hit their mark."

During the always-too-short preparation period, the director has selected locations, discussed the script and a succession of rewrites with producer and writer, examined props, met with the cinematographer or technical director to debate the style and "look" of the project, selected stars and supporting cast, examined and approved sets and discussed their dressing, studied and reorganized the production schedule, sweated over script deletions required by budget, approved the cast's wardrobe, makeup, and hairdressing, plus a hundred other tasks that overflowed each preparation day and made it appear that the project would never get off the ground.

By the time directors begin production, they know each scene practically by heart. They understand their characters in depth. They have explored their characters' childhoods; their manner of speaking, walking, and dressing; their fears; the kind of cars they drive; their relationships with other characters; perhaps even their sexual proclivities.

Before the project begins, the art director gives the director floor plans of all major sets plus photographs or sketches of key locations that the director will use in blocking the action. Inevitably, during preparation, the director begins to visualize how many key scenes will play. That visualization will change in some degree as shooting begins and the characters acquire dimension.

The final blocking of a day's work (or reexamination of an earlier blocking) usually is accomplished the night before shooting. With floor plans and script rewrites in hand, the director sets out to define the actions and business of characters in each scene. Directors, particularly those who are inexperienced, should ask themselves such questions as:

1. Do I clearly understand the writer's intentions? Reexamine the scene in terms of its place in the plot progression. What scene precedes it, what scene follows? What will be the transitions? Is this a key scene or one of relative unimportance?

2. When was the last time these characters were together? What happened then? What was their relationship then? Has anything happened since then that might affect their relationship?

3. What were these characters doing immediately before the scene begins? Will their prior activities affect this scene?

4. What normal activities would the characters be engaged in at this time of day in this setting? (You are searching for logical *business* for the scene.)

5. What will be the characters' emotional state? Will it change during the scene? How will their emotional states affect their movements and business during the scene?

6. Where will the focus of interest (emphasis) lie at each moment? How will I direct audience attention there, with appropriate impact?

7. What will be the audience's position in this scene? How can I generate suspense (tension) and build it to a climax?

8. Will the scene create any special camera or lighting problems? Have I discussed them with my cinematographer or technical director? Generally, what kind of camera coverage will the scene require?

9. Will the dialog fit comfortably into the actors' mouths? (Although dialog should have been polished earlier, television's last-minute urgency often makes such questioning necessary.) Will the dialog need to be adjusted?

10. How much time should I allow myself for the shooting of this scene? Shooting time varies in direct proportion to a scene's importance in the script. Other factors include night shooting on location, weather hazards, and such physical problems as stunts, helicopters, and animals.

After working out the tentative blocking, the director then makes a rough sketch, usually on the blank (left) script page facing the dialog. Examples of such sketches appear in Figure 6.4. Some directors prepare simple, crude diagrams; others draw detailed sketches coded with one specific color for each actor, tracing his or her movement through the scene, another color for the camera, and still another enumerating properties and other physical requirements.

Because the day's work has been carefully prepared, the director walks onto the stage with confidence. Rehearsal procedures vary enormously, but the following pattern is used most frequently in film.

While the crew watches, the director walks through a scene with the actors, trying out blockings tentatively, making sure that the pattern is comfortable for the cast and that it makes the necessary dramatic statement. Because a director has blocked a scene in a certain way on paper doesn't mean it will necessarily end up that way on film. Sometimes homework provides the perfect blocking for a scene; sometimes the blocking requires adjustment; sometimes it goes out the window. Actors often come up with excellent ideas that will change or modify a scene's movement or business. Secure with preblocked concepts, the director can remain flexible, objectively assessing suggested variations rather than clinging defensively to a fixed blocking. If proposed changes are not appropriate, the director must be firm, following original instincts and returning to the original blocking.

Onstage blocking usually begins with a brief discussion of a scene's content followed by tentative explorations of its staging. Actors walk through various patterns, feeling their way, allowing the scene's emotional thrust to guide them, responding to suggestions from the director.

Director Walter Grauman notes that when actors follow their instincts, working out movement and business for themselves, their blocking frequently turns out almost identical to the blocking he had devised at home the previous night. Some actors feel more secure working out a blocking for themselves; others are delighted to have a director guide them. *All* actors need the security of a director who has studied the scene and is ready with answers should problems develop. Chapter 5 has additional notes on rehearsal techniques.

Once everyone has agreed on the pattern of movement, the actors' positions are marked on the floor and the director discusses coverage with the cinematographer. Experienced directors are seldom arbitrary; they encourage input from their cinematographers, who often make major contributions to a scene's visual effectiveness.

Now director and actors leave the set, getting out of the way of the camera crew and electricians. For lighting purposes, stand-ins take the place of the principal cast. Off in a corner, the director and cast go over the scene's dialog. They discuss the dramatic spine of the scene, reexamining relationships and the emotions of the characters, rewriting words or phrases that seem awkward.

On some videotaped shows (rarely in film), the director schedules cast readings prior to the start of production. Seated around a table with the producer and the writer, the cast tentatively reads through the material, exploring its content. Sometimes several days are devoted to such readings. When scenes don't work,

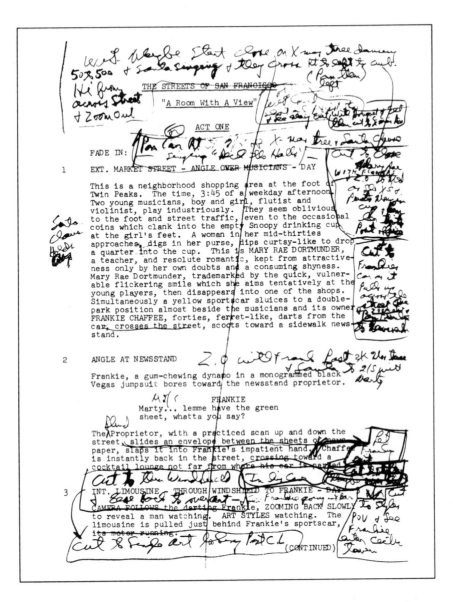

THE STREETS OF SAN FRANCISCO

"A Room With A View"

ACT ONE

FADE IN:

1   EXT. MARKET STREET - ANGLE OVER MUSICIANS - DAY

This is a neighborhood shopping area at the foot of
Twin Peaks. The time, 3:45 of a weekday afternoon.
Two young musicians, boy and girl, flutist and
violinist, play industriously. They seem oblivious
to the foot and street traffic, even to the occasional
coins which clank into the empty Snoopy drinking cup
at the girl's feet. A woman in her mid-thirties
approaches, digs in her purse, dips curtsy-like to drop
a quarter into the cup. This is MARY RAE DORTMUNDER,
a teacher, and resolute romantic, kept from attractiveness only by her own doubts and a consuming shyness.
Mary Rae Dortmunder, trademarked by the quick, vulnerable flickering smile which she aims tentatively at the
young players, then disappears into one of the shops.
Simultaneously a yellow sportscar sluices to a double-
park position almost beside the musicians and its owner
FRANKIE CHAFFEE, forties, ferret-like, darts from the
car, crosses the street, scoots toward a sidewalk newsstand.

2   ANGLE AT NEWSSTAND

Frankie, a gum-chewing dynamo in a monogrammed black
Vegas jumpsuit bores toward the newsstand proprietor.

                        FRANKIE
            Marty... lemme have the green
            sheet, whatta you say?

The Proprietor, with a practiced scan up and down the
street, slides an envelope between the sheets of newspaper, slaps it into Frankie's impatient hand. Chaffee
is instantly back in the street, crossing toward a
cocktail lounge not far from where his car is parked.

3   INT. LIMOUSINE - THROUGH WINDSHIELD TO FRANKIE - DAY
    CAMERA FOLLOWS the darting Frankie, ZOOMING BACK SLOWLY
    to reveal a man watching. ART STYLES watching. The
    limousine is pulled just behind Frankie's sportscar,
    its motor running.

                                            (CONTINUED)

FIGURE 6.4   These notes and sketches from Walter Grauman's shooting script of an
episode of "The Streets of San Francisco" demonstrate a director's attention to
detail and the complexities of staging. The sketches shown here were drawn on
the back of the first page of script. (From "A Room with a View" written by Del
Reisman and Cliff Gould. © 1973 QM Productions. Used with permission of The
Taft Entertainment Company and Walter E. Grauman.)

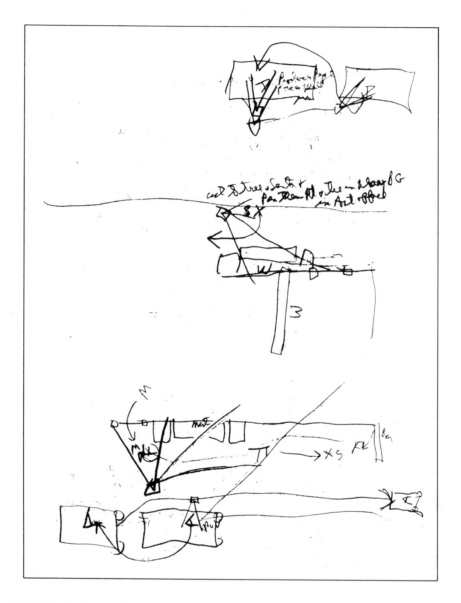

FIGURE 6.4   *(continued)*

they are returned to the writer for reexamination. When speeches seem uncomfortable, the writer fixes them. Problems concerning characters and relationships are ironed out now, removing pressure from the production schedule.

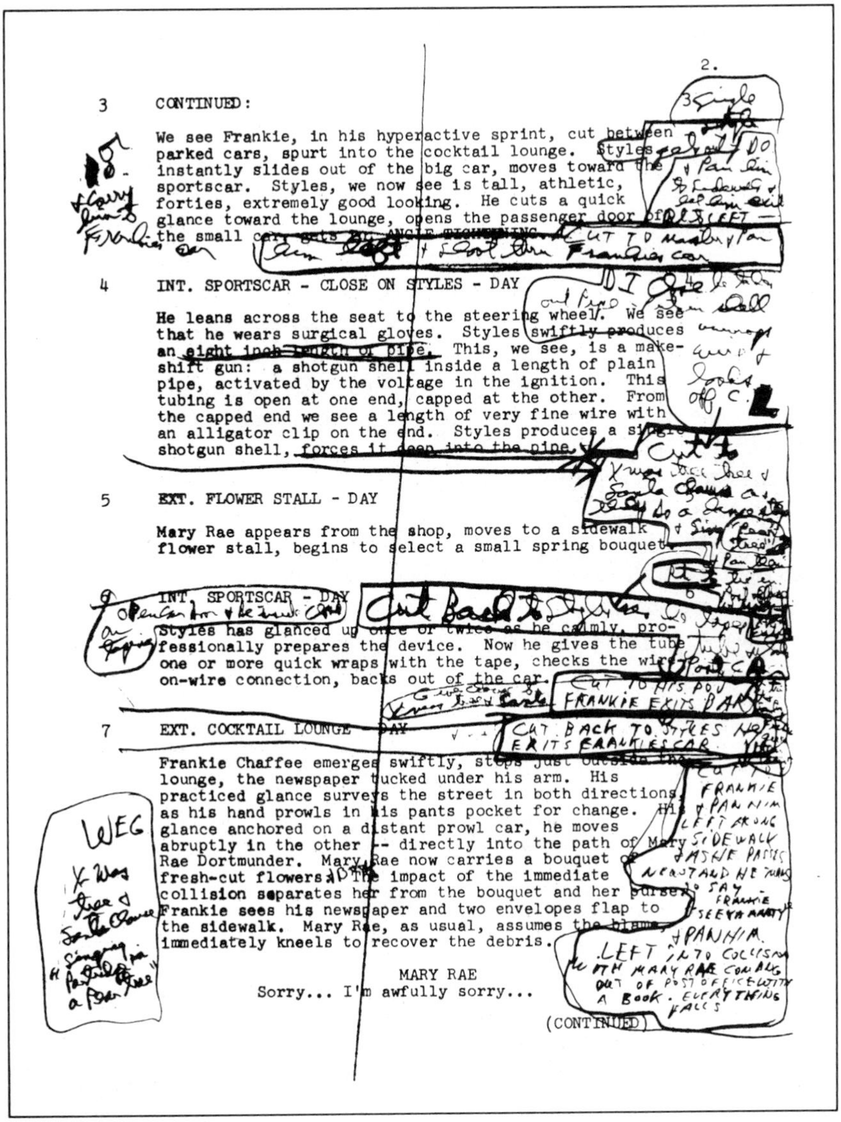

2.

3 CONTINUED:

We see Frankie, in his hyperactive sprint, cut between parked cars, spurt into the cocktail lounge. Styles instantly slides out of the big car, moves toward the sportscar. Styles, we now see is tall, athletic, forties, extremely good looking. He cuts a quick glance toward the lounge, opens the passenger door of the small car, gets in. ANGLE TIGHTENING.

4 INT. SPORTSCAR - CLOSE ON STYLES - DAY

He leans across the seat to the steering wheel. We see that he wears surgical gloves. Styles swiftly produces an eight inch length of pipe. This, we see, is a make-shift gun: a shotgun shell inside a length of plain pipe, activated by the voltage in the ignition. This tubing is open at one end, capped at the other. From the capped end we see a length of very fine wire with an alligator clip on the end. Styles produces a single shotgun shell, forces it deep into the pipe.

5 EXT. FLOWER STALL - DAY

Mary Rae appears from the shop, moves to a sidewalk flower stall, begins to select a small spring bouquet.

6 INT. SPORTSCAR - DAY

Styles has glanced up once or twice as he calmly, professionally prepares the device. Now he gives the tube one or more quick wraps with the tape, checks the wire-on-wire connection, backs out of the car.

7 EXT. COCKTAIL LOUNGE - DAY

Frankie Chaffee emerges swiftly, steps just outside the lounge, the newspaper tucked under his arm. His practiced glance surveys the street in both directions as his hand prowls in his pants pocket for change. His glance anchored on a distant prowl car, he moves abruptly in the other -- directly into the path of Mary Rae Dortmunder. Mary Rae now carries a bouquet of fresh-cut flowers. The impact of the immediate collision separates her from the bouquet and her purse. Frankie sees his newspaper and two envelopes flap to the sidewalk. Mary Rae, as usual, assumes the blame, immediately kneels to recover the debris.

MARY RAE
Sorry... I'm awfully sorry...

(CONTINUED)

FIGURE 6.4 *(continued)*

Once the set has been lighted, the director and cast begin final rehearsal—with camera. Now the director must decide when the actors are "into" a scene, when optimum performance has been reached as well as perfect coordination between camera crew and cast. Traditionally, directors do not ask actors to give their top performance in rehearsal; most actors would refrain anyway, because they try to save their best efforts for tape or film. They dislike reaching a dramatic peak too early, fearing that they will be unable to achieve it again. Some camera-wise professionals are notorious for saving their best performances for closeups, thereby hoping to ensure that those closeups will be included in the final cut.

An experienced director who notices a problem with a performance or a line reading does not embarrass the actor by making a correction in front of the cast and crew. Instead, the director takes the actor aside and quietly discusses the problem. Likewise, if the director is unhappy with a member of the camera crew—a dolly has been too fast or a pan too slow—the director takes it up quietly with the cinematographer, who then discusses it with the crew member.

Once everyone is ready for a "take," final lighting adjustments are made. Makeup people and hairdressers hurry into the set to make certain that the cast's appearance is camera perfect.

The assistant director rolls film. When sound has attained recording speed, the director calls, "Action," and the filming begins. There may be several takes before the director feels the scene is acceptable and advises the camera assistant to print it. Usually, the director then repositions the camera for editorial coverage, breaking the master angle into two-shots, closeups, or whatever angles may be necessary for appropriate dramatic emphasis.

CHAPTER HIGHLIGHTS

— Director Elia Kazan's words, "Directing finally consists of turning Psychology into Behavior," establish the theme for this chapter. Although audiences generally believe that actors' movements in a scene arise spontaneously, professionals recognize that such actions grow from a script, from a director's vision, and from painstaking rehearsal. The director's manipulation of actors on a stage is called *blocking*.

— In general, actions reveal characters' thoughts or emotions more clearly than dialog does. In many scenes, dialog provides a meaningless accompaniment to the more significant content, revealed through actions. Action also demands greater spectator participation because its meaning must be interpreted.

— Specific character activities such as combing hair or mixing drinks are called *business*. While business contributes significantly to the reality of scenes, it should also reveal facets of the character who performs it or contribute to the dynamics of a scene.

— Part of a director's job is to reveal characters' thoughts or emotions through actions. Symbols often provide an answer. For example, the actions of a young woman in relation to a stuffed animal given her by her boyfriend will reveal her feelings about him. Thus, when she trashes the animal, she is, in a sense, trashing him.

— When characters move in a scene, usually there is some *internal motivation*. Such motivations fall into three categories: involvement, disinvolvement, and emotional turbulence. *Involvement* moves a character toward another; it is either an aggressive or a seeking action, motivated by such emotions as anger or love. Its forward moves sometimes arise out of need for an object or information. *Disinvolvement* suggests retreat, movement away from another character, and generally is triggered by such emotions as fear, hatred, or anger. Characters often retreat into business. *Emotional turbulence* suggests nondirected movement and is triggered by anxiety or nervousness. Nondirected movement also is motivated by drugs, alcohol, or psychosis.

— Some character moves are generated by considerations *outside* the character. The script accounts for most such moves, requiring actors to answer telephones, kiss, kill, dress, or undress. These are movements necessary to the simple telling of a story. The second external reason for character movement is to provide dramatic emphasis.

— The theater provides emphasis through staging. Stage directors emphasize actors by moving them downstage, centerstage, or facing the audience, or by contrasting them with other actors through costume, lighting, or stage position. Many of these stage techniques may be adapted to videotape or film with the camera replacing the live audience.

— The relative positions of characters within a frame often symbolize their emotional relationships. Characters far apart emotionally often are staged apart, the external mirroring the internal.

— In staging a scene, directors usually block actors before they block camera. The reason: What happens *in front of* the camera is more significant to audiences than the camera work.

— Sometimes directors place camera considerations first, such as when the director needs to orient the audience, usually at the start of a show or sequence; to use the camera to make a plot point; to establish a mood or atmosphere; and to obtain the photographic basis for animation or optical effects.

— Meticulous preparation is a hallmark of a successful director. Preparation culminates on the night before shooting. The director finalizes each scene in the schedule, blocking (or reblocking) the action, and examining all aspects of the day's work.

— On the soundstage, the director may suggest a pattern of movement or the actors may prefer to find their own blocking, allowing the scene's emotional thrust to steer them. In either event, the director should be prepared.

PROJECTS FOR ASPIRING DIRECTORS

1. Nineteen-year-old Stephanie fears that she is dying of cancer. What business can you invent for her that will reveal her fears? She is alone in her apartment.

2. Write a telephone conversation in which Stephanie tells her overprotective mother how well she is doing at college, pretending gaiety in spite of her fears. Find business for her to perform during the phone conversation that will reveal the lie in her words. We don't have to hear her mother.

3. Find business that will enhance the dramatic values in these scenes:
   a. A young man tries to get a shy young woman to confess her feelings for him.
   b. A murderer intimidates her boyfriend, swearing him to secrecy concerning her crime.
   c. An 8-year-old tries to conceal from his mother a vase he has broken. (You can play this for comedy if you want.)

4. Watch a soap opera. Make a list of (a) involving moves, (b) disinvolving moves, and (c) noninvolving moves. See if you can find an example of a character retreating into business.

5. Stage a scene on paper between Cinderella and her stepmother. Cinderella wants to go to the ball. Her stepmother refuses to let her. In your staging, be conscious of involving or disinvolving moves and symbolic relationships (physical positions representing emotional positions). You don't have to write the dialog. The scene will have 3 beats: (a) Cindy pleads, stepmother refuses, (b) Cindy gets an idea (to telephone her fairy godmother) and tries to conceal her excitement, (c) her stepmother, suddenly suspicious, questions Cindy about her plans. The scene should play in the kitchen. Draw a rough diagram of the kitchen set.

6. Block the actions of characters in the following short scene, but this time draw the most significant moments as frames in a storyboard: a series of sketches (frames), each in screen proportion—three units high by four units wide. Each frame depicts key moments in a scene *as the camera views them.* Artistic talent is not required. Figure 6.5 is a typical storyboard pattern.

   This is a challenging project so take time to think about your staging. What does the setting look like? What moves or actions will your characters take within the setting? What props might be present that would provide revealing business? Your storyboard should present not less than ten key moments in the scene. And consider where you will place your camera at these moments.

   After you have created the scene in your imagination (and you can certainly make notes or preliminary sketches), then draw the frames with a

FIGURE 6.5   Sample storyboard frames.

pen or felt-tipped marker. After you have completed each frame, write on the back of the paper a description of the action the frame depicts. Note: The word **filter** refers to the thin, metallic quality of a telephone voice. See glossary.

The luxurious office of the president of Bronwood Pharmaceuticals. AMY ASHLEY, 29, shakes two aspirin from a bottle and washes them down with a glass of water. Her business partner, JOHN LYNCH, 31, enters.

AMY
Did you forget how to knock?

JOHN
Do you knock when you come into my office?
Look, we've got a problem.

> AMY

It'll have to wait.

> JOHN

I'm sorry, this can't wait.

> AMY

Well, I'm sorry, too, but I've got a headache.
Some other time, okay?

> JOHN

You're going to have a worse headache when I
get through.
> (flips intercom switch)

> AMY

What the hell are you doing?

> JOHN

Making sure no one can hear us in the outer office.

> AMY

That's pretty melodramatic. Did Erika hire
a detective and have us followed?

> JOHN
> (hands report to her)

This just came in from the lab. Our last shipment of
diahexene was polluted.

> AMY

Toxic?

> JOHN

The chemist said it'll attack the nerve centers. The
most pleasant thing that can happen is blindness.

> AMY

How ... how much did we ship out?

> JOHN

Only four cases. Enough to blind a hundred
people.

> AMY

Where'd they go?

> JOHN

Two back east. One to Canada. The other here in
town. I've got a call in to County Medical. We'll get
the tracers out before lunch.

                              AMY

No.

                              JOHN

It's the fastest way, Amy. County'll put it right on
the telex. And our tracers will. . . .

                              AMY

I said no. Cancel your call to County Medical. Cancel
the tracers.

                              JOHN

Are you crazy? The stuff's poison.
                        (pause)
What do you want me to do?

                              AMY

Make a few discreet phone calls. Get the stuff back
quietly. No fuss. No panic. No one has to know <u>why</u>
we're calling it back.

                              JOHN

Suppose we don't get it all? Once the stuff's out of
here, we lose control. Amy, for God's sake, let me
notify the authorities!

                              AMY

No, I'm sorry, hon. When you offered to let me sit in
this chair, you agreed to let me make the final decisions. It's worked pretty well so far, hasn't it?

                              JOHN

So far. But now . . . you're playing Russian roulette.

                              AMY

I'm gambling. We've gambled before, haven't we?

                              JOHN

I won't do it.

                              AMY

Yes, you will. What about the chemist? Does he know?

                              JOHN

He only knows the stuff's been polluted. He doesn't
know it's gone out.

                              AMY

Good. Then it's settled.

                              JOHN

It's <u>not</u> settled. I own as much stock in this company as you do. I'm not going to risk blinding people
when we can stop it. This time you're overruled!

                              AMY

If you call County Medical, so help me God I'll tell
Erika our whole story. I'll smash your marriage just
like you're smashing this company.

> JOHN

I can't believe you.

> AMY

What's it going to be, John? Half owner of a multi-million-dollar company with fringe benefits . . . or instant disaster? Well?

The telephone rings. AMY picks it up.

> SECRETARY'S VOICE
> (filter)

Dr. Feldman on line one.

> AMY

Thanks.
> (punches button)

Hi, George.

> FELDMAN'S VOICE
> (filter)

Hello, Amy. Is it all right if we change your appointment to tomorrow? I've got an emergency.

> AMY

I was hoping to come in today. I've had this headache. . . .

> FELDMAN'S VOICE
> (filter)

Surprised to hear that. Especially after that shot I gave you Monday. Diahexene is supposed to work miracles.

> AMY

What . . . what did you say?

> FELDMAN'S VOICE
> (filter)

Diahexene. We just got a shipment in from your company. Matter of fact, you're the first person I've treated with it.

AMY hangs up the phone slowly.

> JOHN

What's wrong?

> AMY
> (after a moment)

Tell . . . tell County Medical to put the big news on their telex.

JOHN stares at her a moment, puzzled, then hurries out. AMY shakes her head in shock and bewilderment.

CHAPTER

7

STAGING THE CAMERA

The camera has three basic functions. The first is obvious: *It covers the action*. It tells the story; it reveals what's happening. If this function is ineffective, the audience won't stick around to watch the other two.

The second function echoes a major discussion from the preceding chapter. The camera calls attention to the significant; it minimizes or eliminates the unimportant. And it creates various degrees of impact for significant elements. Thus, the *camera creates appropriate emphasis*. The discussion of camera emphasis complements the examination of actor emphasis in Chapter 6.

Less obvious than the first two functions is that the *camera contributes to mood and atmosphere*. Every aspect of camera usage makes some definable dramatic statement, obvious or subtle, that helps to shape an audience's emotional response.

The following discussion won't examine the mechanics of camera operation. As indicated in the Preface, this book is far from a production manual. (If I discussed production aspects of television and film, this volume would double or triple its present length.) This chapter explores the camera's three primary functions:

— COVERING THE ACTION: discussing the various types of camera coverage, with a detailed examination of master angles and a specific demonstration of how one dramatic scene might be photographed

One must compose images as the old masters did their canvases,

with the same preoccupation with effect and expression.

Marcel Carné[1]

— CREATING APPROPRIATE EMPHASIS: exploring the various factors that provide impact for significant moments, including closeness, separation, positioning, duration, and visual composition

— CREATING MOOD AND ATMOSPHERE: the subtle manipulation of audience emotions and the inferences created by camera position, movement, lenses, and filters

COVERING THE ACTION

The simplest method of photographing a dramatic scene would be to place a camera some distance from the cast and record all of the action in a fixed wide angle. This was how pioneer directors first photographed silent films. It was as if

[1]Quoted by Louis Giannetti in *Understanding Movies*, 2nd ed. (Englewood Cliffs, N.J.: Prentice-Hall, 1976), p. 48.

the camera were recording a stage play, anchored to a seat in the fourth or fifth row and photographing the action through a **proscenium arch**. Actors moved laterally across the frame, seldom if ever upstage or downstage; inevitably they were framed head to toe or wider. When David Wark Griffith tried to move his camera closer, the producer screamed. Audiences would never be satisfied with half an actor![2]

Creativity prevailed. Griffith discovered that closer angles not only reveal more of facial expressions and detail but they also carry far greater impact. Moreover, by moving the camera from the fourth row and placing it "on the stage," within the scene itself, actions immediately became more dynamic and camera angles more interesting. The audience became *part* of the unfolding drama rather than merely a spectator.

During this century's first decades, pioneers developed most of the production patterns and techniques today's directors use: dolly and panning shots, **split screens**, **wipes**, and dissolves. As Hollywood grew to become the world's film capital, its filmmaking techniques became oriented to mass production. As yearly output grew, certain filmmaking patterns became standard. Most scenes began with a wide shot. If the producer had spent $10,000 for a set, by thunder, he wanted to make certain the audience would see it! Gradually, as tensions built, camera angles would move closer until, at the scene's climax, action usually was played in closeups.

That was Hollywood's pattern for covering dramatic action in the teens, twenties, and thirties; it is still valid for most scenes today.

The Spectrum of Coverage

Photographing the action, telling a story with a camera, may be achieved in an endless variety of ways, depending on the director's individual style and the dramatic requirements of a scene. Some directors, especially those who learned their craft in the early days of live television, avoid frequent cutting between angles. They prefer a fluid, moving camera that tracks the action within the set.

Other directors shoot a single **master angle**, that is, a comprehensive shot, usually wide, that covers the action of the entire scene. Then they break up the scene into closer angles: **medium shots**, **over-shoulder shots**, and **closeups**. The editor then builds a progressive structure, splicing together and intercutting the various angles for maximum dramatic effectiveness. The additional angles are called *coverage*. In network television where time strictures are necessary, some coverage is essential, allowing the editor to trim final footages to the exact time required.

Camera coverage also provides flexibility in strengthening dramatic values. It

[2]As cited by Arthur Knight in *The Liveliest Art* (New York: New American Library, 1979), p. 24.

permits an editor to tighten scenes that play too slowly or to stretch moments that play magnificently, to eliminate bad line readings or unnecessary dialog, and to favor an excellent actor and minimize a bad one. Such coverage is primarily used, however, for dramatic emphasis, which I discussed in Chapter 6 and will return to in a few pages.

Camera Progression

The next time you go to a stage play, be aware of what happens to your attention. As you take a seat and begin to leaf through your program, you notice that the man next to you has a terrible cough. Involuntarily, you draw away from him. The woman in front of you wears a bouffant hairstyle, almost obscuring the stage. You lean to your left to see better. A young couple two rows down is cuddling and kissing. You try not to watch them. Moments later the lights dim, the curtain rises, and the play begins.

Gradually, as the actors establish their characters and their problems, your interest grows. You forget the man with the cough, the bouffant hairdo, and the cuddling couple. While a small part of your consciousness remains in the audience, aware that you are watching a performance, the larger part is moving closer to the stage ("camera dollying in").

At first you're absorbing the stage setting and figuring out characters' identities and the nature of their relationships. As you become more deeply involved, you fasten your attention first on one actor and then another. As concentration heightens, you lock your attention more closely onto faces. In a sense, your attention has moved from a full shot to a medium shot to individual closeups. When the curtain descends at the end of the act, you reluctantly "dolly back" to your seat in the audience. As the house lights come up, you again become aware of the coughing man and the others. You had totally forgotten them.

By studying your reaction to stage plays and other dramatic forms, you can recognize the psychological validity of the camera progression established in Hollywood so long ago. By creating a pattern of increasingly closer angles, the director parallels the consciousness of the attentive viewer.

Were you to watch the eyes of the people seated around you in the theater, you would see them flick back and forth between actors, fastening their attention on where the greatest drama is taking place from moment to moment to moment. When a wife tells her husband that she is leaving him for another, for example, the audience automatically looks to him to see his reaction. Directors and editors try to parallel this instinctive audience pattern: *to show the audience what it wants to see.*

Realistically, directors have two goals when cutting between characters: (1) showing the audience what it wants to see and (2) showing the audience what the director wants them to see. In any scene the focus of interest changes many times. Directors and editors try to anticipate where the audience will fasten its attention. Dramatic experience and instincts guide them. Directors have their own vision of how a scene should play and where audience attention should be focused. They attempt to maximize significant moments and to minimize dull ones. Occasionally, for sound dramatic reasons (often in so-called thrillers), directors deliberately *prevent* the audience from seeing what it wants to. Therefore, to some degree, directors and editors manipulate the audience by telling it where to look.

The Setup

Each time the director moves the camera to a new position to photograph the action from a different angle, the new position is called a *camera setup*, or simply a *setup*. Each new camera setup requires a change in lighting. Because lighting can consume a considerable amount of time, varying with the size of the set and the complexity of the staging—and each minute of production time costs hundreds of dollars—each setup is expensive. Unnecessary setups are fiscally irresponsible. For feature films with extravagant budgets, economy in the number of setups is not a major consideration. But for many television programs where dollars are limited, the number of setups per day becomes a significant directorial concern.

Can a scene be covered in two setups? Three setups? Four? In planning a day's work, TV directors must think in such terms. Sometimes by eliminating setups in unimportant scenes, a director will gain the luxury of additional coverage for a major scene or the time needed to shoot an additional location, giving the show a handsomer look.

In staging a show, then, the director must consider two factors: how to make it good (dramatically effective) and how to make it "cheap," that is, within the anticipated budget. Although the budget is primarily the concern of the unit manager and the producer, any director who wants to work again must respect its limitations.

The Master Angle

Traditionally, most scenes begin with an **establishing shot**, or master angle (also called a *master shot* or simply a *master*), primarily for reasons of orientation. Audiences need to see where the action is taking place, what characters appear there, and what they are doing. Some directors play an entire scene in such a **full shot**, the camera remaining stationary throughout. Others start wide and then, as tensions build, dolly (or zoom) to a closer angle. Directors who maintain a comprehensive master shot for the entire scene almost never intend that such an angle appear in its entirety in the final cut. Playing an entire scene in a wide angle, especially a significant or lengthy scene, would vastly diminish its impact. These directors usually intend to shoot closer angles that will intercut with the wide master angle, thereby providing appropriate dramatic emphasis.

Master shots create the patterns to be matched in closer angles. They establish the scene's mood and tempo and the physical moves and actions of its characters, although the director has a small degree of latitude in making changes later in closer angles. If those changes contradict the action patterns established in the master shot, the coverage will not "cut" with the master. For example, if a woman paces agitatedly in the master but does not pace in a medium shot, the editor cannot intercut the two angles; there will be a mismatch of action.

Pickup Shots What if an actor blows a line near the end of a difficult, complicated four-minute master? Must the director go back and reshoot the entire master angle again? Some directors, especially in feature films, do just that. But when time and money are critical, most directors simply pick up the action a few lines before the break point. That is, they keep all of the master angle previously shot and photograph the remainder in a new pickup shot. They know that the scene will necessarily be edited, that closer angles will be used for much of the central material, and that these closer angles will certainly bridge the break between the original, incomplete master shot and the new pickup master. If you recall, director Sara used a pickup shot in Chapter 1 when time was critical.

The wide master angle usually appears only occasionally in the final edited version of a scene: at the start (for orientation), perhaps later to accommodate character moves, and possibly at the conclusion. Thus, by picking up the master angle, the director has no fear of shortchanging the editor. And pickup shots can save vast amounts of valuable time.

Starting Close Beginning scene after scene with a wide, orienting angle can become boring; the repetitious pattern seems to lack imagination. Good directors try to find variations that will begin a scene dynamically and still orient audiences reasonably soon. Frequently it becomes effective to begin on the face of an actor as he or she utters a trenchant line or on a close shot of a prop or a significant piece of business. One overused beginning to a scene is a ringing telephone. As a hand enters the shot, the camera pulls back (or tilts up) to reveal the person answering. Starting a scene on hands pouring a drink, a phonograph record in a dead groove, or a close reflection in a mirror are all patterns we have seen repeatedly. Close opening shots are frequently effective following a scene that ends wide; the contrast in angles provides a visually refreshing change.

When a scene begins in a close angle, the director needs to orient the audience soon, especially if the scene lasts more than a minute or two. Such orientation may be achieved by dollying back, cutting to a wider angle, or panning the actor across the set, thereby allowing viewers to discover where the action is taking place.

In one scene in *Children of a Lesser God*, director Randa Haines used *eleven* shots before establishing a master angle. She began close, revealing the row of deaf students in a panning shot; she connected their teacher (actor William Hurt) with them by means of a **point of view (POV) shot**; she used a tie-in shot that demonstrated the physical relationship between students and teacher. But over a minute elapsed before she employed a comprehensive angle revealing all of the scene's elements. The point: You don't always have to begin a scene with a wide angle.

Masters without Coverage   There are several occasions when an entire scene can play in a single master angle without coverage of any kind: (a) when the scene is relatively unimportant, (b) when the scene is short, (c) when the scene requires no special emphasis and contains movement, and (d) when a single angle is more effective than many.

As an example of (a) and (b), a scene of a woman driving up to a house, getting out of her car and entering probably won't need coverage. Two exceptions come to mind: (1) if the action consumes too much time and the director wants to give the editor coverage that may shorten the playing time and (2) if the scene is more than an ordinary entrance. (See "Editing to Diminish" later in this chapter for more on the first exception.) For example, if the woman is extremely nervous or afraid to go into the house, then closer angles will help the director to establish her state of mind.

(c) Scenes that require no special emphasis may play in a single angle if there is sufficient physical movement. For example, when the camera trucks with two people walking through a park, no coverage is needed usually because the continually changing background in the **trucking shot** prevents the scene from becoming static. If the park scene contains significant emotional content, however, the director will probably want to provide closer angles—perhaps two **tracking shots** in closeup as they walk.

(d) Occasionally, when script and actors are excellent and the angle is a close two-shot, the quality of the moment is so special, so enormously affecting, that coverage would only depreciate it. Insecure directors probably would cover such a scene with closer angles. But if the master angle is sufficiently rich, most editors would leave it alone.

Some inexperienced directors add camera angles, apparently "just for the hell of it." Whenever a director or editor cuts to a new angle, there should be a legitimate dramatic reason, not on a whim, not just for visual variety, not just to add pace. Remember, pace must come from the content of the scene and the way that it's played. Frequent cutting back and forth won't speed up a leaden scene.

The same principle applies to movement by actors. Actors should move because of plot needs or because of some internal motivation. Movement for movement's sake often creates scenes that are confused and confusing. We see them occasionally in afternoon soaps, when rehearsal is minimal and the director feels the need to "keep a scene moving."

Directors are frequently under pressure, especially in television. To avoid time-consuming coverage, they sometimes stage a two-character scene so that both actors move down into camera, facing the camera. Such a dynamic angle contains such emphasis and visual interest with both faces full into camera that it will sustain nicely for a brief interchange. For example, a husband cannot face his wife and moves guiltily away from her (into camera). Angry, seeking answers, she pursues him. He cannot face her and continues to face camera. Finally, at the end of the scene, the husband turns and exits. Camera either pans him to the door or holds on the wife's face as she reacts. We have played an entire short scene in a single angle, without coverage of any kind.

Multiple Master Angles   Sometimes if a scene is long or the staging complicated, directors shoot more than one master angle. Each new master usually accompanies a significant change of direction in the action. Chapter 5 described such new directions as **beats** and called them scenes within scenes, sometimes begun by the entrance or exit of a character or by a new direction in subject matter.

In the classroom scene described earlier, the teacher initially sits at a desk facing his deaf students. Attempting to teach the students to speak aloud, the instructor asks the purpose of speech. One student (male) answers by signing "to pick up hearing girls." Grinning, the teacher takes two students of each sex to a small table where they sit and begin a mock "pickup" situation. Because the scene has changed direction, the director employs a new master angle here, a group shot of the five characters at the small table.

Directors of feature films occasionally use multiple masters, but directors of series TV or low-budget features do so rarely. The reason is that multiple masters are expensive, consuming considerable lighting time. Standard coverage in a scene—going from a wide master angle to medium shots, over-shoulder shots, or closeups—consumes far less time than beginning over, restaging and relighting the set for a new master angle. Watch television. You'll discover that most scenes in most shows are staged with but a single master angle.

Traveling Master Shots   Some directors feel that there is something old fashioned about a stationary, wide angle master shot. They prefer a master that adjusts itself to the movement of actors, that emphasizes key moments and, perhaps, simplifies coverage. Often such staging is beautiful to watch, a carefully choreographed ballet of actors and camera.

Here's a typical example. Director Paul Wendkos staged a three-page master angle for a TV episode in which the camera began its coverage on the door of a den. As Actor A entered, the camera pulled back with him until he encountered Actor B. The actors moved so that this angle became an over-shoulder shot, tightening as emotions became intense. After a brief, angry interchange, Actor A strode to the window, camera panning with him and moving closer. (Actor B was now offstage.) They exchanged a page of dialog and then Actor A returned to Actor B (camera again panning), grabbing him, threatening him. Camera held the close two-shot for this final interchange, and then Actor A exited, the camera remaining on Actor B and moving in to a closeup as he shook his head, bewildered. We heard the door slam as Actor A exited. Wendkos needed only two additional pieces of coverage to complete his scene: the complementary (reverse angle) over-shoulder shot when the two actors first spoke and the complementary closeup of Actor B when Actor A stood at the window.

Although the above example may seem complicated at first glance, many directors stage traveling master shots that are considerably more complicated. The camera, mounted on a crab dolly capable of movement in any direction (including up and down), does more than merely cover the action; it plays the most significant moments in close angles and contributes to the scene's mood and atmosphere. Such masters require careful rehearsal with each camera position carefully marked on the floor. They can also be time consuming.

Scenes with No Master Angle On rare occasions some scenes play with no master angle at all because of their special nature. They consist of a **montage** of angles that build a composite picture.

Recall the famous shower scene in *Psycho* in which Alfred Hitchcock shot seventy (!) camera setups to build forty-five seconds of shocking action, a mosaic of flash cuts: Janet Leigh in the shower: upturned face, water splashing down, an intruder with knife, slashing, mouth screaming, knife slashing, horror on face, blood running down drain, hand pulling at shower curtain, knife slashing, hand clutching at tile wall, nude body falling, blood down drain, face on tile floor, eyes open. Such a scene plays dynamically with no master angle; it is more effective for its staccato, jarring juxtapositioning of images, horrifying in their cumulative impact.

Generally, directors omit master angles when scenes or sequences either have great emotional intensity (such as the *Psycho* scene) or such size and scope that a single master angle would be relatively meaningless. The Odessa steps sequence in Eisenstein's *Battleship Potemkin* is an example of the latter category. (Eisenstein may have shot a master angle, but it is not apparent in the edited film.) Each element in the shower scene and Odessa steps sequence is filmed in an individual angle. Like a piece in a jigsaw puzzle, it only becomes meaningful when placed in context with pieces that surround it. Spatial relationships in the Odessa steps sequence are sometimes suggested by POV shots, as when the student looks off in horror and sees the baby carriage rolling down the steps. Such a spatial relationship was also suggested in the scene in *Children of a Lesser God* when the student glanced away and the director cut to a POV shot of the teacher at his desk.

Staging a Scene

Let's stage a simple, two-person scene and consider some of the ways in which a director might use a camera to cover the action.

The Plot: David McKenna's girlfriend, Eileen, announces that she is leaving him. She has been promoted to an executive-level job in another city and has decided to go there alone. As she packs, David reminds her of the good times they've shared, asking her to reconsider. She coldly refuses. David's anger flares and he tells her to get out of his house. She starts to pack a photograph of David and herself but he takes it, tears it

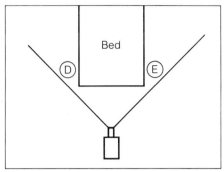

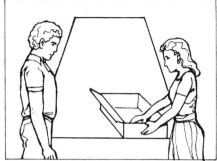

FIGURE 7.1 The camera photographs David and Eileen in profile, from a static side angle, with minimum impact.

FIGURE 7.2 What the camera in Figure 7.1 would see: Separated by a bed, the figures would probably be smaller (with less impact) than indicated here.

in half, and gives her back the half in which only she appears. He grins wryly as she slams shut the suitcase and storms from the room.

Simplicity in Staging The most economical way of covering the action of this scene (though probably not the most effective) is in a single camera angle. Examine Figure 7.1, which arbitrarily places the bed in the center of the set with Eileen on one side packing and David facing her from the other side.

What are the positive aspects of such a staging? Well, the physical separation of David and Eileen symbolizes their emotional separation. Perhaps at one point he walks around the bed to her as he asks her to stay. But she pulls away, takes her suitcase, and leaves. Now the separation is complete. Camera dollies in to a closeup of David to register his reaction.

What are the negative aspects of such a staging? First, the camera shoots both actors in profile for most of the scene, which provides little impact. Second, the separation requires a wide angle, which (again) diminishes the impact. Third, the angle presents a balanced, symmetrical, static composition, lacking in visual dynamics. Glance at Figure 7.2. Not very exciting, is it?

Suppose I change things around and place the bed in the *foreground*. Eileen enters through a door in the upstage wall, takes a suitcase from under the bed, and begins packing. David comes up behind her. Now both actors *face the camera* (see Figure 7.3). Eileen ignores his pleas, continues to pack hurriedly, crossing to dresser from time to time, camera panning with her, the motion helping to keep the picture visually interesting. The audience sees both characters in full face and in depth for most of the scene, which makes the picture dynamic.

 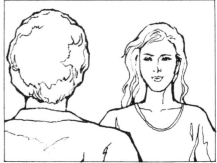

FIGURE 7.3   With David behind Eileen, both facing camera, impact would be vastly increased. Dotted lines indicate a closer shot, achieved when the camera zooms or dollies closer.

FIGURE 7.4   The reverse of the closer shot shown in Figure 7.3: what the camera would photograph when Eileen turns to face David.

Some directors use this kind of simplified staging when they need to gain time. It's functional, especially if a scene is short or not of great dramatic significance.

If time permits, a director might expand the camera coverage slightly by allowing Eileen to turn upstage at a critical point to face David. Now the scene will require one additional angle, a complementary over-shoulder shot, camera facing Eileen (see Figure 7.4). With this added angle, the film or tape editor will be able to cut back and forth between the two shots, favoring first David and then Eileen, as dramatic emphasis changes.

A Typical Staging   If given more time and if the scene is critically important to the script, a director might attempt a slightly more ambitious staging that requires additional camera coverage. Examine Figure 7.5.

The letter *D* indicates David's position as the scene begins; *E* indicates Eileen's entrance and her initial move to David, who is repairing a cabinet. The camera is positioned in a master angle that will photograph the entire scene. Note that from this basic camera position a director could either start the scene close on David and widen to include the door as Eileen enters, or start close on the door (adjusting David's opening position slightly) as Eileen enters and dolly back to include David when she moves to him. Each of these shots carries implications. Beginning on David implies faintly that it's "his" scene, played from his general perspective. Beginning on Eileen's entrance implies the reverse.

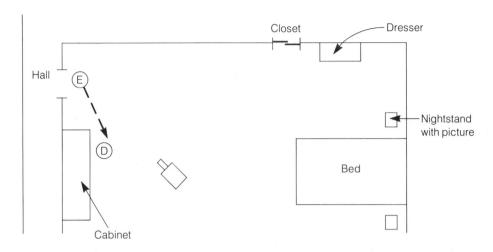

FIGURE 7.5 David and Eileen's bedroom. The diagram indicates camera's position as Eileen enters and moves to David.

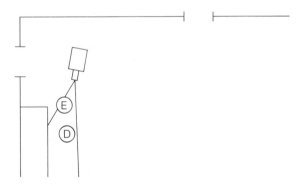

FIGURE 7.6 The reverse angle (from Figure 7.5) used when Eileen confronts David.

A director could play the first confrontation in the master angle, but to see David's reaction to Eileen's announcement, the director would need a second setup, a reverse of the master angle, as indicated in Figure 7.6. The master angle plus the reverse provide loose over-shoulder shots as indicated in Figures 7.7 and 7.8.

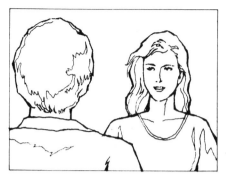

FIGURE 7.7   What the camera photographs in Figure 7.5: an over-shoulder
shot favoring Eileen.

FIGURE 7.8   The reverse of Figure 7.7: an
over-shoulder shot favoring David.

Crossing the Line   Notice that in Figures 7.7 and 7.8 David is positioned on the left side of the frame and Eileen is on the right. If after photographing the first setup, a director accidentally moves the camera across the invisible line that bisects the two actors (that is, if the camera in Figure 7.6 were located in the doorway) and photographs the second setup from that camera position, there would be a disconcerting *jump* when the two angles were edited together. David would abruptly switch position from the left to the right side of the frame, and Eileen would do the reverse. Such mismatching of images, **crossing the line**, is awkward and disorienting—the mark of an amateur director. So long as the camera angles remain on the same side of that magic line running through the two players, their positions will remain constant from cut to cut. See Figure 7.9 and 7.10.

Developing the Scene   Let's assume that after a few lines of dialog Eileen impatiently crosses to the closet and pulls out a small suitcase that she places on the bed. During the remaining action, she repeatedly crosses to the dresser, taking clothing from drawers and arranging it hurriedly in the suitcase. Picture the action in your mind as the camera in its master angle (as shown in Figure 7.11) pans with Eileen as she crosses first to the closet and then to the bed. When she moves to the dresser to get her clothing, she moves upstage in the shot.

When David pursues her, moving to the bed, he enters the master angle, which had panned away from him and stayed with Eileen when she crossed to the closet. The camera now holds another two-shot. The camera operator may have to adjust slightly in order to accommodate a clean over-shoulder shot. (See Figure 7.12.)

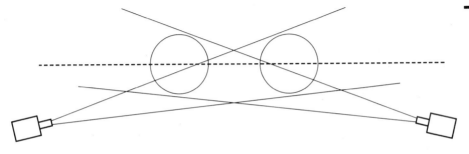

FIGURE 7.9 The two setups here demonstrate the correct way of photographing over-shoulder shots.

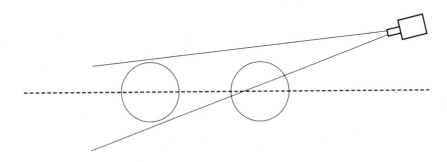

FIGURE 7.10 The camera has crossed the imaginary line that bisects the characters. Now setup 2 will not intercut well with setup 1 because characters will change positions in the frame.

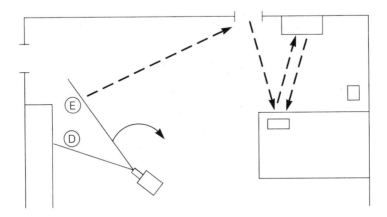

FIGURE 7.11 Camera pans right as Eileen crosses to the closet, takes out the suitcase, and carries it to the bed. She moves back and forth from closet to bed, carrying and packing clothes.

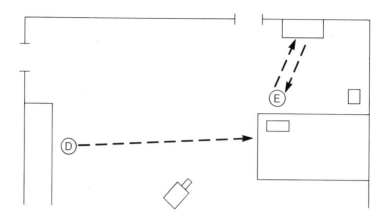

FIGURE 7.12   David crosses from his original position and enters the camera's master angle, a loose over-shoulder shot.

In this position, as in the initial angle by the cabinet, the camera is angled past David toward Eileen, who packs busily, almost ignoring him. We need one final angle, one that favors David as he stands at the corner of the bed. It will be the reverse of the master angle, shooting past Eileen toward David. We begin this last setup with David still in his position by the cabinet and bring him into the two-shot. The editor may be able to use this extra coverage, intercutting it with Eileen's action. Figure 7.13 shows this third camera position. The camera shoots past Eileen in profile toward David, who now faces almost directly into camera as he pleads with her.

Finally, when Eileen picks up the picture from the nightstand beside the bed, her movement can be photographed from the master angle (Figure 7.12), panning her to the nightstand and back to the suitcase.

If a director wants to play the final confrontation between Eileen and David in closeups rather than complementary over-shoulder shots, he or she may do so "merely" by changing lenses. This gives the scene the feeling of five setups rather than three, with relatively little loss of time. I placed the word *merely* in quotation marks because the lighting director may want to make a few adjustments to accommodate the closer angles.

The staging of this scene, while relatively simple, was arbitrary. It could be blocked in a hundred different ways, with dozens of camera variations of each. The set was designed specifically to accommodate the action and to permit effective but economical camera coverage. Similarly, action can be tailored to accommodate a set. When production companies work in actual locations, a phenomenon happening more and more frequently in both television and feature

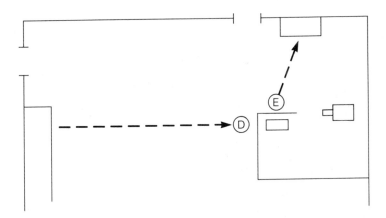

FIGURE 7.13   Camera is now in a reverse position, shooting toward David as he crosses from the cabinet to the bed.

films, directors must adapt their staging to whatever settings they encounter. In most productions, directors and art directors work closely together, shaping both the set and its dressing to specific staging requirements.

Creative Planning   Let's consider the thought processes that went into designing the set for the Eileen-David scene so that you can understand the integration necessary among scenery, properties, and physical action.

In a sense, the tail wagged the dog. The action of packing a suitcase seemed essential to the thrust of the scene; it added immediacy and urgency. Therefore the closet (location for the suitcase), dresser, and bed had to be relatively near one another to keep the action from becoming unwieldy. I needed to find some occupation for David (business) at the top of the scene. Yes, he could have been reading a newspaper or watching television, but both are passive occupations. Either would have created the impression that Eileen was out working while David lounged lazily at home. Better by far to put him to work painting or repairing a cabinet.

Because of the need to be economical in terms of setups, I placed the door in a spot where Eileen could enter and walk immediately into an over-shoulder shot. Had a director been shooting in an actual location, he or she would have placed David wherever necessary to achieve that same camera position. Similarly, the closet and bed were placed in proximity so that a camera could conveniently pan Eileen into her second position (packing), maintaining the economy of one master angle.

FIGURE 7.14   A triangular three-shot. Additional angles, two-shots, would maintain same relative positions of performers: Terry remaining on the left, Vanessa on the right. Tim, who stands at the apex of the triangle in the three-shot, now faces left to Terry and right to Vanessa. Directions refer to *camera* left or *camera* right.

Three-Shots

Camera coverage of three characters requires more care in planning than coverage of two because the possibility of disorientation becomes more likely. Recall that the director of a two-shot creates an imaginary line that bisects the characters. So long as the camera remains on one side of that invisible line, the characters maintain their same relative positions in the frame, from cut to cut.

The same principle applies to the camera coverage of three characters. Figure 7.14 shows how the camera in its master angle would view the positions of three characters. When a director breaks a three-shot into two-shots or singles, he or she must again respect the invisible lines that bisect the characters. Note that three characters will always form some kind of triangle—unless you place them in a line, which is visually static and therefore generally undesirable.

FIGURE 7.15   In an over-shoulder angle, Tim faces left to Terry.

If you were the director, how would you cover the actors in Figure 7.14? (That is, how would you break the three-shot into two-shots?) To avoid confusing the audience, the characters in each leg of the triangle must maintain their same relative positions (see Figures 7.15–7.18). Note that Terry always remains on screen left when facing Tim or Vanessa. Vanessa always remains on screen right when facing Tim or Terry. But Tim, who stands at the apex of the triangle, will always face *right* when playing to Vanessa and *left* when playing to Terry.

In closeups, the directions of characters' *looks* conform to these patterns. Thus, Tim will look right to Vanessa and left to Terry. In actual practice, the director may **cheat** the characters' looks toward the camera to give the audience a more frontal view without disorientation.

If the characters change positions relative to one another or if the camera changes position relative to the three characters, an entirely new triangle is estab-

FIGURE 7.16   In a reverse of Figure 7.15, Terry faces right to Tim.

lished—with new physical relationships. Then the director must carefully re-examine the camera coverage and adjust it to the new positions in order to avoid disorienting the audience.

Direction of Movement

Audience expectancy grows from simple logic. When the cavalry chases Indians over a hill, both head in the same direction. When directors divide their camera coverage into separate shots—as they necessarily must when cavalry and Indians are separated by considerable distance—audience expectancy logically remains the same. Both will be heading in the same screen direction: right to left or left to right (see Figure 7.19).

Removed from any story context, such intercut shots carry no implications regarding who is chasing whom. Indians could be chasing cavalry; cavalry could be chasing Indians. But the very fact that the shots are intercut unmistakably

FIGURE 7.17   In over-shoulder angles between Tim and Vanessa, Tim faces right, Vanessa faces left, maintaining the same relative positions as in the original triangular three-shot.

implies *relationship*. The two elements somehow are acting upon each other. Visual clues such as the Indians repeatedly glancing backward or the cavalry firing their guns in a forward direction would, of course, clarify their relative positions.

Examine the implications if just one of the elements changes directions. Suppose that the cavalry races from left to right and the Indians from right to left. What would be your assumption? It would depend on story context, but generally there are two possibilities. The cavalry and Indians are either running *away* from each other (perhaps after a confrontation) or heading *toward* each other (the confrontation still to come).

Practical Applications   Writers and directors frequently use the implications in the direction of character movement in order to build suspense. It is late at night. A young secretary has been working late. She locks the office and begins walking home, left to right. On another street, a psychotic killer moves along in the

FIGURE 7.18   In a reverse of Figure 7.17, Vanessa faces left to Tim.

shadows. He walks *right to left*! Clearly, the implication of these intercut camera angles is that the killer will intercept the secretary. Instantly the audience begins to worry.

Patterns of **screen direction** create geographies in the spectator's mind. If the young secretary has repeatedly left her office to go home and each time has walked from left to right, viewers create a mental picture of her home located somewhere to the right of her office. If after establishing such a geographical relationship, the director then violates it, allowing the girl to walk toward her home in a right-to-left progression, the audience would certainly become disoriented.

A similar principle governs the flight of airplanes or the direction of travel of buses or trains. Audiences tend to superimpose a subconscious map of the United States (or whatever country) over the action. Thus, planes flying from Los Angeles to New York should travel from left to right; trains traveling from Washington,

FIGURE 7.19 When the cavalry charges, its screen direction must remain constant from shot to shot (left to right or right to left); otherwise, audiences may become disoriented. (From *Stagecoach*. Used with permission of Caidin Film Company.)

D.C., to Denver should travel from right to left. When they travel north or south, you're on your own!

Changing Directions If the cavalry has been pursuing Indians in a right-to-left progression and the director for good reason needs to change directions, the pursuit cannot be arbitrarily reversed without creating audience confusion.

The director's problem may be solved in two ways. The director can *change the progression within a shot.* If, for example, the Indians have been riding left to right and suddenly confront a sheer cliff wall, they must stop, wheel their horses and take off on another tangent, right to left. In another cut, the cavalry riders also reach the cliff, spot the hoofprints in the soft earth, and change directions, following the new progression taken by the Indians. Also, when the camera changes position, crossing the line of progression, the cavalry or Indians change direction within the shot. Such a transition is easily accomplished when the camera is mounted in a helicopter that flies from one side of the racing cavalry to the other.

The director can also change the direction of movement by *placing a direction-less shot between the two progressions*. If the Indians ride from left to right in cut A and then ride directly *into camera* in cut B, they may ride from right to left in cut C without causing confusion in the audience.

Certain directors (such as John Ford) have occasionally violated these "rules" with no great degree of audience confusion, proving again that in any creative medium principles cannot be arbitrary.

Complementary Shots

Another of the conventions that audiences have accepted as a normal and necessary part of camera treatment is the use of **complementary angles**. When David and Eileen confronted each other earlier, camera coverage consisted of two matching over-shoulder shots, one favoring David, one favoring Eileen. Later in the scene, camera coverage consisted of closeups: one of David, one of Eileen.

Because close shots are used primarily for emphasis—to give special impact to an action, reaction, or line of dialog—isn't it possible to highlight a moment without the necessity of playing closeups on *both* characters? Yes, of course it is possible. We see such coverage occasionally in both television and motion pictures. But directors use matching shots for three reasons: audience expectancy, dramatic progression, and balance.

Audience Expectancy Since the 1920s, movie scenes usually have been photographed in complementary angles. When live television gained popularity in the late 1940s, it adopted some of Hollywood's film techniques. When TV drama made the transition to film a few years later, it completed the pattern, embracing the tools, facilities, and techniques that the film industry had developed.

"It has always been done that way" is not a valid reason for continuing any practice. Yet, for television directors working within narrow budgets, with little time for experimentation, there is a certain security in the crisp, professional, symmetrical look afforded by complementary coverage. They know that they are providing audiences with a visually dramatic pattern they will feel comfortable with. But symmetry is the least important of reasons for using complementary angles.

Dramatic Progression Most well-written scenes grow through increasing levels of conflict and audience arousal. Such growth usually results from an interplay between characters, each feeding the other, emotion building on emotion. With more than one character in a scene, it's seldom possible to play a big moment for one without involving big moments for others. Ted tells his father that he's quitting college. When his father tries to reason with him, Ted's jaw tightens. His father becomes angry. Ted forces down his own anger, finally explodes. His father threatens financial reprisal. Ted storms out, slamming the door. Such dramatic peaks are seldom onesided. Their growth can effectively be recorded through increasingly close complementary angles.

Balance Every camera angle carries implications. When in a two-person scene the angles are close on one and medium on the other, audiences infer that the scene is played from the general perspective of the closeup character. It would become his or her scene. In the dramatic confrontation between Ted and his father, the director could play Ted in close angles and his father in wider shots. This would suggest that Ted's dialog, actions, and reactions are more significant than his father's. Shots of his father would become (almost literally) Ted's point of view.

Such unequal coverage, were it to remain consistent throughout a scene, would have a second—and necessarily negative—effect. It would create far greater impact for the closeup character than was possible for the distant character. Thus, if Ted's father made a jarring pronouncement, its impact would be diminished simply because it wasn't played in a close angle. As indicated before, scenes usually build in tension between characters; both contribute. If the buildup appears one-sided, unequal in intensity, the scene limps along on one strong leg and one weak one. Thus, for a scene to build to a powerful climax, most directors fashion a complementary camera pattern with characters assuming approximately equal size and occupying similar (complementary) screen positions. To complete the pattern, focal length of lenses in both angles should be the same.

However, rules cannot be inflexible. There are occasions when—for reasons of mood, character, or style—some directors deliberately elect not to use complementary angles. But for most directors most of the time in most dramatic situations, such angles facilitate the buildup of dramatic intensity.

CREATING APPROPRIATE EMPHASIS

In any dramatic scene, certain moments obviously are more significant than others. At those moments, certain characters or actions require the audience's concentrated attention as well as the greatest degree of impact that a director can provide.

We will examine six methods by which the camera can direct audience attention and create varying degrees of emphasis: closeness, separation, positioning, duration, variation, and visual composition.

Closeness

Chapter 6 explored stage directors' techniques for calling the audience's attention to specific characters or actions at specific moments. In one such technique, as characters moved downstage, that is, closer to the audience, they gained importance. In film and television, when a camera replaces the audience, its distance from the actors (or size of the image) will, to a great extent, determine the degree of impact. A closeup provides greater impact than a medium shot, a medium shot more impact than a full shot. (See Figure 6.2.)

What if the director cuts to a close shot of an unimportant character, a mere spectator to the action? Would such a cut create enormous impact? Of course not. It would simply bring that character to the audience's attention. Impact is inextricably related to a character's dramatic significance at any particular moment. Mere closeness is not sufficient. If the moment has created dramatic tension and a character directly contributes to that tension, then a closeup will effectively heighten its impact. Thus, closeups tend to *increase* or *intensify* whatever dramatic values already exist.

Close angles serve additional dramatic functions. They *clarify* in that they reveal detail that could not be seen in wider angles. A tear in the corner of an eye would be lost in a medium shot. Knuckles tightly clenched or an ingenue nervously biting her lip might go unseen in a full shot. Closeups may reveal thoughts or feelings, internal values, or such emotional subtleties as darkening despair in a character's eyes.

Part of the fascination of drama is the degree to which spectators contribute to such emotions. A classic example of such involvement is the famous Lev Kuleshov experiment in the 1920s with film footage of the actor Ivan Mozhukin. Kuleshov snipped a neutral, expressionless closeup of Mozhukin from one of his feature films and broke it arbitrarily into three pieces. He carefully spliced three other shots between the closeups: a steaming bowl of soup, a little girl playing with a teddy bear, and the corpse of an old woman lying in her coffin. Remember, the actor's expression remained unchanging throughout the sequence. When audiences viewed the strip of film, they applauded Mozhukin's virtuosity as an actor, remarking how hungry he looked when viewing the bowl of soup, how gentle and loving he appeared when watching his daughter at play, how sad he looked at the death of his mother. Although such was not his purpose, Kuleshov vividly demonstrated audiences' ability to project their own thoughts or emotions into what they watch. (Would a spectator who despised soup have read Mozhukin's expression as repugnance when he viewed the steaming bowl?)

Many film and tape editors, aware of such audience projection and lacking necessary footage, sometimes "steal" closeups of an actor from one scene in a picture and insert it into another. Such deceptions are rarely detected; viewers see in the actor's face the emotions or reactions they expect.

Close shots also tend to *create empathy* between an audience and a character. Repeated closeups create the illusion that the story is being told from that char-

FIGURE 7.20 When three actors are covered in a two-shot and a single, the performer in the single shot gains emphasis.

acter's essential point of view. They create a feeling of physical closeness, of relationship, so that the spectator may more easily identify with a character.

Separation

Camera emphasis uses another principle adapted from the theater whenever it separates one character from a group, thereby focusing attention on that character. Earlier I discussed breaking a triangular staging—three characters facing one another—into over-shoulder shots. In such a pattern, each character assumes a degree of impact exactly equal to that of the other characters. But what would be the implication if the director divided that triangle into a single shot and a two-shot? Note (in Figure 7.20) that the isolated character suddenly becomes important. Whether in the theater or in film or TV, separating one character from a group implies that that character is special. The director can heighten the degree of that specialness by moving the camera closer.

Whenever characters share a frame, the emphasis, logically, is divided. In the two-shot in Figure 7.20, the character nearer the camera, being larger, assumes more importance than the background character. Thus, in a scene among three characters, the one occupying a frame alone assumes primary importance, the figure in the foreground in the two-shot assumes secondary importance, and the background figure assumes the least importance.

When four characters appear in a scene, the director can photograph them in a pair of two-shots without giving emphasis to any specific character. But when the director divides camera coverage into a three-shot and a single, the character in the single instantly becomes significant.

Positioning

In another parallel from the theater, whenever a character presents full face to camera, impact is greater than if that character appears in profile. In film and TV, profile shots are generally unsatisfactory. They rob us of the chance to look into a character's eyes. Occasionally, students in my directing class cut to a closeup of an actor in profile and then wonder why the shot doesn't have the power usually associated with closeups. To be sure, closeness provides impact—but only if the actor faces camera. Thus, my students achieve a degree of impact by using a close camera angle and then dissipate the energy of the closeup by playing the actor in profile.

When two characters share a frame, with both identical in screen size, the one more nearly facing the camera draws the greater degree of attention. When two characters share a frame in profile, emphasis is divided equally; one will have the same impact as the other (see Figure 7.21).

As the camera arcs (moves in a curved path) around to favor one character, that character gradually assumes more impact while the second character's impact simultaneously diminishes. Once the camera has moved to a position where it completely favors one character, the shot becomes the traditional over-shoulder shot.

Directors often use over-shoulder angles when they want to convey a sense of relationship between characters. Visualize a love scene played entirely in closeups. True, closeups increase impact, but consider that each character is separated from the other, isolated, confined within a frame. Such fragmentation of images is sterile. Visualize the same scene in tight over-shoulder angles in which viewers see the physical as well as emotional closeness of the two lovers, together within a single frame. The difference is unmistakable. Over-shoulder shots also place the favored character in the Golden Mean of the frame, that spot some believe to be the aesthetically perfect focus of attention.

Duration

It's a common experience: Moments with a loved one at an exhilarating party fly by quickly. Other moments, boring or distasteful, drag on for hours. Time seems to be elastic, changing with the event, stretching in anguish, contracting in pleasure.

Film or video time, like real time, distorts to fit the quality of the moment. Although in real life such distortions are primarily psychological, in television and film they become tools of the sensitive director. In real life such distortions relate to criteria of pleasure, boredom, or pain, but in TV and film they depend primarily on the significance of the dramatic event and its need for emphasis.

One of the Soviet Union's great director-theorists, Sergei Eisenstein, discovered making *The Battleship Potemkin* that it is possible to gain greater emphasis for a dramatic moment simply by *prolonging* it. In a sequence in which a rebellious sailor smashes a dinner plate, Eisenstein photographed the action from a number

FIGURE 7.21 When two characters share a frame equally, each assumes the same degree of emphasis as the other.

of different angles. Then, in editing those angles together, he repeated a few frames of film at each splice, thus stretching the duration of the action.

Eisenstein's method for prolonging a moment is used only occasionally today. In *Poltergeist*, a chair slides across the kitchen floor, supposedly manipulated by spirits from the Great Beyond. Director Tobe Hooper photographed the action from two angles, each the reverse of the other. When the editor spliced those angles together, he prolonged the chair's movement in exactly the same way that Eisenstein did, by playing the middle section of the action twice. Such overlapping action made the kitchen appear longer than it actually was, but audiences didn't notice.

In *All That Jazz*, director Bob Fosse filmed Roy Scheider (actually a stunt double) balancing on a high wire in a circus tent. When Scheider fell into the net beneath, Fosse photographed the action from four angles, thereby allowing the editor to extend the duration of the fall, giving it importance.

Modern directors have other ways of achieving emphasis through duration. Among these are slow motion and editing. The latter may be used both to extend significant moments and to shorten or eliminate dull ones.

Slow Motion The most easily recognized method of prolonging a dramatic moment is to play it in slow motion. In film, **slow motion** is achieved by photographing action at speeds greater than the normal sound speed of twenty-four frames per second. When the film is later projected at the normal rate, actions are slowed proportionately; moments are prolonged, usually achieving greater dramatic weight. In video, slow motion may be achieved simply by slowing the replay of tape upon which an action has been recorded, an effect seen most frequently in sports broadcasts.

In drama this type of emphasis plays most effectively with broad or violent action: a criminal leaping to his death from a twelfth-story window in an explosion of shattered glass; the final scene of *Bonnie and Clyde*, prolonging the shocking moment when the two criminals are stitched with machine gun bullets.

Slow motion also adds an emotional factor to its emphasis: a sense of grace, an almost poetic fluidity. Terrifying as it was, *Bonnie and Clyde*'s final moment conveyed a feeling of pas de deux, a grotesque ballet choreographed by the devil. In the film *The World of Henry Orient*, director George Roy Hill used slow motion primarily for its balletic qualities, transforming a romp through the streets by two teenage girls leaping over fire hydrants and swinging on lampposts into visual poetry.

Editing to Prolong Significant moments may be expanded simply by adding reaction cuts. Ideally, such a technique heightens dramatic tension rather than merely stretching time. Returning to Eileen's departure from David, let's change circumstances and suggest that the scene builds to a confession that Eileen is carrying David's child. The director or the editor decides that such a momentous revelation must play in extreme closeup for maximum impact. David's reaction must also play in a tight closeup because his reaction may be more evocative than the announcement itself. Because of the moment's significance, many directors and editors would cut to Eileen as she tries to repress her deep emotions and again to David as he stares unbelievingly. In real life such a moment might take a second or less. Television and film are able to enrich the drama by prolonging the tension. On tape or film, such a moment might last three or four seconds.

Of course, we have all seen such editorial techniques abused. Directors and editors sometimes try to build a moment that simply isn't there, to extract drama from a scene where little drama exists. They prolong; they do not enrich. The result frequently becomes laughable.

David's initial reaction to Eileen's announcement should be noted well. A reaction can be far more powerful in evoking spectators' emotions than the action or statement that motivates it. Reaction from a character becomes a catalyst for the spectator.

In the early days of television I produced an NBC series that showcased young professional actors. My partners and I had written a scene in which a psychotic French artist threatened to kill his young model. As we began rehearsal, the actress cast as the model came to us in tears. Her partner in the scene had most of the dialog; she had little. It wasn't fair!

We assured her that it was really her scene. She had most of the reactions, and reactions are far more important than dialog. She nodded, accepting but disbelieving. When she saw the completed show, she was ecstatic. The NBC director, a veteran, had recognized instantly that the scene's drama was on her face, in her growing hysteria. And that's where he focused his cameras for most of the scene.

In most dramatic situations, characters and spectators echo and reflect each other. Both react to the same stimulus. Through identification, we as the viewers *become* the reactor. In the model's scene, her tears evoked our own. Her fear sparked our own—and *magnified* our own.

Prolonging a dramatic moment is frequently used for purposes of building suspense. In a small Western town a robber casually enters a bank. In individual cuts the robber notes each of his cohorts, who have timed their arrival to coincide with his. One is at the counter filling out a deposit slip. Another stands in line at a teller's window. A third has been hired as a janitor; he sweeps the floor near the vault. The last waits outside the bank, holding the horses at the hitching rail.

The leader's eyes flick around the room. Again director or editor cuts to each as the gang members exchange glances with their leader. Now each moves into position in their well-rehearsed robbery scheme. *Again* director or editor cuts to each of them, setting the stage, delaying the moment when violence will erupt, stretching seconds into minutes to build anticipation and apprehension.

Prolonging Action Significant moments can also be extended through the actions of characters. In *Jagged Edge*, someone sent incriminating notes to defense attorney Glenn Close; they were typed on an old Corona typewriter. When she discovered such a typewriter in the closet of her lover, the man she had defended, she was shattered. With shaking hands she typed a phrase to see if the letter *t* corrresponded to the faulty letter in the notes. She typed "He is innocen. . . ." And then she hesitated, prolonging our anxiety, building the moment. When, finally, she typed the letter *t*, it proved that the defendant, her lover, had been guilty after all.

Examples of prolonging a dramatic moment through the actions of characters occur in almost every movie or TV show. Watch for them. (See also "Delaying the Revelation" in Chapter 8.)

Just as moments may be prolonged to achieve greater dramatic emphasis, the reverse is also true. Moments of little or no dramatic weight may be shortened or eliminated.

Editing to Diminish One of director D. W. Griffith's discoveries was that a scene didn't always have to play from beginning to end. Frequently beginnings of scenes are littered with meaningless niceties. Griffith often found it effective to jump into the middle of a scene, getting right to the heart of a dramatic situation. Similarly, he discovered that concluding moments may sometimes be amputated to achieve dramatic economy, dispensing with cluttering handshakes and goodbyes.

Modern directors borrow from Griffith's concept, cutting away the fat, eliminating the dispensable or the boring to quicken pace and build spectators' interest. A director named Jack English introduced me to the concept of a **timing cut** on 20th Century Fox's very first television series, "My Friend, Flicka." The script called for the boy, Ken, to receive a tongue lashing from his father on the porch of their ranch house. Then Ken was to walk to the barn and pour out his woes to his horse, Flicka. The barn was located about a hundred feet from the house. To watch the boy traverse that distance would consume thirty seconds of dull film. How did English handle it? At the conclusion of the tongue lashing, Ken exited the two-shot and the camera held on his father who stared regretfully after him. The next shot began inside the barn on a closeup of the horse and then panned to the barn door as Ken entered. Audiences never missed what certainly would have been tedious footage. Without an opportunity to study the distances involved, they accepted the timing. Timing cuts eliminate dull footage by cutting around it, jumping ahead to the next shot in a dramatic sequence, and thus propelling a story forward. In commercials where every second must contribute, timing cuts are a way of life.

Another form of timing cut is the **cutaway**. When an action plays too long, the director or editor simply cuts away to (say) someone watching and then cuts back to the action, which now has been completed. In *Fletch*, director Michael Ritchie photographed Chevy Chase climbing onto the roof of a real estate office and then breaking a window and lowering himself through it. To speed the action, he cut away occasionally to a vicious dog watching. During each cutaway many feet of tedious footage were eliminated.

Variations of the timing cut are used repeatedly by TV and film directors. Dramatic time is elastic. It expands to enhance the significant or the entertaining. It contracts to eliminate or minimize the boring. Wouldn't it be exhilarating to transfer such techniques to everyday life?

Variation

We have looked at four ways to increase dramatic emphasis on a character or an action. We have not yet considered the fact that emphasis necessarily changes during the progression of a scene.

Editing The **editing** of videotape or film seems the most obvious technique for changing emphasis within a scene. Yet surprisingly, many within the entertainment industry (including some directors) do not fully understand why editors cut from actor to actor. Usually, such cutting is intended to maximize the drama, to show the audience where the greatest interest lies. But interest necessarily changes from moment to moment, thereby requiring cuts back and forth to accommodate the changing values.

By showing the audience only what they want it to see, directors manipulate the audience. Because they filmed their scenes in a certain way, guided by their vision of the potential drama in them, directors edit their videotape or film to conform to that vision and, perhaps, to enrich it. On some occasions, usually in so-called suspense sequences, directors deliberately prevent audiences from seeing what they want, thereby building apprehension.

When Eileen tells her boyfriend that she is leaving him, most editors would play the announcement in her closeup so that we will receive the full impact. Then editors would probably change emphasis, cutting to David's face as he registers shock and then back to Eileen as her face softens in regret. "Being on the right face at the right time" is the way some directors phrase it, directing audience attention to where the maximum drama is taking place.

But editing also, usually, shows audiences what they want most to see. Study your reaction to actors upon a theatrical stage—or, in real life, to people arguing at the scene of an automobile accident. As they speak, your interest goes from one to the other; your eyes flick back and forth from face to face. But your eyes are not always on the face of the person speaking. Frequently they go to the face of someone reacting to a powerful statement or, on the stage, to an actor performing a surreptitious piece of business (for example, putting poison in someone's drink). Your eyes follow your interest, fastening on wherever the most significant dramatic action takes place.

Ultimately, then, the final version of most shows is a composite: what the audience wants to see and what the director wants it to see.

Selective Focus   One frequently seen device for changing emphasis is to **rack focus** (also known as pulling focus), that is, changing the camera focus from foreground objects or characters to background objects or characters. Viewer attention understandably follows the focus, attracted to what is most easily distinguished. Such changes in focus parallel the action of our eyes in everyday life; as we focus on a single person in a room, all others (without our being aware of it) automatically go out of focus.

Through the years a few motion pictures have been photographed with **deep focus**, that is, with both foreground and background sharp. Such overall sharpness is achieved through use of an extremely small aperture, or **f-stop**, in the camera lens and correspondingly higher levels of light. *Citizen Kane* and *The Magnificent Ambersons*, both directed by Orson Welles and photographed by Gregg Toland, are celebrated examples (see Figure 7.22).

Deep focus photography has been heralded by some critics as contributing significantly to a film's reality.[3] In real life no one tells us where to look; we fasten our attention wherever we like. There are no cuts from face to face or predesigned changes of focus from one friend to another. Similarly, claim these critics, deep

[3]Specifically, André Bazin in *What Is Cinema?* volume I (Berkeley: University of California Press, 1967), pp. 33–37.

FIGURE 7.22   In *Citizen Kane*, director Orson Welles used deep focus photography, presenting both foreground and background in needle-sharp focus. (Used with permission of RKO Pictures, Inc.)

focus photography allows a spectator the freedom to direct attention without manipulation.

In most pictures, whether tape or film, camera focus tends to be needle sharp on some characters and fuzzy on others. This **selective focus** remains a time-honored device for directing and controlling emphasis. As shown in Figure 7.23, when focus is sharp on the man, viewers direct their attention to him. When focus changes to the woman, viewers transfer attention to her, usually without any awareness that they are being manipulated.

Visual Composition

Just as theater directors use elements of their sets to create emphasis for characters (see Chapter 6), TV and film directors use elements of visual composition to achieve the same goal. The intersecting lines of doorways and windows attract

FIGURE 7.23 Selective focus offers directors a method of directing emphasis. With two actors, audience attention goes to the actor in focus. Attention shifts as the focus changes.

FIGURE 7.24   Directors often use lines created by the environment to bring emphasis to characters, as in this scene from *Midnight Cowboy*. (A UA release © 1969 Jerome Hellman Productions, Inc. Used with permission.)

and focus audience attention on the character within the structural **frame** (see Figures 3.16 and 3.21). In Figure 7.22, the strong **vertical lines** of two columns drive the viewer's attention downward to the figure of Citizen Kane. Notice two other compositional forces at work here: The rapt attention of the character at camera left creates a line that directs our gaze to Kane. Finally, Kane stands in strong contrast to his background, a composition that also compels audience attention.

In Figure 7.24, director John Schlesinger has used his New York City location to create a framing that surrounds and confines Jon Voight and Dustin Hoffman, locking our attention within the frame. The powerful diagonal line of the wooden beam also drives attention inward. Notice that the setting does more than compel our attention; it creates dramatic tension; it compresses the characters within its claustrophobic walls; it contributes significantly to the scene's mood and atmosphere.

CREATING MOOD AND ATMOSPHERE

Every story has its own flavor, personality, style, ambience. When reading a script, a director becomes aware of such characteristics often from the very first scene. They will color all major creative choices: selection of actors, music, editing, scenery, locations, wardrobe, properties, and hairstyles. They will also color the style of acting and even the nature of characters' relationships. They will shape the way in which the director uses a camera.

Is it a hard-hitting contemporary yarn played in the back alleys of a large city, calling for a gritty, grainy, naturalistic texture? Is it a nineteenth-century love story suggesting sepia tones and the soft, shimmery image created by fog filters? During preparation, most directors develop an image in their minds, a "look" that they feel is appropriate for the project. The intelligent director discusses that look in preproduction with the cinematographer or technical director, seeking his or her creative input.

The camera contributes to the creation of mood and atmosphere through *position*, *movement*, and *lenses and filters*.

Camera Position

Where a camera is placed in relation to actors creates certain inferences in the spectator. Many directors position their cameras instinctively. That "instinct" is based on knowledge accumulated through study and experience, knowledge overtly concerned with the mechanics of telling a story. But other factors are also at work. The relationship between camera and characters in a scene creates emotional overtones, making subtle, almost subliminal statements about the characters' relationships with one another and with their environment, about their states of mind and their emotional stances.

Usually a camera is placed at eye level because that is our normal way of looking at the world. When we look at our friends, we look them in the face, eye to eye. But there are many occasions when a director intentionally places a camera above or below eye level.

High Angle Shots   When we look down on someone, figuratively as well as literally, we place him or her in an inferior position. He or she becomes subordinate, recessive, smaller than we. An insecure teacher has a violent argument with a girl in his class. In protest, all of the students get up and leave. The teacher sinks down in his chair, stunned, ashamed, sick at what he has brought about. A director photographing that scene might play the final moment in a wide, high angle shot, aware that the high angle would make the teacher appear diminished, a small, pitiful figure surrounded by empty chairs, alone in an empty room. The angle does more than cover the action. It makes a comment. It instills emotion.

A little boy breaks a valuable antique and his mother threatens to punish him. A high camera position provides a down angle on the boy, diminishing him, showing the spectator how small and insignificant he feels, evoking sympathy.

High angle shots are ideal for establishing a locale. They orient the viewer to a scene's geography and the physical relationship of its elements. They provide an overview.

A seldom used variation of the high angle shot is the **overhead shot** in which a camera is positioned directly above the actors rather than angling down at them. Now the camera assumes a detached, godlike role, watching impersonally from high above as actors move through their paces. The overhead shot gained attention in the early 1930s when director Busby Berkeley used it to photograph elaborate dance numbers. The high camera emphasized patterns rather than people as dancers performed intricately choreographed maneuvers.

The overhead shot is occasionally used for startling or shocking effect. Director Walter Grauman devised such a shot in a television movie starring Edward G. Robinson. The camera began close on the actor's face in what appeared to be a normal closeup. Then the camera suddenly zoomed back to an overhead position, revealing that Robinson was strapped to a bed in a mental hospital.

Low Angle Shots   We look up to people we respect, who occupy a higher position in society, who tower above us intellectually or professionally. Similarly, when a director places the camera below eye level, looking up at a character, that character assumes a position of dominance, of strength, of importance. Perhaps such feeling originated when we were children staring up at the gigantic adult figures who dominated our world.

In the classroom scene earlier, another director might play the argument between teacher and student in two angles: a high shot angling down on the student to make her appear diminutive and a **low angle** on the teacher to make him dominant. At the scene's conclusion, a low angle shot on the teacher would make a totally opposite statement from the one described earlier, leaving him in a position of power.

Because some characters are dominant and others recessive in almost every scene, a few directors believe that a camera should almost never be placed at eye level. Others, a misguided minority, prefer to shoot scene after scene from a low camera position, claiming that such shots are more "dynamic," that low angles automatically add theatricality. Realistically, such low angles seem striking because all characters tower above the camera, creating a melodramatic world of diagonal lines and dominating fantasy figures. Theatricality works in scenes that require it. Scenes of horror or terror, a chase sequence, a gunfight, and any action that leans toward melodrama can all be enhanced through occasional use of low or bizarre camera angles. But such theatricality in normal, honest dramatic scenes becomes an overstatement that renders serious dramatic values absurd.

When photographing dance numbers, directors often use low angles because they tend to accentuate and exaggerate legs and feet. Such shots feature skillful footwork and at the same time convey a certain sensual appeal.

Subjective Shots In most cases, the camera records action dispassionately; it is **objective**. Occasionally, however, the camera becomes a participant; its angle becomes the perspective of a character. The most usual form of such participation is known as a *point of view (POV) shot*. Frequently, such a shot is preceded by a closeup as a character stares at some off-camera action. The subsequent angle then becomes that character's point of view, the camera looking at the world through that character's eyes.

An entire motion picture once used **subjective camera** treatment. The 1946 film *Lady in the Lake* played its story from the literal point of view of its protagonist, actor Robert Montgomery. The only time the audience saw him was when he glanced in a mirror.

Usually, POV shots depict a character's actual angle of vision. If a character glances up at a tower, the POV shot necessarily becomes an up angle. Often, however, because of directorial or audience needs, POV shots are cheated. If specific visual information is needed, for example, the **cheated POV** may be moved closer so that viewers can read the headline of a newspaper. During the chase sequence in *The French Connection*, director William Friedkin cheated the driver's forward POV, attaching his camera to the car's bumper. The lower angle gave a far greater sensation of speed as the pavement whizzed by only inches below, yet audiences accepted the shot as the driver's perspective.

The POV shot represents the most easily understood as well as the most commonly practiced example of subjective camera. In a broader sense, however, the entire stylistic treatment of a film or videotape can represent and reflect a character's emotional state. The most celebrated example is the German expressionistic film *The Cabinet of Dr. Caligari*, in which the distorted sets and grotesque actions represented the perspective of the story's narrator, a madman.

We have seen examples of this broader concept of subjective camera dozens of times. When a central character becomes depressed, the entire ambience changes. The sky becomes gray. Accompanying music sours. Colors seem muted. The lighting key is low. When characters become elated, the film's atmosphere echoes and enhances that elation through blue skies and brilliant sunshine, sparkling music, upbeat settings, and a smiling world. How many burial scenes have you watched on television or in the movies in which mourners held umbrellas to protect them from the rain?

Film theorists use the term **pathetic fallacy** (from *pathos*) when the external world, usually the elements of nature, reflect the inner state of a character. Thus in burial scenes the darkening skies reflect the characters' grief; the rain externalizes the tears they are shedding or withholding.

Flashbacks sometimes—but not always—demonstrate subjectivity in that they glimpse the past through a character's eyes. Thus, characters and incidents in flashbacks may be distorted or reshaped by recollection. If an old woman remembers her childhood with joy, for example, flashback scenes may color parents, brothers, and sisters in an unrealistically positive way, the scenes photographed through fog filters, with rosy hues predominant and the lighting high key, suggesting happiness. When a criminal recalls his grim teenage years, flashbacks may reconstruct them harshly, exaggerating the cruelty of others, ignoring strict reality because the flashback has been filtered through memory, reshaped by time and emotion.

Note the difference between the two types of subjective camera. When a man becomes drunk, his world distorts tipsily. On the one hand, if the editor cuts to the character's literal POV, viewers watch his room tilt or grow fuzzy *through his eyes, as he sees it*. If, on the other hand, the audience watches him stagger down a street (objective camera) and the entire picture tilts or blurs or weaves, the picture is *suggesting* his mental condition rather than depicting his point of view literally. Viewers infer his mental state through the entire stylistic treatment, including camera work, music, sound, special effects, and whatever the directorial imagination may invent.

Tilted or Canted Angles You stand on the front porch of your home staring at the horizon, reassured by its implications of serenity, stability, security. But what if the horizon suddenly tilted? What would be your feelings? Shock? Panic? Dizziness?

To make their camera angles effective, directors frequently rely on basic human emotions and psychological responses. The tilted shot is a prime example, suggesting a world that has gone awry, in which normal standards of security have disappeared.

Directors search for appropriate ways to suggest a character's mental or emotional state. A **canted angle** (also called a **Dutch angle**), suggesting **disequilibrium**, might be appropriate for the drunken gentleman described above, for a madman whose world has come unhinged, or for someone on an LSD trip. It fits comfortably with "tangerine trees and marmalade skies."[4] Such an angle might even comment appropriately on a love scene if the lovers' emotions swell to the bursting point, so ecstatic that the horizon itself stands on end.

Tilted or canted shots work effectively in montage sequences, image superimposed on image, composing a kaleidoscope of visual concepts. Most of the time, however, they simply call attention to themselves. They are bizarre and unnerving. Use them sparingly.

Hand-Held Shots Often without realizing it, our minds make associations. When a character in a script is named Ethel, that character takes on resonances of everyone we have ever known, seen, or read about named Ethel. We automati-

---

[4]From the Beatles' song "Lucy in the Sky with Diamonds," said to suggest the LSD experience.

cally transfer feelings from those earlier associations. Similarly, when a television or movie program is tinted sepia, we recall yellowing tintypes; the program connotes the nineteenth century, which is exactly what its director intended.

Hand-held shots are often associated with news and special events coverage. When network or local TV station personnel carry minicams to the scene of a fire, they grab shots as best they can, camera moving awkwardly, unevenly, jostled by passersby. When a director selects a hand-held camera for a sequence, it is with the knowledge that such coverage will create such associations in the viewers' minds. A **hand-held camera** creates a feeling of participation in exciting action, of shots taken on the run. It provides a gritty, documentary look as opposed to the conventional "slick and glossy" appearance often associated with Hollywood filmmaking.

Camera Movement

You should understand by now that in well-directed drama little happens without a reason. Each element in a script exists to serve a plot or a character goal. Every move or piece of business performed by actors ideally is motivated by a thought or emotion.

In a similar fashion, movement by the camera should be neither arbitrary nor capricious. Each time a director pans, tilts, dollies, or trucks a camera, there should be a definable dramatic reason. In some cases, those reasons will touch on concepts we examined earlier.

The Dolly As interest in a scene increases, the audience fastens its attention more and more closely on the faces of the players. In early Hollywood films, such growing involvement was paralleled by a series of editorial cuts, each angle closer to the action than the last.

The forward **dolly** accomplishes the identical purpose but in a more fluid way. Although dramatically motivated cuts (those that follow viewer interest) are almost undetectable, the slow forward dolly is still less apparent. It is almost invisible because in most scenes it exactly mirrors the action of the viewer's mind. Conversely, dollying out represents a disengaging of attention, distancing the viewer from the scene.

As the camera dollies toward characters, their impact increases with their screen size. *Increasing emphasis* constitutes the primary reason for most dollying in. If the dollying camera is in a low position, its approach will gradually increase the degree of the up angle so that a character will simultaneously gain *impact* and become more *dominant*.

Although a scene's increasing tension is the primary reason for dollying in, such a move may also be physically motivated. When a dolly accompanies a character (or other physical action), the character seems to propel the camera's forward movement. Such a dolly appears both graceful and motivated.

By definition, dolly shots move forward and backward. **Zoom shots** also move forward or backward within a lens field. While a dolly physically travels within a setting, changing perspectives as it moves, the zoom changes the focal length of its lens, foreshortening space, moving within a single plane. Zoom lenses provide great convenience for directors, but they have been overused in recent years.

Trucking and Tracking Shots   When a camera trucks, the movement is usually lateral (sideways) and almost always accompanies the movement of characters. The traditional trucking shot is a raking side angle, moving with characters as they walk down a path. Inventive directors often allow lampposts, passersby, or other atmospheric elements to pass between camera and characters, thus contributing to the ambience.

When a camera precedes characters, retreating as they advance, such a variation is labeled a tracking shot. Sometimes a tracking shot follows characters from behind, accentuating their destination or the direction in which they are heading. In actual practice, the terms *trucking* and *tracking* are often used interchangeably. In script language, a camera *tracks* a character but *trucks with* a character. Because both trucking and tracking shots are propelled by the movement of characters, there is dramatic motivation behind them.

For many directors, the moving camera does more than merely follow the movements of principal characters, mechanically covering the action. The concept "style reflects content" is recognized in the work of sophisticated professionals, and the moving camera frequently contributes to the style of each. For Germany's Max Ophuls (*La Ronde, Lola Montes*), it became a poetic expression of a romantic world. Long, languorous trucking shots up stairs, down corridors, through opera house lobbies commented on his characters and the mood they generated, underscoring and embellishing that mood. Contrast such graceful camera movement with a different directorial treatment, a staccato series of quick cuts from room to room as a character moves through them. Note the difference in mood generated by each treatment.

Crane Shots   When a camera is mounted on the end of a **crane**, a long steel arm, it suddenly becomes capable of vast, sweeping moves, usually up or down. Recall the opening shot of many Westerns, with camera high in the air, angling down on a small Western town, effectively orienting the audience. A lone rider appears at the far end of the street. As he approaches, the camera cranes slowly down so that, as he directs his horse to the hitching rail in the foreground, the camera is low, close to the ground, looking up at this mythic hero, emphasizing his strength.

The high camera is objective, dispassionate, godlike. It makes little comment. As it descends, quickly or slowly, it becomes part of the unfolding drama, a participant, creating greater audience involvement. There is drama, even theatricality, in the great, majestic sweeping movement of the crane arm. At the conclusion of the Western, the hero rides off into the desert. The camera cranes upward, making him appear a small, lonely figure and exaggerating the vastness of the desert that surrounds him.

Panning and Tilting Previous examples of camera moves (trucking, dollying, craning) all involved the physical movement of both the camera and its base or pedestal. The camera changed position in relation to either actors or scenery. In panning and tilting, the camera base remains stationary and only the head moves. In a **panning shot**, the camera head rotates from side to side, its field travelling horizontally, from right to left or from left to right. In a **tilt**, the head moves upward or downward, the camera field traveling vertically.

Directors use panning and tilting most frequently to follow character movement. The camera pans with characters as they cross a room. It tilts upward as a character climbs a ladder. Sometimes directors use a pan as an orientation shot. Imagine an opening sequence, perhaps under titles, as the camera pans slowly across acres of farmland (a panorama, from which the word *pan* derives), ultimately picking up a sports car speeding along a dusty road.

Another reason for panning is to follow the direction of a character's glance, replacing the more conventional POV shot. For instance, a teenage girl looks at a window and screams. The camera slowly pans to the window to reveal a monstrous face leering at her. The slowness of the pan deliberately teases the audience, building apprehension, delaying the revelation. The director would probably cut rather than pan back to the girl because a second pan might seem labored.

A quick pan offers shock value. In the example, the trauma would have been increased if the camera had unexpectedly *whipped* to the window in a sudden, stunning moment. The director of *Poltergeist* occasionally used such quick pans for their nerve-scraping effect.

Lenses and Filters

Lenses and filters affect the quality of an image as it is focused on film or videotape. Lenses control the optical properties of an image; **filters** modify that image either by making aesthetic corrections or refinements, by affecting the color balance, or by changing the exposure.

Lenses Lenses may be divided into three categories: normal, wide angle, and long. Because normal (middle range) lenses are relatively distortion free, they are used most often by most directors for most scenes. Their angle approximates the angle of the human eye.

Both **wide angle lenses** and **long lenses** distort the photographic image; the wider or longer the lens, the greater the degree of distortion. Directors intentionally use such distortion to meet aesthetic or practical requirements of a dramatic situation.

Because by definition a wide angle lens (also called a **short lens**) provides a comprehensive view, it is often used for orientation shots. Such orientation should not include panning, however, because extremely wide angle lenses spherically distort the field of vision, making the field appear to bend at the center. Wide angles are ideal for dollying because they minimize camera vibration and seem to provide great depth of focus.

Directors frequently face situations where sets or locations are not as large as desired. One property of the wide angle lens is that it enlarges objects in the immediate foreground and reduces background objects, distorting space, creating the illusion of greater distance than actually exists. Thus, a corridor actually twenty feet long might appear double or triple that length when photographed with a wide angle lens. Should characters walk down the corridor toward camera, the director must exercise care to prevent them from coming too close because objects in the immediate foreground will appear distorted and grotesque. If a director wants facial distortion for comedic or horrific effect, however, then such a lens is appropriate. Also, because wide angle lenses make distances appear greater than they actually are, movement toward or away from camera appears faster than normal.

The reverse is true of long lenses, which appear to compress space, to diminish distance. Characters moving toward camera seem to make little progress. Director Mike Nichols effectively used such a lens in *The Graduate* when Dustin Hoffman raced to stop a wedding, to prevent his loved one from marrying another. The long lens created the effect of a nightmare—he ran and ran but appeared to get nowhere.

Long lenses, like binoculars or a telescope, enlarge images; the longer the lens, the greater the degree of magnification. They are used for obtaining close shots from a distance. Because long lenses have a narrow depth of field, they lend themselves to selective focus. When directors want a character sharply defined but with foreground and background indistinct, they will inevitably select a long lens.

Filters The normal, properly focused, properly exposed picture is crisp and sharp and renders color or black and white values in relatively realistic fashion. The quality of the image may be intentionally altered by special gelatin or glass filters, which fall into two categories: technical and aesthetic. In the technical category are **neutral density (ND) filters**, which reduce the quantity of light transmitted through a lens, and *polarizing filters*, which control reflections or glare or darken skies for night effects.

In the aesthetic category lies a spectrum of filters that diffuse, distort, or otherwise change the image in order to achieve a visual effect that will influence the audience's emotional response.

Star filters have been used extensively in commercials as well as in some television and motion picture drama. The star filter transforms highlights into star-shaped lens flares or halations, creating an effect that is bright, glamorous, upbeat, glittery. Such an effect might enrich a wedding day, a posh Bel Air party, or a romp through wet streets by a couple in love.

Diffusion filters soften an image. They have been used since Hollywood's earliest days to add beauty to glamor queens, simultaneously concealing lines, wrinkles, or other skin imperfections. Early cameramen placed a scrap of silk stocking over their lenses to achieve the same gauzy effect. Diffusion filters have been used in outdoor locations to create a pastoral mood of tranquility, transforming the sometimes harsh lines of nature into the shimmery ambience of Impressionist paintings. Diffusion filters have partially given way to fog filters.

Fog filters were originally used to create the effect of—guess what? That's right, *fog*! Such shots usually were enhanced by a few wisps of smoke from a smoke pot just off camera. If you watch television, you have seen fog filters used many, many times. They create a soft, hazy brightness in a scene, especially noticeable when the camera is directed at a light source such as a window. The picture becomes idealized—misty and luminous and vastly flattering to the doe-eyed ladies who inhabit television commercials.

Although commercials were first to expand the uses of fog filters, television programming quickly followed suit. Fog filters seem appropriate for flashback sequences. The noticeable difference in picture quality places such sequences apart from scenes happening in the present. Fog filters seem appropriate for fantasy. The hazy quality suggests another dimension. They seem appropriate for dream sequences. Some directors use them recklessly in comedies and in action adventure shows. One of their most effective uses was in the motion picture *All That Jazz*. Director Bob Fosse used fog filters for a number of stylized hallucinatory sequences in which protagonist Roy Scheider conversed with a glamorous woman later to be revealed as Death.

Polarizing filters reduce glare and reflective flare. Usually they are paired in a lens mount so that one may be rotated for varying degrees of effectiveness. Polarizing filters tend to darken skies, rendering them a deeper blue, and are sometimes used to approximate a night effect. They also seem to intensify colors.

Color correction filters are used ocasionally to alter the color values in a scene, usually very subtly. Because color is one of a director's most valuable tools in eliciting emotional response from an audience, correction filters can prove extremely useful, especially in the hands of a sensitive cinematographer. In nightmare sequences, for example, a saturation of color might contribute to the sense of unreality; it would make proceedings seem strange, bizarre, tinged with insanity. In more realistic photography, a subtle addition of blue (of which audiences would be consciously unaware) might contribute to feelings of serenity. With film the processing laboratory can shift color values in postproduction.

In addition to color correction filters, many areas of color control are available throughout the production process: in scenery, lighting, wardrobe, and set dressing. Discussion of color motifs with a production designer or an art director during preparation can prove enormously helpful to the director.

This book is not designed to examine audience response to color, but I suggest that you not overlook such a significant area. Many art books discuss the subject in depth.

Color can be selected equally for dramatic effects and subtle ones, to convey a feeling of conflict, of tranquility, evil or well-being, happiness or sadness, intimacy or remoteness, warmth or coolness, masculinity or femininity, and so on. It is close to music in its effects upon the sensibilities of the audience.[5]

CHAPTER HIGHLIGHTS

— A director's use of the camera falls into three major categories: (1) *covering the action* (revealing what is happening, telling the story), (2) *creating emphasis* (providing impact for significant elements), (3) contributing to a drama's *mood and atmosphere*.

— The most accepted pattern of covering the action emerged in early Hollywood films with the camera starting wide and moving closer. Beginning a scene in a wide angle provided the audience with orientation. Then, as tensions grew, camera angles came closer until, at a scene's climax, the action played in closeups. The pattern remains valid today.

— Camera usage and style vary from director to director. Some prefer a fluid, moving camera; others shoot a single master angle that covers action of the entire scene and then break the scene into closer angles that an editor can assemble later. Each camera position is called a *setup*.

— Master angles usually are wide. They show the physical relationships between characters and movement that cannot be seen in closer angles. Master shots sometimes start wide and then move closer. Sometimes they start close and then move wider. Occasionally, scenes are shot with no master angle at all.

— Because certain moments in drama are more significant than others, directors use their cameras to emphasize those moments. They achieve such emphasis through closeness, separation, positioning, duration, variation, and visual composition.

— Close angles do not create significance; they increase or intensify whatever dramatic values already exist. They also clarify, allowing audiences to glimpse details and sense emotions they would be unaware of in wider angles. Finally, closeups create a feeling of intimacy, allowing sympathy between the audience and a character.

---

[5]Eliot Elisophon, *Color Photography* (New York: Viking Press, 1961), p. 12.

— A character facing camera achieves greater impact than characters in profile. Over-shoulder shots do not have the impact of closeups, but they create a strong feeling of *relationship* between characters.

— Emphasis may also be achieved by prolonging a moment. Slow motion is one method. Significant moments may also be prolonged through editing, expanding key actions or reactions by adding additional cuts. Similarly, editing can shorten or eliminate dull moments.

— Emphasis changes many times in most dramatic scenes. Cutting between angles is the most common method for directing audience attention from character to character. Another method is changing (or racking) focus. Audiences give attention to what is most easily viewed. When a foreground character appears sharp and the background character is out of focus, attention centers on the foreground character.

— Camera may contribute to a drama's mood and atmosphere through position, movement, and lenses and filters.

— Viewers make inferences based on camera placement. A high angle shot looks down on characters, figuratively as well as literally. A low angle shot looks up at characters, making them appear powerful and dominant.

— Although the camera usually provides an objective view of action, occasionally it becomes subjective. In the most usual form of subjective photography, the POV (point of view) shot, the camera literally assumes a character's angle of vision. A second form of subjective camera is demonstrated when the entire treatment of subject matter (lighting, sets, music, and so on) reflects a character's state of mind—the external mirroring the internal.

— Tilted or canted angles suggest disequilibrium, the world of a madman or a drunk. Hand-held shots are associated in viewers' minds with news and special events coverage. Directors take advantage of this association to create a gritty, documentary look for appropriate sequences.

— Every camera move should have a dramatic motivation. Forward and backward movements by a dolly, for example, usually echo the viewer's increasing or decreasing involvement. Trucking and tracking shots usually are impelled by the movement of characters.

— Lenses and filters affect the quality of an image. Lenses control the optical qualities; filters modify the image by making aesthetic corrections, affecting the color balance, or changing the exposure.

— Both wide angle and long lenses distort the image; the wider or longer the lens, the greater the degree of distortion. Wide angle lenses enlarge objects in the immediate foreground and diminish background objects, thereby creating the illusion of increased depth. Long lenses reduce depth, flattening the photographic field.

Star filters transform highlights into star-shaped **flares**, creating an effect that is bright, upbeat, glittery. Diffusion filters soften an image, concealing imperfections, idealizing subject matter. Fog filters, originally used to create the illusion of fog, today have almost replaced diffusion filters. They create a soft, hazy, luminous image.

PROJECTS FOR ASPIRING DIRECTORS

1. Take paper and a pencil to the next movie you see. Make notes on camera coverage. When did the director begin a scene wide? When did he or she begin in a close angle? In scenes that began close, how soon did the director provide orientation? How was it provided? Did any scenes lack orientation?

2. Were there any scenes in the movie without coverage of any kind? What was the nature of the scene? Did you miss coverage? Why or why not?

3. In the scene in Chapter 6's "Projects for Aspiring Directors," mark the moments that you would play in closeups or extreme closeups. Why did you select these moments? (Caution: Don't get too close too soon. Allow for progression.)

4. Is there any moment in the preceding scene that you would double cut? That is, a moment that you would prolong by intercutting closeups?

5. In *Final Curtain* in Chapter 6, mark the moments that you would play in closeups and the moments that you would prolong through double cutting.

6. Watch an hour or two of television. Make notes on how many commercials use fog filters. Why did the director use fog filters in these commercials? Did any of the commercials use star filters? Which? Why did the director use them?

8

KEEPING THEM IN SUSPENSE

The term **suspense** usually is applied to a genre of television shows or motion pictures designed to generate anxiety or apprehension. Such "thrillers" traditionally attempt to terrorize audiences by placing protagonists in sinister, frequently life-threatening situations. Exploitative feature films sometimes intensify shock values by adding scenes of gratuitous violence, bloodshed, and mutilation.

Because elements of the thriller occur so frequently in television and film, this chapter will examine them in considerable detail. But suspense has a far broader application than providing thrills. It is an essential part of all successful shows, the foundation of every dramatic scene, as necessary an ingredient as actors or dialog. Through an understanding of the nature of suspense, a director can deepen audience involvement and heighten a project's impact.

The four major parts of this chapter will examine:

— THE NATURE OF FEAR : studying its basis in the unknown and the unseen

— HITCHCOCK'S TECHNIQUES for generating suspense: delaying the revelation of what audiences want to see, scraping their nerves, and providing them with information that will provoke worry

The art of creating suspense is also the art of involving the audience, so that the viewer is actually a participant in the film. In this area of the spectacle, filmmaking is not a dual interplay between the director and his picture, but a three-way game in which the audience, too, is required to play.

François Truffaut[1]

— SCRIPT ASPECTS OF SUSPENSE : examining the victim, the antagonist, and the need for believability

— THE ESSENCE OF SUSPENSE : shattering many misconceptions

THE NATURE OF FEAR

All successful drama stirs us fundamentally. It touches basic hungers; it awakens our collective unconscious. When a particular show unaccountably makes us cry, often it is because we have been moved on a very primitive level, beneath all of the scaly, protective layers that make us "civilized."

---

[1]François Truffaut, *Hitchcock* (New York: Simon & Schuster, 1967), p. 11.

When directors seek to create fear in an audience, they look for its primordial roots. What subject makes all people afraid, no matter what their culture, no matter where they live? Many such deep-seated terrors go back to a time when we were all primitives: when we were children.

The Unknown

If you were like most youngsters, you were afraid of the dark. Why? Because you couldn't see what might be lurking there. Your imagination supplied terrifying possibilities. The Boogeyman. The ghost of Uncle Arthur. A grotesque monster. When, finally, you turned on a light, the room became reassuringly normal. Fright disappeared. Moments later, with the lights out, the Unknown again manifested itself and darkness conjured up the hobgoblins of imagination.

What else were we afraid of as children? Graveyards. Coffins. Ghosts. Skeletons. All symbols of the dead. What is death? For many of us, it is (again) the Unknown.

When I was young I used to shiver at thunderstorms. The sudden, stroboscopic flashes of light were warnings of explosions to come, deep-throated rumblings and crashings that set the walls of our small home (and me) trembling. Some primitive cultures worshipped thunder as a god. The phenomenon is awesome and powerful. And it, too, represents the Unknown.

Fear of the unknown also confronts us on the most mundane levels. When we visit a dentist, most of us become tense, afraid of possible pain. The moment the dentist describes exactly what to expect—even when it's going to hurt—much of the fear vanishes.

For a director, fear of the unknown is a valuable tool when fashioning projects that seek to create terror. One of the earliest television series I produced was called "The Untouchables." Its violent, fear-oriented stories concerned activities of mobsters in the late 1920s and early 1930s. Significantly, five out of six scenes in every episode played at *night*. Had anyone applied simple logic, it would have been obvious that the lengths of those nights were absurd. The reason for so many night sequences, of course, had nothing to do with logic. Scenes of unexpected violence, where fear constitutes the essential subtext, play more excitingly in a darkened, shadowed world. In such a world, our childhood fears return to choke us. Have you ever watched a ghost story that played in the *daytime*? Has any hero in any story ever visited a haunted house at *noon*?

The Unseen

There's a special kind of dread of things unseen, unrecognizable, unassessable, whose grotesqueness can only be guessed at. In *Jaws*, the sea monster swam concealed in the ocean depths. It is significant that in a story whose primary character was a great white shark, audiences never actually saw the shark until the film was almost two-thirds over. The opening sequence featured the shark's moving point of view under the water. Certain throbbing, Stravinski-like music became the shark's theme. Whenever we heard it, we trembled at the monster's presence. From time to time we saw results of the shark's depredations. But it wasn't until scene 178 (out of a total of 255) that we actually glimpsed the sea monster. The ocean, like the night, protectively concealed his presence, magnifying him, increasing our terror. Like the night, the ocean itself became frightening; the most innocent seascape might erupt into stomach-wrenching horror.

Using the Camera to Conceal   Revealing only a shadow or a portion of someone's anatomy is the most obvious way by which a camera may conceal the identity of a character. A **claustrophobic frame**, or angle is another. Ordinarily when a protagonist enters a new locale, the director provides a wide angle for orientation. Sometimes, however, directors deliberately prevent orientation. They fasten their camera angles close on a protagonist's face, placing blinders on the audience, not allowing them to see who or what may lurk in the vicinity. Such shots generate claustrophobia, an uneasy frustration at being confined. Sustained for a period of time, they build genuine apprehension.

Let's watch as Rachel approaches the decrepit old mansion and stares at it nervously. The angle changes. Now we're watching Rachel through a small upstairs window. As she glances up, camera moves back quickly, as if to avoid being seen. We suspect (and worry) that this angle may be someone's point of view. Our suspicion is confirmed a moment later when we see a gnarled hand pull a curtain across the window, blocking our view.

Now Rachel goes to the front door and opens it. The huge door squeaks, scraping our nerves in the tradition of such clichés. Once inside the mansion, Rachel walks slowly from room to room, *camera remaining close*, tracking her. Because the camera is restricted to a head-and-shoulders closeup, we wear blinders. Like a child in a darkened bedroom, we cannot see what is there. We grow apprehensive, anxious to survey the off-screen space, aware that Rachel may be in danger, and eager to *see* that danger before it descends. The longer our view is restricted, the more disquieted we become. When, finally, Rachel turns a corner and screams, we're jolted by a shocking closeup of the maniac/monster/fiend.

Note that in this example, the director used both kinds of camera concealment. When someone watched from an upstairs window, the observer's *identity* was kept hidden, revealing only his gnarled hand. When the camera stayed claustrophobically close on poor Rachel, the villain's *whereabouts* remained a secret, denying the audience the chance to look around and assess the dangers.

The suspense values of a claustrophobic angle are twofold: (1) Things unseen are more frightening than things seen, and (2) delaying the moment of revelation also builds spectators' anticipation. (More about delaying the revelation in a moment.)

The Viewer's Imagination   When thrillers enter the realm of fantasy, directors confront a new set of problems. Recall the low-budget "shockers" you may have seen on late-night television in which special effects trickery or costume design appeared inept and amateurish and loathsome monsters emerged as nothing more frightening than a man in a green rubber suit. When a show becomes laughable, the director inevitably and rightfully inherits much of the blame.

Readers of best-sellers often complain that a motion picture or miniseries falls short of expectations. The reason, usually, is that a novel depends on the reader's *imagination* to supply its visuals: scenery, costumes, faces. So abundant is the reservoir of the human imagination that dramatic artists have difficulty competing. Most prefer not to try.

In so-called thrillers, the more familiar audiences become with antagonists, the less fear those villains engender. Imagine a diabolical fiend, a sadistic murderer-rapist who talks volubly with everyone, revealing in detail his innermost hopes and dreams. Compare him with an identical figure who never speaks a word. Now our imaginations go to work as we wonder what vile, unspeakable passions motivate his actions. Although the first characterization might represent a more original approach, the second promises to generate far more terror.

When the director conceals or partially conceals a figure, the imagination adds dimension, transmuting the unseen into something far more terrifying than anything a special effects artist could devise.

HITCHCOCK'S TECHNIQUES

One of the acknowledged masters of screen suspense, director Alfred Hitchcock, used a variety of techniques to create apprehension and shock. Among these were delaying showing the audience what they wanted to see, building a succession of incidents or creating elements whose only purpose was to scrape audience nerves raw, and providing spectators with more knowledge than a protagonist possessed, thus allowing them to worry.

Delaying the Revelation

In almost all of his films, Hitchcock capitalized on the fear of the unseen and the unknown, prolonging anxiety by putting off showing an audience what they wanted to see. Let's examine two examples from one of his less-celebrated films, *Under Capricorn*, the first using camera and editing and the other simple staging in order to postpone a revelation.

The opening scenes created an aura of mystery concerning the character played by Ingrid Bergman, wife of the Australian businessman played by Joseph Cotton. The audience had never seen her and was anxious to discover the nature of her problem. At a dinner party, Cotton's business colleagues started to ask about her and then, embarrassed, changed the subject. Moments later, we heard her footsteps on the staircase. Anticipation built as the footsteps came closer. All eyes fastened on the closed double doors leading to the hall. The footsteps stopped outside. The double doors started to slide open. But before we could glimpse Bergman, Hitchcock cut to a group shot of the dinner guests as they reacted. Now, certainly, we expected Hitchcock to cut to their point of view. But again the director denied showing us what we wanted to see. Instead, he slowly panned the faces of party guests as each reacted in dismay. When, finally, the director cut to Bergman, he again delayed the revelation, starting close on her feet and slowly tilting up. By the time he arrived at her closeup, it seemed anticlimactic. The revelation of Ingrid Bergman in a drunken stupor unfortunately did not justify all of Hitchcock's delaying tactics. The director had built our sense of anticipation to such a high point that what followed was a letdown.

Here's a more successful example from the same picture. Bergman, in bed, suddenly became aware of *something* beneath the covers. Instead of throwing back the bedclothes and immediately exposing it, she arose cautiously from the bed, staring in mute horror at a spot on the far side. She crossed slowly all the way around the bed, camera panning with her, delaying, delaying, delaying showing us the cause of her horror. When, after no less than twenty seconds, she reached the far side of the bed, she suddenly whipped back the covers, and Hitchcock cut to an extreme closeup of a hideous shrunken head there. The shock was extreme this time, justifying all of Hitchcock's delaying tactics.

In many suspense mysteries, directors conceal a killer's identity by showing only a shadow or a pair of legs. The plot purpose is to delay the revelation, but such concealment has a rich fringe benefit: It simultaneously creates an aura of fear because it stimulates the viewer's imagination to create a far more terrifying "heavy" than if he or she had been revealed.

Directors sometimes use a slow panning action to build anticipation or apprehension. For example, a baby lies sleeping in a crib. Off screen, a window opens surreptitiously. When the camera pans slowly, slowly away from the child, we wait anxiously to see what the camera will reveal. The slow pan deliberately teases us, delaying the revelation.

Scraping Nerves

A jolting moment does not happen accidentally. Usually it is meticulously crafted by author and director. Shocking moments grow out of a succession of minor abrasions that make spectators nervous and uneasy. Notice that the element of **progression**, so intrinsically a part of drama, again manifests itself. When half a dozen minor irritations rub our nerves raw, we are off balance, edgy, apprehensive. Any major shock will now cause us to leap from our seats.

In a newspaper interview with Hitchcock concerning *Psycho*, the reporter asked why an abrasive motorcycle policeman wearing bizarre, mirrored sunglasses had appeared early in the film. Hitchcock's reply: for one reason only— to make people nervous. The director occasionally used canted angles for the same reason. At suspenseful moments he would throw in a tilted angle unexpect-edly, jarringly, to build audience apprehension.

Abrasions or irritations may arise from *characters*, *incidents*, or *atmospheric elements*. They may even arise from *editing techniques*. Verna Field, discussing her editing of *Jaws*, revealed that most editors develop a rhythm by which they cut film elements together. In *Jaws* she deliberately violated that rhythm, always cutting a moment before or after her normal instincts would dictate, thereby hoping to create uneasiness in spectators.

Because of its great capacity to evoke emotion, music heads the list of atmo-spheric elements capable of creating spectator disquiet. Audience shock during the stunning shower sequence in *Psycho* came primarily, of course, from the avalanche of brutal visual images, but the accompanying music contributed pow-erfully. High-pitched, almost discordant violins created in tone and quality the sound of a woman screaming, so nightmarish, so alarming in quality that it sandpapered nerve endings.

Probably the most familiar of atmospheric elements designed to make us nervous are thunder and lightning. Despite the fact that they have become ste-reotypical in ghost stories, so powerful is their ability to evoke primitive fears that they continue to disquiet us.

Wind is also effective in creating disquiet through a variety of disturbing noises: shutters or screen doors banging, creakings and groanings of wooden structures, tree limbs clattering against a window pane. Recall the effective use of wind in *Poltergeist* when it climaxed with the uprooting of a tree. Note that the wind cannot be seen and therefore the director must devise some visual manifes-tation of it—such as a lantern or a hinged sign that can blow back and forth, tree branches that can swing, or leaves or dust that can be blown about—to make the viewer visually conscious of the wind's presence. When an exterior light source (moonlight or a street lamp) throws shadows on an interior wall, supplementing the wind sound, uneasiness is transmitted from outdoors to indoors.

Suspense Versus Surprise

One of the ways Hitchcock made audiences nervous was to tell them more than he told his characters. By knowing in advance that the villain was waiting to pounce, the audience could anticipate violence, agonizing for the unsuspecting heroine, unable to warn her of her impending doom, and stew in ever-increasing apprehension as she approached the fateful moment.

As with other directors of the genre, Hitchcock preferred suspense to sur-prise. He described the difference through the classic example of a bomb ticking beneath a desk as two executives talk. If the audience is unaware of the bomb, they will be stunned at the explosion. But how much more satisfying, claimed Hitchcock, to let the audience in on the secret. Aware of the bomb ticking beneath the desk, aware that it will explode in fifteen minutes, anxious to warn the victims-to-be of their danger, the audience can participate in the scene. "In the first case we have given the public fifteen seconds of surprise at the moment of the explosion. In the second case we have provided them with fifteen minutes of suspense. The conclusion is that whenever possible the public must be informed."[2]

SCRIPT ASPECTS OF SUSPENSE

In drama as in other art forms, rules cannot be arbitrary. They do not apply one hundred times out of one hundred. Yet certain characteristics have manifested themselves so frequently in successful fear-inducing drama that we cannot ignore them. They concern the victim, the antagonist, and the believability of the production.

Here's a theorem not found in any physics textbook: The degree of suspense generated in any story varies directly with the *vulnerability* of the victim and the *fearsomeness* of the pursuer.

The Victim

It is night. Frail, thirteen-year-old Beth, searching for her younger sister on an abandoned ocean pier, approaches an ancient building. Below, pilings creak loudly as if in warning. Beth shivers, swallows hard, and enters. Inside, a shadow moves wetly.

Beth, *stop*! Don't go in there! As audience, we sense danger. We're uneasy, nervous, apprehensive: prime components of suspense. Yet when I first wrote Beth's story as part of a magazine article, the opening paragraphs (considerably expanded from that above) created neither involvement nor worry. I rewrote them half a dozen times before realizing what the problem was. I had ignored one of the elementary principles of suspense.

Attempting to create teenage vernacular, I had given Beth the thought, "Why did the little creep keep coming here?" congratulating myself on having captured the typical sibling animus. Beth's character went downhill from there. Nowhere in that ill-fated scene did she do anything to make herself remotely sympathetic. When she reluctantly entered the old building, it appeared she was acting out of duty rather than concern.

Audiences have to *care* about the victim. If they don't care, they won't worry. The more they care, the more they will worry. It seems so obvious, doesn't it? (See "A Person with a Problem" in Chapter 4.)

With new words implying that Beth loved her sister, was actually afraid for her, was forcing down her fears in order to find her, the story vastly improved. For the first time, the reader became involved and started to worry.

---

[2]Truffaut, *Hitchcock*, p. 52.

Perhaps you have a picture in your mind of what Beth looks like: thirteen years old, skinny, wide frightened eyes, freckles. What if I changed her into a lady wrestler, a muscular, leathery woman of thirty-five or forty years? Think about it. Would you worry as much?

What if I performed transsexual surgery and changed her into a man? Would there be any difference in apprehension between a scared teenage girl and a seasoned, gun-carrying policeman?

The answer is obvious. We tend to worry more about vulnerable protagonists than those who can easily protect themselves. We worry more about the female of the species because of some distorted cultural notion that women are the weaker sex. Do you remember the classic suspense film *Wait Until Dark*? Audience apprehension was excruciating because the frail heroine, Audrey Hepburn, was pitted against three cruel and sadistic killers. And she was blind.

In *Jaws*, Roy Scheider played the role of a police chief, theoretically a strong authority figure. But the director, Steven Spielberg, wisely built in a character weakness, a way to make him vulnerable. His enemy was a twenty-foot great white shark—and the police chief was acutely afraid of the water!

Jaws exploited vulnerability throughout. Do you recall the opening sequence? A beach party at night. A girl decided to go swimming. Alone. We became conscious of the unseen Menace swimming in ocean darkness. The girl shed her clothing, making her seem even more vulnerable. We held our breath, afraid to look.

Later the shark appeared in an ocean inlet where the police chief's own son swam. Our apprehension increased, for now the potential victim was someone we knew, someone we cared about. Someone extremely vulnerable: His hand was bleeding (blood attracts sharks), he was only a child, and because he was in the shark's milieu, he was not fully in control.

Although vulnerability begins with the character defined by the script, it can be enhanced and exploited by the director—for example, in casting. The choice of Audrey Hepburn for the blind heroine in *Wait Until Dark* was brilliant. Would audiences have worried quite as much about an older, tougher, more authoritative woman? By selecting actors whose physical appearances suggest vulnerability, directors can ensure audience sympathy for their victims and vastly increase the suspense potential of their projects.

The Antagonist

Vulnerability is only half of the suspense equation. If the hero(ine) flees from a truly formidable antagonist, the director can build effective suspense. When the attackers/pursuers come from outer space, capable of concealing themselves in your best friend's viscera or masquerading as your fiancé (as in *Alien* and *Invasion of the Body Snatchers*), paranoia is set aquiver. When they leave their graves (*Poltergeist*) to steal our children or terrorize our families, the viewer reaches for a tranquilizer (see Figure 8.1).

Remember *The Exorcist* and *The Omen* and *Rosemary's Baby*? There is

FIGURE 8.1   In *Poltergeist*, a typical American family is rendered acutely vulnerable
by forces from beyond the grave. (An MGM release © 1982 Metro-Goldwyn-Mayer
Film Co. and SLM Entertainment Ltd. Used with permission.)

no more fearsome adversary than the devil or his emissaries: dedicated to evil,
powerful, crafty. Hitchcock used to say, "The better the villain, the better the
picture!"[3]

Thus, to create fear-inducing suspense the director must involve the audience
emotionally, make the protagonist vulnerable, and make the attacker or pursuer
fearsome. An apparent contradiction to such requirements is presented on television each Saturday morning in the multiplicity of kids' cartoons. These hyperki-
netic programs display the most fearsome antagonists imaginable—ghosts and
living skeletons and green, scaly monsters—as well as heros and heroines who
usually are young (and thus vulnerable). Yet these shows generate surprisingly
little fear or suspense.

The contradiction is apparently reinforced by the many classic thrillers that
featured antagonists who were normal human beings: neither demon nor sea
monster nor alien from outer space.

---

[3]Truffaut, p. 141.

To explain this contradiction, let's examine the third factor contributing to the suspense thriller—believability.

Believability

Believability cannot be computed mathematically; it depends largely on the abilities of the director and the writers to create the illusion of reality.

Jaws could have been ludicrous: a comic strip sea monster attacking swimmers. Yet the production team was so skillful that they made us believe it. I was fortunate enough to read five different drafts of the script as well as the novel from which the scripts were adapted—which perhaps explains my preoccupation with *Jaws*. The most significant development that occurred between the first script and the fifth is that the characters assumed dimension; they became more honest, less melodramatic. It is revealing that director Steven Spielberg remained a guiding, contributing force during the shaping of the *Jaws* screenplay. As a matter of fact, he wrote the third version himself.

The final script examined for the first time the minutiae of a police chief's daily routine: downing an Alka Seltzer for his queasy stomach, showing a secretary how to set up files, typing a drowning report, reporting a stolen bicycle—all unimportant to the plot but vital in converting spectators into believers. For the first time the audience explored the relationship between the chief and his wife and discovered that they loved each other: certainly not innovative material, yet essential in terms of making viewers care. Altogether, the director and the writers seduced the audience into believing a bizarre concept by surrounding it with believable people.

Saturday morning cartoons offer little suspense because we cannot accept them as reality. And because we cannot identify (other than superficially) with their one-dimensional characters, we remain spectators rather than participants.

In real life when your sister or your father loses a job, it's a serious matter. If someone should threaten to *kill* them, it would become mind boggling. In a TV show or movie, if the hero becomes as real as a member of your family, you're going to worry when that person is threatened. When pressures build, your worry changes first to apprehension and then to terror.

THE ESSENCE OF SUSPENSE

Answer the following questions true or false:

1. Suspense films necessarily deal with elements of fear and shock.

2. A well-written, well-directed show does not need suspense.

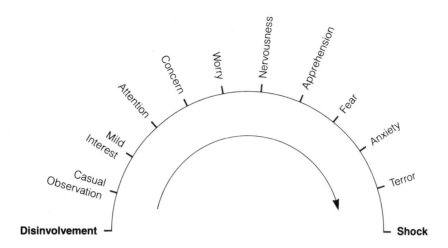

FIGURE 8.2 Every drama creates audience responses described somewhere on this continuum. Thrillers provoke those responses at the far right; human interest shows fall somewhere in the center; those at the far left usually fail.

3. A show either has suspense or it doesn't. There's no in-between.

If you answered false to all of the questions, give yourself a perfect score. Let's clarify a few misconceptions about the nature of suspense. When asked by François Truffaut, "Is suspense necessarily related to fear?" Hitchcock replied, "Not at all."[4] Suspense does not deal solely in shock and terror. What we have discussed in this chapter concerns only a single genre in the dramatic universe, the thriller. In the minds of the general audience, the word *suspense* equates with *thriller.* "You're going to see a suspense film? Better take a tranquilizer to steady your nerves!"

Every well-designed dramatic program—whether on TV, the stage, or the movie screen—contains suspense. Will the lost child find her way home? Will the husband desert the wife who loves him? Will the executive sacrifice her moral values in order to gain promotion? Every story in which someone has a problem (and that's 98.8 percent of all stories ever written) engenders worry. And worry and suspense are virtually synonymous.

Every drama ever written evokes reactions that fall somewhere on the continuum in Figure 8.2. Between the beginning of a show and its final curtain, the action will move back and forth, generating mild interest at some moments and dark apprehension at others. Any show whose viewers remain for an extended time in the area of disinvolvement or casual observation is in serious trouble.

---

[4]Truffaut, p. 50.

Most successful human interest drama generates concern and worry, and perhaps moments of apprehension and fear. In watching the award-winning movie, *Kramer Versus Kramer*, for example, we worry first that Ted's need to care for his small son will destroy the career that means so much to him. After Ted changes his values, treasuring the newfound relationship with Billy, finding joy and fulfillment in his role as father, he loses his job—and our concern deepens. It turns to genuine apprehension when former wife, Joanna, appears and threatens to take Billy away from him. Our apprehension turns to anguish when the court rules in favor of Joanna. Is *Kramer Versus Kramer* a suspense film? Unquestionably. Audience fears soar higher when that father–son relationship is threatened than during most blood-and-guts horror films. Ted Kramer becomes acutely vulnerable because what he loves most in this world may be taken from him. And Joanna becomes a "fearsome monster" by threatening to wrest Billy away.

Think of films you have seen recently. Did they make you worry? Then they were "suspense" films. Westerns, soap operas, science fiction, children's stories all make us care about a hero and fear an antagonist. When written, produced, and directed effectively, all are suspense films.

In addition to using the fundamental dramatic techniques discussed in earlier chapters, directors can heighten suspense by exploiting the specific elements that make an audience worry. The basis for that worry is contained in the threat posed by the antagonist. Any device, action, or visual image that reminds the audience of that threat will maintain or build dramatic tension.

In Hitchcock's *Strangers on a Train*, the character played by Farley Granger is tormented by guilt feelings concerning the death of his wife. In one scene where he sits in his den, we can read the titles of books on a shelf behind him. Hitchcock has placed prominently among them a book titled *Murderer*, commenting on the man's inner tension and reminding the audience. In another scene, Granger meets surreptitiously with Robert Walker, who plays the actual murderer. At the sound of police sirens, the two duck behind a gate. Hitchcock again reminds us of Granger's guilty fears by framing camera angles through the gate, now suggesting jail bars, and by casting on the wall behind them shadows that also resemble jail bars. In each instance, the director found ways to underscore Farley Granger's problem, to externalize his inner tensions, thus deepening audience concern.

CHAPTER HIGHLIGHTS

— The label *suspense* often is applied to shows that generate anxiety or fear. But the word has a far wider application. It is the foundation of all dramatic action.

— So-called thrillers achieve their success through exploitation of our most basic fears. Children (and many adults) are afraid of the dark because it conceals the *unknown*, which lies at the heart of most fears. Graveyards, skeletons, ghosts, and coffins—frightening to many of us—are symbols of death, the greatest unknown of all. The unknown and the unseen are tools for directors.

— Concealing antagonists usually makes them more fearsome; the viewer's imagination magnifies them, increasing their power. Directors sometimes conceal the identities of villains by revealing only a portion of their bodies. Directors also sometimes conceal the antagonist by keeping the camera angle close on the protagonist (a claustrophobic frame).

— Director Alfred Hitchcock used these principles repeatedly in his films, delaying the revelation of what an audience wanted to see. He also devised scenes with the sole intent of making an audience nervous, thus making the moment of shock more traumatic. His suspense techniques included providing the audience with more information than he gave his protagonist, thus allowing the audience to worry.

— Many fear-inducing techniques begin at the script stage. The degree of suspense generated in a story varies directly with the vulnerability of the victim and the fearsomeness of the attacker. Vulnerability usually manifests itself through age, sex, or physical disability.

— The fearsomeness of the pursuer or attacker is demonstrated in such films as *Jaws*, *Poltergeist*, *Alien*, and *The Exorcist*, featuring sea monsters, ghosts, aliens from deep space, and demons from hell.

— The final factor in the suspense equation is believability. The more audiences accept the drama as reality, the more terrified they will become.

— A major point: Every well-designed TV show, stage play, and feature film contain suspense; it is not limited to thrillers. Suspense is manifested whenever audiences worry about a protagonist.

PROJECTS FOR ASPIRING DIRECTORS

1. Let's examine the Cinderella story in terms of suspense.
    a. Name two elements that create suspense in the Cinderella story. (You may find more.)
    b. Is Cindy vulnerable? Are her antagonists fearsome? Does the story contain antagonists other than the stepmother and stepsisters?

2. Let's convert the Cinderella story into a thriller. How can we make her situation life threatening? Despicable as they are, the stepmother and stepsisters probably don't have sufficient motivation to do away with Cindy, so we may have to create a new, more desperate antagonist. (Consider: Could there be a character whose life will be devastated if the prince marries?) Yes, such a change will alter our story somewhat. Don't worry about it; the author won't sue.

3. Now that you have a formidable antagonist, devise a final, climactic sequence in which Cindy tries to make her way home from the ball. It is after midnight. In the woods she is threatened by the antagonist's sinister forces.

4. Draw a storyboard of not more than ten frames (see "Projects for Aspiring Directors" in Chapter 6) showing how you would play such a sequence for maximum suspense. Use your imagination. Try to think in terms of both *staging* and *camera*.

5. Key question: Is it better to show the audience where Cindy's danger lies or to play the revelation for shock value? Is there any way to obtain both values?

DIRECTING
NONFICTION

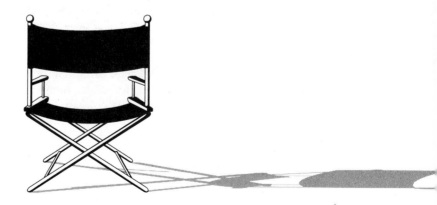

NOTES ON THE MULTIPLE-CAMERA OPERATION[1]

Some directors find it harrowing. Others thrive on it. Some, especially those trained in film, call a live TV operation heart stopping, filled with potential disaster; if something goes wrong, you're stuck; you can't reshoot it; your mistakes are broadcast to the world. Ironically, other directors enjoy live directing for exactly the same reason; they are stimulated by the sense of danger.

The challenge of directing a multiple-camera operation arises from the need to focus attention in a dozen different directions: the script with its typewritten and penciled instructions, multiple camera monitors revealing different perspectives of (usually) the same subject matter, a **preview monitor** and an on-the-air monitor, an auxiliary monitor for film or videotape, perhaps a **nemo** (**remote**) monitor, the audio engineer's balance of sound elements, the clock that shows you ahead of schedule or behind, the frantic whisperings of the floor manager on your **program line** (**PL**), scrawled notes from a nervous producer, the needs of your technical director, and helpful suggestions from the associate director (if your budget can afford one). On top of everything else, you have to go to the bathroom.

If such an operation sounds Herculean, it is. A major multiple-camera production is not for the faint of heart. Yet several factors make such an operation practicable. Chief among these is *preparation*. If you have studied your script, mulled its contents, and thoroughly rehearsed your program, you should know it almost by heart. With familiarity comes security. (Alas, directors seldom get the amount of rehearsal they would like. Rehearsal time with cameras can be expensive.) When directors are sufficiently secure, their dependence upon a script diminishes.

Inexperienced or insecure directors tend to keep their eyes glued to the script, afraid they will miss a note or camera direction. The pros keep their eyes on the camera monitors, glancing only occasionally at their scripts, studying the action, getting into the program rhythm, watching for magic moments, poised to make knife-sharp cuts from performer to performer.

When professionals prepare their scripts, do they include hundreds of notes and commands? No, most good directors write only what is necessary, avoiding clutter. Some use different colors for different aspects of the operation—for example, red for videotape, green for projected timings, black for camera shots.

Good directors have no time for wasted motion. They don't give three **preparatory commands** when one will do. I have heard students say (so help me) "Ready to roll and record, ready music, ready to fade up on camera one, ready

---

[1]The material in Part 3 deals with a multiple-camera videotape or live operation unless stated otherwise.

to open floor mikes, ready to cue talent." What's wrong with "All right, every-body. Stand by"? Or an old-fashioned countdown? "Five, four, three, two, one." I've heard one top director cue the opening of an Emmy awards show with only two crisp words, "Hit it!"

NBC news director Jay Roper prepared the following excellent essay on di-rectorial commands for television production classes at California State Univer-sity, Northridge.

DIRECTORIAL COMMANDS[2]

It is important for beginning directors to develop the means of communicating effectively with their crews. A TV crew may be a terrific group of professionals, but the one thing they lack is the ability to read minds. Therefore, it is up to you as director to be clear and concise with your instructions. The following guide-lines may be helpful.

A director gives two types of commands: *preparatory* and **executional**.

Let us say that your friend is headed for the bookstore and you would like her to buy notebook paper that will fit into your three-ring binder. You are explicit in your instructions, saying, "Please get me some lined, three-hole punched notebook paper, preferably about 100 pages." Your request is comparable to a director's preparatory command. You have explained exactly what it is that you need.

Later in the day you watch a track team practicing for the 100 meter dash. The starter raises his gun and shouts "On your marks . . . get set . . ." and then, when all are ready, he fires his pistol to start the race. The act of firing the gun is comparable to the executional command in the control room. It is at that precise moment that the action set up in the preparatory command takes place.

If the preparatory command has been clear and precise, there is really no need to repeat it as part of the executional command. Often there just isn't time. For example, the preparatory command "Ready, camera 2" should be followed by the executional command "Take!"—not "Take Camera 2." What else would the TD take? If you have been clear, then you can be concise. Most executional commands are "stand alone" commands. Others may need some reinforcing. (See the list of commands at end of this essay.)

It is important that directors mark their scripts carefully, inserting just the information necessary to get the job done. Novice directors tend to write prepa-ratory commands as well as executional commands in their scripts. This, frankly, is a mistake. Directing a program is complex enough without having to interpret

---

[2]Used by permission of Jay Roper.

all those marks in the heat of battle. If you have written "Roll playback tape" in your script, the preparatory command would be given just prior to rolling that tape. Thus, "Stand by to roll playback tape." And, "Roll" or "Roll it!"

Remember that your crew is equally under pressure and can suffer from "buck fever," responding to what they *think* they hear. With that in mind, it's a good idea to avoid embedding the executional command in the preparatory command. For example, the executional command "Ready to take camera 2" contains within it the executional command "Take camera 2." If you just use the phrase "Ready, two," you will avoid flirting with Murphy's Law.

A variation in preparatory commands occurs when you change from *cutting* to the next shot to *dissolving* [or some other effect]. Now you must be extremely careful how you prepare such a command. Many directors use the words *set* or *prepare* in such a situation. For example, "Set a dissolve to camera 2" or "Prepare a wipe to Aux[iliary] One." These words usually act as a trigger, advising the TD that something new is coming and that an action other than pushing a button will be needed.

Inappropriately timed commands can sometimes trap directors. Once in a control room during a live NBC production, the director gave the command, "Stand by, playback tape." At that point, distracted by a noise, he turned and bellowed, "Quiet!" Sure enough, with tension in the control room as high as it was, the TD rolled the playback tape. Remember, once you have given a preparatory command, the next thing—and the *only* thing—that you should voice is the command of execution.

Which leads to a discussion about timing your commands. If you have rolled a tape which runs three minutes in length, it would not be appropriate to give the next preparatory command as soon as you get into the tape. You should wait until the tape is almost finished. Generally, a preparatory command should cause a crew to poise for action. If the preparatory command is "Ready to cue talent," the stage manager should have an arm poised to throw the cue. If the preparatory command is "Ready to zoom," the camera person should have his or her hand on the zoom crank. If the preparatory command is given too early, a quick response will be lost when the executional command is given.

Normally, preparatory commands are given in the order in which events are to occur. If you have given too many preparatory commands—or given them too early—your crew may become confused. Sometimes, however, preparatory commands may be given out of sequence in order to *avoid* confusion. Let us say that you plan on coming out of a 30-second tape to a camera alone and then going into a complicated effect. I suggest that you set up the difficult effect first and then give the preparatory command. The preparatory command might be "On effects, set a matte of camera 3 over camera 2. That will follow stand by camera 1 in the clear." Thus, you will be starting on camera 1 first and then later going to the effect of camera 3 matted over camera 2. You have helped your TD by setting up the more difficult process first.

Another item for consideration: how to address your commands. If you say "Camera 2, pan left," the camera operator will certainly comply. But most of us respond to our names more quickly than we respond to our positions. If you were to say "Ralph, pan left," you would probably get a quicker response. In production classes where students rotate positions, they sometimes fail to make the instantaneous connection between the command and their position. Such a delay can destroy the split-second effectiveness of editing. It's a good idea to have your TD label each camera monitor with the name of the person operating that camera. Then, as you watch your monitors, you will be able to address each camera operator by name. Using an individual's name also removes the stigma of being just a cog in a machine and will make your crew member more of a team player.

Command List

The following are commands commonly used by directors of live TV or videotape programs. Most directors develop their own "patter" as they gain experience. I suggest that finger snapping be avoided.

— *Stand by* or *Ready*—a preparatory command implying a quick or simple action. Examples: "Ready, 1," "Stand by, music," "Ready to cue talent."

— *Prepare to* or *Set a*—a preparatory command implying a more complicated action that requires a setup on the switcher. Example: "Set a dissolve to camera 3," "Prepare to matte camera 4 over camera 1," "Prepare to fade audio and video to black."

The following commands are executional and stand alone.

— *Matte* (sometimes called *Insert*)

— *Take*

— *Dissolve*

— *Cue*

— *Roll*

— *Fade up* (Fade out)

— *Black* (take the show to "crash black")

— *Cut* (stop everything)

— *Boom up* (Raise the arm on the microphone boom)

— *Lose effects* (go to the background camera alone, without the key)

— *Change* (to a projectionist to change slides)

— *Switch* (to a TD using a **multiplexer** during a slide sequence, with slides all loaded in the same projector)

— *Crawl* (to a character generator operator to achieve the crawl motion)

— *Next page* (to a character generator operator to change to the next page of copy)

The following commands are executional, but they do not stand alone. They generally include a video or audio source. *All* camera directions should be preceded by reference to the camera or to the camera operator.

— *Lose camera 2* (lose the effect coming from Camera 2; sometimes stated "Go to Camera 3 *in the clear*"; for example, camera 2 matted over camera 3, or a split screen wipe between camera 2 and camera 3

— *Preview a matte of camera 3 over camera 2*

— *Fade music* (if the production requires two audio sources, identify each in commands)

— **Undercut** *camera 1* (change the "background" to another video source)

— **Overcut** *camera 3* (change the "foreground matte" to another video source)

— *Camera 1, loosen/tighten*

— *Camera 2, tilt down/up*

— *Camera 3, pan left/right*

— *Camera 4, truck right/left*

9

QUESTIONS AND ANSWERS: THE INTERVIEW

Interviews seldom stand alone; that is, they are usually part of a larger program.[1] For example, interviews sometimes appear as elements in news or sports programs, homemaker shows, documentaries, game shows . . . in short, as components of almost every television program type, with the exception of drama. They enable us to meet a wide variety of people, to learn from their reservoir of knowledge, to be entertained by their stories, perhaps to become emotionally involved with them.

In preparing an interview, the director must begin with an understanding of the program's basic appeal. What aspects of the interview will attract viewers? Will they tune in to learn? For self-improvement? Or to be entertained? After pinpointing the interview's most appealing aspects, the director may then enhance or highlight those aspects through staging, set decor, camera, and editorial treatment.

---

[1]Exceptions include "Face the Nation" and Barbara Walters specials, which are 100 percent interview.

*This instrument can teach, it can illuminate; yes, it can even inspire.
But it can do so only to the extent that humans are determined to use it
to those ends. Otherwise, it is merely lights and wires in a box.*

Edward R. Murrow[2]

In this chapter we will examine:

— THE INTERVIEWER: a director's perspective on the three types of interviewers most often seen on TV

— SHOT SELECTION: angles most often used in interviews and other shows, their descriptions and reasons for their use

— THE INTERVIEW SPECTRUM: a cross section of programs in which interviews appear; the problems they pose for directors

— THE MARTHA MCCULLER SHOW, a daytime homemaker program. A practical demonstration of how a local station director puts such a show together

[2]Quoted by Fred W. Friendly in *Due to Circumstances Beyond Our Control* (New York: Random House, 1967).

THE INTERVIEWER

Many people believe that a director's function is solely to stage the action and call camera shots. To them, such concerns as the interviewer's style or personality or the guest's subject matter are inappropriate, outside the director's purview.

Such attitudes, of course, are totally incorrect. The director's goal is to build a program to its highest potential, seeking the greatest possible audience involvement. Accordingly, the director's province must be the entire show—all of its elements, including camera, music, lighting, scenery, editing, staging, performers, and performances. The interviewer's style of presentation is central to the success of an interview show and, therefore, a legitimate directorial concern.

Interviewers generally fall into one of three categories: the Sincere Inquirer, the Personality, and the Provocateur.

The Sincere Inquirer

Most interviewers take a secondary or background position, emphasizing their guest's unique knowledge or expertise. Such Sincere Inquirers allow their guests to star. They are creative listeners, attentive, encouraging, and supportive, subtly leading their guests into conversational paths that have been prearranged. Dick Cavett typifies the Sincere Inquirer.

Usually, a production assistant has met with the guest prior to the telecast and conducted a preinterview. Preinterviews are designed to assess a body of information and to select specific facts, stories, and anecdotes as well as a theme for the broadcast. Such items may be chosen for their humor, human interest, or whatever characteristics make them unique or audience worthy. When the production assistant later relays this information to the host, the host prepares notes, perhaps on unobtrusive three-by-five cards, for use during the broadcast. If the host meets a guest immediately before the broadcast, it is to extend a warm greeting rather than to labor through an exploration of possible discussion areas. In the televised interview, host and guest may now explore conversational areas with purpose and direction, yet spontaneously, with no sense of having traveled this road before.

Frequently, guests pinpoint certain taboo areas: topics they prefer not to discuss. This is often true of show business celebrities, many of whom rule out discussion of controversial aspects of their private lives. If the host expects to conduct a pleasant, upbeat interview, or wants ever to interview this celebrity again, he or she must respect the guest's wishes and avoid discussion of taboo subjects. Frequently showbiz celebrities exact a "price" for their appearances: a plug for their latest film or television series. Sometimes such appearances include film clips, which the director must view prior to broadcast. If language or actions in the film clip seem questionable, the director must have the material approved by station management.

With the multitude of details and time strictures in preproduction, a director is often tempted to take the guest's word that material is inoffensive. A word of advice: Don't take anyone's word! When directing a live show for NBC, I once accepted the star's solemn assurance that film she had brought back from Europe ("home movies") would pose no problems. Innocently I bypassed my directorial responsibilities and instructed our film department to run the material on cue. When aired, the film seemed harmless enough—until a Folies Bergère sequence appeared, featuring a number of bare-breasted showgirls! We quickly cut away from the film—and the hostess giggled in some embarrassment. The station management didn't giggle at all.

If you believe that in a two-shot the camera-right figure receives primary audience attention (as many directors do), then you should place the guest at camera right.[3] Because the Sincere Inquirer defers to the guest, realizing that the interview will succeed or fail based upon the guest's effectiveness, the director should favor the guest in camera treatment. The host's introduction may occur at some distance from the guest, in another portion of the set. Following the introduction, the director may cut to the guest (assuming a two-camera operation), allowing the host/interviewer to enter this new shot and sit down beside the guest. Or the director may simply pan the host to the guest and hold this two-shot for the start of the interview. It is important to provide the audience with a closeup of the guest early in the interview, perhaps at the moment of introduction. Beyond the obvious function of showing the audience what the guest looks like, a closeup allows the audience to feel "close," establishing a sense of personal involvement.

In basic interview coverage, camera angles always will cross to achieve the impact of full face rather than profile. The camera on the left photographs the performer on the right; the camera on the right photographs the performer on the left. Note cameras 1 and 3 in Figure 9.1. If the camera on the left photographed the performer on the left, it would provide unsatisfying profile shots.

Notice that when interviewer and guest are seated on a couch or in chairs, a medium two-shot fits the three-by-four TV **aspect ratio** almost exactly. When they sit farther apart, the camera must widen to accommodate both figures. This wider angle is inferior for two reasons. First, because the performers are farther away, the picture loses detail and facial expressions. Viewers lose emotional closeness, becoming spectators rather than participants. Second, the center of the frame in the wider angle is empty; the most valuable part of the camera frame features an expanse of blank wall or empty couch. In staging action, experienced directors always try to place their performers within the aspect ratio of a camera frame.

---

[3]In theory, the spectator's eye moves across the frame from left to right.

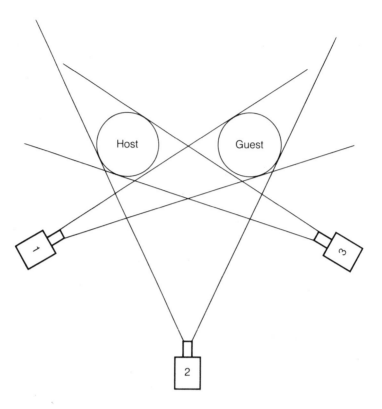

FIGURE 9.1   To photograph full faces rather than profiles during an interview, the right-hand camera (3) photographs the host, the left-hand camera (1) photographs the guest, and the center camera (2) maintains a two-shot.

Because most interviews are conducted from seated positions, it is imperative that cameras photograph the performers *at their level*. Either cameras must pedestal down so that the lens looks performers in the eye, or the couch or chairs must be placed on a small raised platform (a **riser**), bringing guests up to camera level. Risers have a secondary benefit: Because couches or chairs are higher, they make the acts of rising and sitting much easier for host and guest.

The Personality

Unlike the Sincere Inquirer, the Personality seldom if ever assumes a secondary role. Instead, he or she *shares* audience attention. In many cases, because of humor, flamboyance, or personal magnetism, the Personality attracts more interest than the guest. Johnny Carson and David Letterman are examples of such personalities. Note that they, rather than their guests, sit at camera right.

The director's use of cameras for the Personality differs from coverage of the Sincere Inquirer. Instead of favoring the guest, cameras divide attention, attempting to highlight first the guest and then the host as each sparks audience interest. Ideally, such an interview should be covered with three cameras: cameras 1 and 3 on individual shots, the middle camera holding a two-shot (see Figure 9.1).

If the director is limited to two cameras, one should favor the guest and one should favor the host. Repeated cuts from a two-shot to a single and back to a two-shot can become tedious. More importantly, such a pattern minimizes the host's impact by playing him or her constantly in a two-shot. When the host is a celebrity, such treatment will weaken the show and will probably motivate the host to seek a new director. In a two-camera situation, then, the director should live dangerously. After orienting the audience with a two-shot, the director should instruct the wide angle camera to provide closeups of the host. The director then must juggle closeups. Such a situation can cause adrenaline to flow when the dialog quickens, when speeches overlap, when one (or both!) performers decide to stand or move about the set. With competent camera operators, such a situation can be gracefully handled by zooming out to an angle that is wide enough to accommodate the action. Playing a two-camera interview in closeups isn't always "safe." It requires intense concentration from the director. The director can appear inept or incompetent when action unexpectedly boils out of the close angles. But those angles provide energy and impact, creating excitement, bringing the interview to life. Such a result is worth some risk.

One of the advantages of working with the same personality over an extended time is that director and star each come to understand the other's needs. Bobby Quinn, who directs "The Tonight Show," believes that his intuitive sense of Johnny's reactions has been significant in the show's success.

> "I know what triggers him," the director explains. "I know when somebody says that certain something, you're going to get a response. I try to beat him with the camera so I'm there when he says it."
>
> The challenge of capturing that off-the-cuff remark or that sudden arched eyebrow—knowing there is no going back if he misses it—is what keeps Quinn on the edge of his seat in the control booth.[4]

A continuing relationship between star and director has other advantages. The star realizes, for example, that standing suddenly will cause camera problems. Accordingly, the star will "telegraph" his or her movement by first leaning forward and then rising slowly so that the camera operator can follow the movement. When displaying a small object for closeups, the star knows where the

---

[4]Directors Guild of America Newsletter, March 1988, p. 1.

director wants it placed. Most stars are eager for their shows to be successful, so they cooperate with their directors to make interview segments visually effective.

Realistically, the needs of director and star are not always the same. Friction sometimes occurs. It is important for aspiring directors to realize that the qualities that make a star effective as a performer are closely related to personal ego. The more successful the star, the more inflated the ego can become (although this is not always the case). I emphasize again: A skilled director is more than a mere stager of action and manipulator of cameras. A skilled director also displays talent in the often sensitive area of human relationships. Such a director is able to get along with people, even stubborn or unreasonable or childish people. Skilled directors understand how to persuade others to do their bidding, sometimes through humor, sometimes through flattery, sometimes through patient explanation. There are times, too, when directors must be firm. Controlled, polite— but firm.

The Provocateur

Some interviewers use the key entertainment element of conflict to arouse audience interest and elicit information from a guest. Mike Wallace in his early network interviews was a prime example of the **Provocateur** in that he goaded guests into making statements they might not otherwise have made. By stating blunt truths, by refusing to accept cop-outs, by provoking anger and emotionality, Wallace was able to knock aside a guest's defenses and stimulate honest, incensed, sometimes fiery reactions. Barbara Walters also sometimes acts the role of Provocateur.

Provocateurs do not always respect guests' wishes regarding taboo subject matter—although many do. Many state candidly that they will pry into sensitive areas as exhaustively as the guest will allow because these are areas of intense audience interest. Thus, the guest knows what is coming and that he or she is expected to draw the line when the host goes too far.

With the Provocateur, the director must be alert, instantly ready to catch unexpected reactions. Most of us are trained from childhood to conceal emotions; we build protective walls around ourselves to shield us from remarks or actions that might shatter our equanimity. The Provocateur's goal is to shatter the protective wall, to cut through defensive devices, and to elicit honesty. With a camera trained on the guest's face, the director must try to anticipate reactions that betray true feelings, the twitch of a facial muscle, eyes suddenly watering, lips clenching to prevent speech. The director must try to *be* there when a guest suddenly comes apart, erupting emotionally, spilling feelings and honesty all over the studio floor.

Countless books have been written about cameras, lenses, switches, and a director's commands to the crew. It is easy for directors to become so caught up in the mechanics of directing that they lose sight of the single most important broadcast element: the people in front of the cameras. Directing a live or video-taped program can be hectic, frenetic, busy. Under such circumstances, directors sometimes cut back and forth willy-nilly, maintaining visual variety but overlooking what the program is all about. Good directors listen. *Really* listen. They concentrate on the interaction between performers so that their cameras will be on the right face at the right time. Such concentration is doubly essential in interviews conducted by a Provocateur.

SHOT SELECTION IN INTERVIEWS

Although I described the significance of most camera shots earlier, a brief summary is appropriate here. These descriptions apply to interview programs—and to all programs, fiction as well as nonfiction.

Wide Angle; Full Shot; Long Shot	Used primarily for orientation, to show where the action is taking place, who is there, and what they are doing. Wide angles are useful for covering broad movement such as a character walking from one room to another.
Medium Shot; Two-Shot; Three-Shot	Intermediate angles. A medium shot of a character is usually framed at the waist or hips. In an interview, a medium shot covers both characters, providing visual relief from continual close angles.
Medium Closeup; Medium Close Shot	Usually framed at midchest. Provides greater impact than preceding shots do.
Close Shot; Closeup	Usually head and shoulders. Provides considerable emphasis/impact. Also reveals detail and creates empathy between character and audience.
Extreme Close Shot	Usually a head shot, may be even tighter. Rarely used in interview shows. Enormous impact. Useful for demonstrating small objects.
Over-Shoulder Shot	Shoots past head and shoulders of character A toward character B. May be tight or loose. Shot often suggests relationship between characters.
High Angle Shot	Sometimes used for orientation. Tends to look down on characters, figuratively as well as literally. Tends to make them seem weak, inferior.
Low Angle Shot	Looks up at characters, tending to make them seem strong, dominant.

A rule of thumb for interviews or any type of program: The closer the shot, the more impact it creates. Wise directors save closeups for moments that count.

THE INTERVIEW SPECTRUM

Watch a day of television and you will see interviews in almost every program category. In sports broadcasts they are a favorite halftime activity: questioning coaches about first-half glitches, their estimates of opposing team's strengths and weaknesses, and their predictions on the contest's final outcome. Game shows inevitably interview guests before the competition begins, searching for quick audience identification—a "handle" for each contestant. In panel shows, interviews generally consist of probing for information that experts gain in the pursuit of their professions as well as for their opinions on newsworthy issues. Newscasts often include interviews with disaster victims, witnesses to catastrophes, or perhaps local celebrities. Sports commentators frequently question local coaches or ballplayers. On talk shows, the interview comprises the major part of the program, often slanted toward humor, seeking entertainment rather than information. The interview manifests itself briefly in comedy or musical variety programs as a way of allowing audiences to become acquainted with guests. In daytime programming, interviews are one of the major elements of the homemaker show.

Sports

Although audience interest usually derives from the spoken words of the person interviewed, a director often may heighten interest by introducing the element of spectacle (see Chapter 2). Sports figures discussing game strategy in front of a neutral background may be fascinating or incredibly dull depending entirely on the personality of the interviewee and the content of the discussion. The wise director provides a measure of "insurance" by contriving a more colorful backdrop, thereby adding visual appeal to prevent the interview from degenerating into "talking heads." In the 1988 Olympic Games, directors used **chroma key** to place scenic or graphic backgrounds behind their program hosts.

Conducting interviews on the playing field provides color. During halftime (in basketball or football), with teams off the field, there is usually little competition from distracting background action. However, directors must take care that other halftime activities do not compete for spectator attention. A marching band, for example, generates sound, motion, and color, the primary ingredients of spectacle. Briefly clad cheerleaders constitute another element of compelling visual appeal. Appearing in the background of an interview, such diversions can cause restlessness and dissatisfaction. Audiences either lose interest in the interview, irresistibly drawn to the colorful background action, or uneasily divide their attention.

In television newscasts, sports commentators often interview coaches or ball players. When such interviews originate on a playing field rather than in the studio, visual interest is vastly enhanced. Such question-and-answer sessions are frequently staged with the commentator in the studio and the interviewee at the ballpark, responding on camera to the commentator's questions, which he or she hears through an earpiece.

Even when both commentator and guest are at the playing field, spectator interest is weighted heavily in favor of the guest. Usually such interviews are conducted with a single camera. The director frequently begins such an interview with a single shot of the commentator as he or she begins an introduction, widening to a two-shot at the appropriate moment. Beginning close prevents the awkward situation of a guest standing with nothing to do or say, "with egg on his face," while waiting for the commentator to conclude introductory remarks. Once the camera has zoomed out to a two-shot, interest usually fastens on the guest.

After a brief exchange between commentator and guest (played in the two-shot), the camera's movement should follow the pattern of viewer attention. As interest in the guest increases, the camera should zoom slowly in to a close shot and should hold there. By placing the commentator slightly downstage (closer to camera), the director clearly favors the guest, with the camera providing a three-quarter view of the guest's face rather than an ineffectual profile. Occasionally, the camera may pan to the commentator, paralleling viewers' interest in his or her questions and observations. If the discussion is an extended one, the director may want to zoom back to a two-shot from time to time, for orientation and visual relief.

In my television classes in directing drama, there is an almost compulsive tendency among students to fasten their cameras on the face of the character who is speaking. About halfway through the semester they come to understand that dramatic significance often appears on the face of the person *reacting*, who may not speak more than a word or two in the entire scene. In drama, the camera angle ideally reflects audience interest. In other forms of television and film, instructional or entertainment, the same rule applies. Audience interest dictates what appears on the screen—not the source of the words. In a sports interview, some beginning students might pan the camera to the host/commentator every time he or she phrased a question. Such literal camera movement is unnecessary, awkward, and probably dizzying. As in drama, there is no supreme law requiring the audience to see the person speaking. They can hear the question; they know perfectly well what the person looks like. So the camera can remain on the guest's face because that's where viewers' primary interest lies. Yes, occasionally it is visually refreshing to pan to the host as he or she adds commentary or phrases a new question. But such camera moves should be the exception, not the rule.

If it is practical, the sports interview can be further enhanced through the addition of videotape or film. Visuals relating to the guest's message, depicting specific team actions on the playing field, add impact to the interview and increase audience appeal. When videotape or film is not available, still photographs of sports action also can underscore spoken comments and provide effective visual variety.

In sports as in any interview, the choice of guest remains supremely important. A ball player with a sense of humor, vitality, and an infectious grin that compels us to like him is certainly preferable to Barnaby Bland, who speaks in a monosyllabic monotone. As discussed in Chapter 2, audiences are attracted to life. Performers with *aliveness* compel us to watch. In an interview, dynamic personalities command our attention; those who radiate energy create vibrations that do not allow boredom.

Panel Shows

Panel shows satisfy viewer curiosity, emotional or intellectual. They provide information that may help us achieve goals, impress friends, or simply enlarge our knowledge of the world. At times, when conflict divides panelists into armed camps, such shows also provide an element of drama.

The purpose of the panel show directly parallels that of an interview: to extract information from guests or to provoke discussion. Although panel formats vary, the content usually consists of variations of a question-and-answer pattern, which defines the interview. Thus, a panel show *is* an interview show, with a greater number of guests.

Positioning Panel shows generally feature a host, who coordinates the flow of information, plus expert panelists. The placement of panelists within a set depends upon how many there are and what their relationship is, either to the host or to one another. For example, when a panel of investigative reporters questions a political celebrity, the director places the celebrity in a position where he or she may confront the team of interrogators as, for example, on "Face the Nation." The reporters, allied in purpose, are seated together. (See Figure 9.2.) The celebrity, target of the questions, perhaps assuming a defensive stance, sits opposite, either alone or beside a program host. Such positioning of panelists seems almost too obvious for comment, yet the staging principle should be understood because it applies to the composition of any panel show, no matter what the chemistry of its experts or how diverse its elements.

In a situation where a host questions experts, where there is no clear division into separate, allied groups, the director may place the team of experts either facing the host or to either side of him or her. "Washington Week in Review" is an example. Where experts are allied by profession or a commonality of interest, they should be grouped together for easy audience identification. Thus, a host interrogating union members and management about a current work stoppage would place union members on one side of a table facing management personnel on the other.

Identification for the audience may be further facilitated either by placing small identification cards in front of the panelists or by superimposing name and title when each is introduced. When superimposing name and title, the director

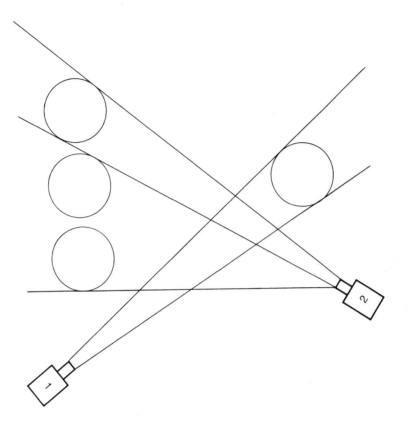

FIGURE 9.2   When a celebrity is interviewed by multiple questioners, one camera covers the celebrity, the other provides either a group shot or single shots of the questioners. A third camera would allow both a group shot of questioners as well as individual closeups.

should occasionally repeat such identification, reminding the audience of panelists' identities, especially when there are more than two guests or when guests are so prominent that their names will attract additional viewers.

Setting The setting for a panel show generally should be simple and utilitarian, reflecting the character of the program and its host and focusing attention on the table and chairs in which panelists sit. The table facilitates the panelists' use of notes, conceals unsightly laps and legs, and provides a base for identification cards.

Sometimes, directors (or producers) create a warmer, more relaxed atmosphere with a den or living room setting, guests seated in sofas and easy chairs. "Wall Street Week" on PBS combines both patterns, with a host querying "regular" experts seated at a table and then moving into a living room setting to meet a special guest expert, seated on a couch. In this new setting, host and regulars also occupy couches and easy chairs. Easy chairs and couches are often difficult to get out of. Placing boards beneath the seats helps performers to rise as does placing couches or chairs on risers.

In some cases, the setting must also provide for concealment of cameras. Where panelists sit at a table conversing with a host or with one another, cameras must be able to move deep into the set to photograph the action (see Figure 9.3). Because there will be some risk of cameras photographing each other, the setting and lighting must provide concealment. Long lenses that allow cameras to shoot from a distance sometimes prevent such an accident.

Some shows surround their panels with gauzelike curtains. Lighted from the front, the curtains reflect whatever patterns have been painted on them. But from the rear (cameras in darkness shooting toward a lighted set), they are sufficiently transparent that cameras can shoot through them with relatively little loss of picture quality. Under such circumstances, the director must make certain that each camera's **tally light** has been covered or extinguished. When cameras are hidden in shadow or behind a transparent drape, glowing tally lights instantly reveal their presence.

Camera Coverage In the basic situation of a celebrity being questioned by the press, camera treatment is fairly simple (see Figure 9.2). Basic two-camera coverage would direct one camera toward the guest with the second recording both three-shots and singles of the press panel. Addition of a third camera would allow the director greater flexibility, permitting tie-in shots past the celebrity to the questioners and the freedom to cut between three-shots of the press panel and individual shots of its members.

Camera treatment of the six-member panel (and host) seated around a table is somewhat more complicated. As indicated in Figure 9.3, camera 2 shoots a master establishing angle as well as a single shot of the host seated at the head of the table. By moving slightly to the left or right, camera 2 can also pick up two-shots, three-shots, and four-shots of the host and panel members (Figure 9.4). Camera 1 shoots the complementary angles, past the host to panel members (Figure 9.5). Guests seated on the camera-left side of the table would be covered by cameras 2 and 3, paralleling the patterns indicated in Figures 9.4 and 9.5.

In photographing the six-member panel, complications quickly arise when panel members turn away from the host to address one another. Cameras 1 and 3 must move downstage—that is, toward camera 2—to shoot complementary angles on the panel members addressing one another (see Figure 9.6). If the angles are not complementary, one of two possible mishaps may occur: The panelists may not appear to be addressing one another, that is, the direction of their looks

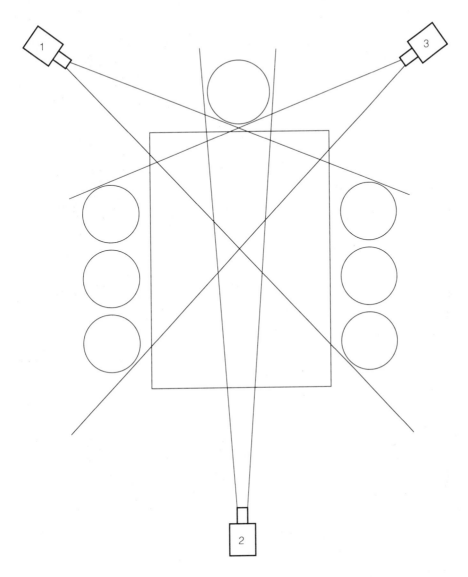

FIGURE 9.3   When a panel is seated around a table and questioned by a host, the center
camera (2) provides orientation shots as well as closeups of the host. Cameras 1 and 3
provide angles on panelists (singles, two-shots, or three-shots).

will seem incorrect. Or the cameras will *cross the line*. Remember that in complementary angles each figure must occupy the same frame position in each angle.
Imagine a line that bisects both figures. So long as both cameras remain on the

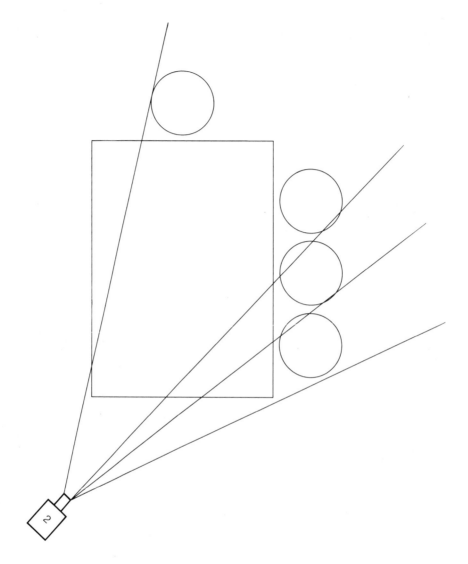

FIGURE 9.4   On a taped show, camera 2 may truck right or left to provide two-shots, three-shots, or four-shots of the host and panelists.

same side of that imaginary line, the figures will remain in their same relative positions (see Figure 9.7). When one of the cameras inadvertently crosses the line, a jump occurs—that is, the figures appear to jump from one side of the

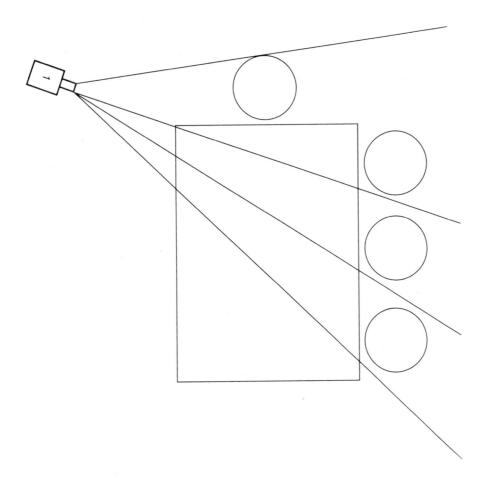

FIGURE 9.5   Cameras 1 or 3 may shoot reverse angles from Figure 9.3. That is, they provide angles past the host toward the panelist(s). From almost identical positions, they may provide closeups of panelists when they direct their attention to the host.

frame to the other (see Figure 9.8). As discussed in Chapter 7, such a cut is usually awkward and disorienting, the mark of an amateur director.

Finding complementary angles quickly, in the midst of sometimes heated discussions among panelists, requires careful coordination among camera operators, director, and technical director. Such coordination is facilitated by experience. Keeping the same key personnel on a program from week to week ensures optimum production values and minimizes the number and severity of a director's ulcers.

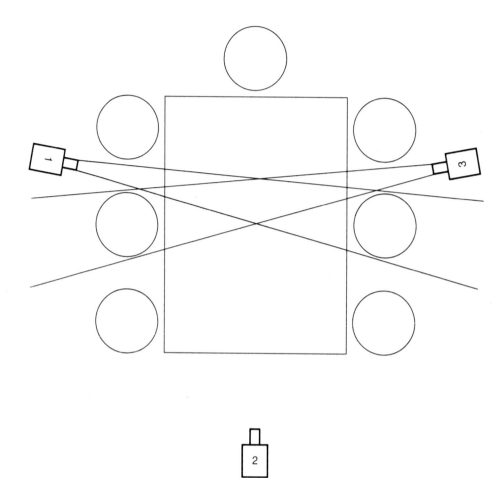

FIGURE 9.6   When panelists discuss issues with one another, cameras 1 and 3 must truck to complementary positions, shooting across the table. Long lenses prevent the cameras from photographing each other. Maintaining appropriate, corresponding looks becomes a formidable problem, so that panelists will appear to be speaking to (looking at) one another.

Newscasts

Interviews in newscasts, like those in **magazine format** shows (such as "60 Minutes" and "20/20") or documentaries, rely for their effectiveness upon immediacy, veracity, and authority. We will look at news interviews in some depth in Chapter 11.

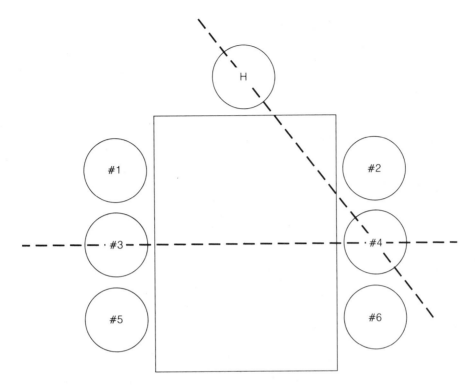

FIGURE 9.7   Performers will change positions in the frame when a camera crosses the
imaginary line that bisects them. So long as both cameras remain on the same side
of the line, both performers will retain their original two-shot positions. When Host
addresses Guest 4, for example, the imaginary line bisects them. When Guest 4
questions Guest 3, the line bisects them.

Game Shows

The interview is but a minor part of game shows, but its purpose is an important
one: to make the viewer root for contestants. In discussing drama earlier, I indicated that in order for a play to be successful, spectators must identify with a
protagonist. The stronger the audience's emotional involvement with a character,
the more powerful will be their rooting interest, their need to see that character
win his or her battles.

The game show uses many elements of drama; it provides an arena of conflict
and a number of sympathetic protagonists. Awareness that a contestant is a
newlywed or a working mother or a newly recruited marine transforms that

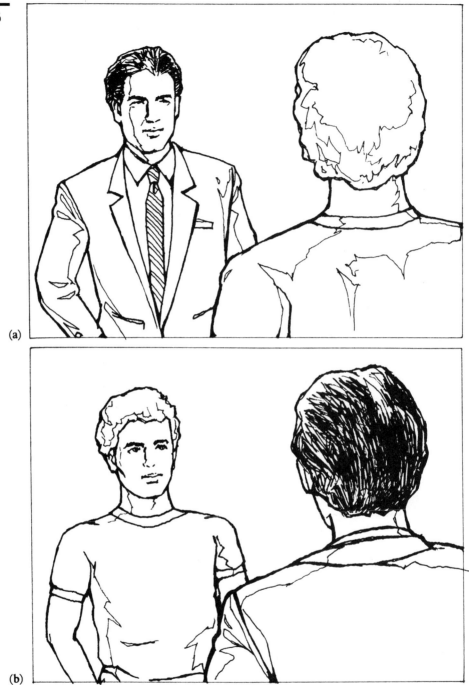

(a)

(b)

FIGURE 9.8 When Host addresses Guest 4 and a camera crosses the imaginary line that bisects them, the Host will appear on camera left in one shot (a) and camera right in another (b).

contestant from a total stranger to an empathetic figure. The director can intensify emotional involvement by providing viewers with close shots of the contestant. Creating physical closeness helps to intensify emotional closeness. By identifying with contestants, the viewer can participate in the game more wholeheartedly.

Realistically, game show viewers divide their attention; part of their consciousness identifies with contestants, the other remains in den or living room, competing with them. Such a splitting of consciousness is characteristic of most audiences; a portion of their minds always remains aware that they are spectators.

Interviews in musical variety or comedy shows occur essentially for the same purpose as the game show interview: to win audience empathy, to promote a feeling of closeness between audience and guest performer. As with the Personality interviewer, close shots of both interviewer and guest allow a sharing of emphasis.

THE MARTHA McCULLER SHOW

You have been assigned to direct "The Martha McCuller Show." As a staff director at a local television station, you're aware that Martha McCuller has hosted homemaker shows at other stations in your city. She's an articulate and attractive woman in her middle thirties and comes to your station with a sponsor (a local bank) for two of her thrice-weekly afternoon programs. Your program director advises you that she has a reputation for being "difficult," but he's sure you will have no problems. (You resolve to talk to the director of her latest show.)

You meet Martha McCuller; you're charmed by her. She's gracious, cordial, and apparently delighted at the prospect of working with you. (Certainly your program director was wrong!) You also meet her production assistant, Terry, a bright, no-nonsense woman in her early twenties. Terry worked with Martha (you're on a first-name basis by now) on her earlier shows. At this first meeting you discuss the program format. Most shows will divide into two major segments that will deal with fashion, makeup, exercise, gardening, interior decoration, cooking, or marital counseling—subject matter of interest primarily to homemakers. Although some of the segments will feature demonstrations (such as fashion modeling, gift wrapping, pruning roses), most will consist of guest interviews. Terry will give you a detailed outline (names of guests, commercial information, and so on) on the morning of each show day.

After Martha leaves, Terry discusses a confidential matter: We must make certain that Martha looks as attractive as possible on each show. Will you please talk to your lighting crew? You assure Terry that your lighting director is a genius in such matters.

Your assurance is more than a "pat on the head" designed to keep a star happy. You know that a major part of any show's success depends upon its star. If audiences relate strongly to a personality, they will watch that personality's show, even if the show itself is less than extraordinary. Wise directors provide tender loving care, trying to make a star feel secure so that he or she will function effectively. They do everything possible to maximize their star's personal appeal. Understanding these facts of life, you discuss Martha's appearance with your lighting director immediately, perhaps ordering videotapes of Martha's earlier shows in order to study the lighting.

The setting for "The Martha McCuller Show" will appear to be her home. In the comfortable living room will be two primary staging arenas: a conversational group (couch, chairs, coffee table) and a desk where Martha will deliver commercials. Adjoining the living room will be a small, latticed patio with wrought iron chairs and a glass table, either for interviews or demonstrations. (See Figure 9.9.)

Later, Terry delivers to you a **cartridge** containing the program theme, a bright, lively melody that effectively sets the mood. You send the name and the publisher of the melody to your station's Music Clearance Department. They will handle legalities, determining whether the song is ASCAP or BMI and whether the station will have to pay fees.

Your program director advises you that you will have one hour for lighting and camera rehearsal. You're delighted; it's more time than you expected. You check with your station's art director. He has already drawn plans for the "Martha McCuller" set. He gives you a copy. Construction begins immediately. The first show will air next Monday.

Early Monday morning you receive a telephone call at home. It's Martha. Will you be a darling and try to help her? She's deeply concerned. There is simply no way she can rehearse the program and talk to her guests while the lighting director and crew are lighting the set. It will be pandemonium, utter chaos.

You swallow hard and try to reassure her. You and the lighting director have already prelighted the set but, realistically, he will have to see each program's specific action before he can fine-tune the lighting. He's eager to make the show look just as rich as possible, and he wants to do justice to the show's star, to make her look as beautiful as possible. Well, Martha understands and appreciates that, but. . . . You ask if it would be possible for her to talk with her guests in the dressing room, away from the lighting activity. You suggest meeting Martha an hour prior to rehearsal so that together you can discuss and plan the physical action of the show. Then she will be free for makeup and to meet her guests. When she's done, she should come to the set and walk through the action, allowing the lighting crew to make final adjustments. You hope this will be comfortable for her.

She's not sure it will be. In your heart you suspect that part of the reason for this early-morning call is apprehension. First-show jitters. You also understand that this is a test of your future working relationship. You try to mix kindness

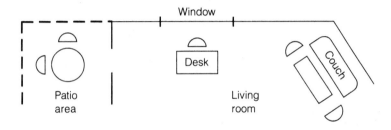

FIGURE 9.9   The setting for "The Martha McCuller Show."

with authority. You assure her there's simply no other solution. Most of the station's shows aren't allotted as much rehearsal time as she's getting. You don't know how she did it; she really must have clout. But the studio is in use immediately prior to her rehearsal/lighting time, so you and she will just have to make the schedule work. Ultimately she accepts your suggestions. You're a darling and she appreciates your help. She'll meet you an hour before rehearsal time.

You sit down, push aside your cold scrambled eggs, and pour yourself another cup of coffee. First crisis averted.

Prior to rehearsal and lighting time, you preview the one-minute film commercial that Martha will use in the show. She will do two commercials, the first live, the second on tape. You note that the taped commercial contains two seconds of so-called **freeze frame** after the announcer's final words, a continuation of picture after the sound portion has concluded. Thus, the commercial has fifty-eight seconds of audio and sixty seconds of picture, allowing you two seconds in which to cut or dissolve to Martha before the tape runs out. You write down the announcer's final words because they will be your cutaway cue.

The booth announcer (the staff announcer on duty) will have a few words at both the opening and the closing of the show, so you make certain that he has his copy and has familiarized himself with it. (He hasn't but he will claim that he has.) Because his announcement is divided into two parts ("KBEX welcomes a new personality to its family of stars, Martha McCuller . . ." and ". . . brought to you by the City-State Bank, with a branch in your neighborhood"), you advise him to wait until he sees the City-State Bank logo (identification) on his booth monitor before reading the second half of his opening.

Prior to rehearsal time, you meet with Martha and Terry and discuss the content of the first show. Martha would like to interview her first guest, a famous author, in the living room and her second guest, an aging movie star, in the patio section. She suggests beginning at the couch with her guest. You suggest that because this is her first show it might be gracious to spend a quiet moment welcoming the audience to her new home, perhaps beginning in the patio and then walking into the living room to greet her guest. Martha applauds the idea.

Martha's notes (based on a preinterview conducted by Terry) will be placed on the coffee table in front of the couch. No preinterview was held with the second guest. Her appearance was arranged by a press agent who sent along a mimeographed list of credits plus suggested discussion topics. Ordinarily, such limited (and predigested) preparation would worry Martha. This star, however, has worked with Martha before. They are old friends, and Martha knows exactly what she wants to discuss. Terry suggests that you ask the prop department to place a pitcher of iced tea and glasses in the living room for Martha to share with her guest. You will be happy to make such arrangements, but in the future would Terry advise you of such needs ahead of time, to prevent possible disappointment?

The first (live) commercial will use another prop, a steel safety deposit box. The property person will place it on Martha's desk after she has begun her first interview. You have read the commercial and blocked the camera shots on paper. You will stage it with Martha and cameras when rehearsal time begins.

When Martha goes to makeup, you walk to the stage where the show will be televised. You spend a few quiet minutes by yourself, before the crew arrives, visualizing the action as you and Martha have discussed it. You make a few tentative markings on the floor and a few notes on the program outline that Terry personally delivered to you this morning.

Later, after you and Martha have walked through the essential moves of the show and indicated where she and her guests will work, the lighting director and crew finish lighting. You take a few minutes to examine Martha's makeup and lighting on camera. You have invited the makeup artist to come to the control booth to study Martha's makeup on camera. You both agree she looks magnificent. Her lighting will come almost entirely from the front, filling in facial imperfections. There will be some modeling, some shadow area, but it will be softened with **fill light**.

The show's opening titles have been videotaped. You have put the program theme on the tape. Titles will be keyed over live shots of Martha moving busily within her living room. At the conclusion of titles, Martha will acknowledge the audience and begin the program. Under titles for this first show, Martha will pour tea for her guest (seated on living room couch) and then will cross to the patio area to greet viewers. Along the way she will pass her desk with its bank **logo** and camera will zoom in to a full screen closeup of the logo. You decide to open on camera 2 with a medium full shot of Martha and guest. But you work backward. That is, you carefully position camera 2 for its zoom to the bank logo and then ask the camera operator to swing around—*without moving the base*—to a shot of Martha and her guest. After the zoom to the bank logo, you will cut to camera 1 framing the patio area as Martha enters.

Once Martha has welcomed her guests (camera moving slowly in), she will exit the patio. Camera 1 will hold and camera 2 will pick her up as she enters the living room and **crosses** to her guest at the couch. As Martha crosses right, camera 1 will be moving into the living room area. As Martha sits (camera 2 in a loose two-shot), she will introduce her guest. As soon as camera 1 is in position, you will cut to a medium closeup of the guest. And that will be the general pattern for the interview: camera 1 favoring the guest and camera 2 favoring Martha, either in over-shoulder shots, medium closeups, or closeups. Because the opening pattern will probably be the busiest of the program, you rehearse it several times, making sure that the camera operators and others are secure.

You have arranged for the couch to be placed upon an unobtrusive six-inch riser so that Martha and guests may sit or stand more easily, so that there will be less separation between a standing Martha and a seated guest, and so that cameras will not produce severe down angles when photographing interviews.

On future shows when Martha interviews two or more guests at once, you will be careful to place them on the same side of her, perhaps with guests on the couch and Martha seated in the upstage chair. If guests were seated on both sides of Martha, you would have a difficult time covering the interview effectively with only two cameras.

At the conclusion of the first interview, camera 1 (wide) will carry Martha to her desk. She will then sit, address the audience directly, camera moving in slowly to a closeup. At one point she will unlock and lock a safety deposit box. In rehearsal you position the box at Martha's right (camera left) so that it will face camera 2, which will provide tight shots. Following the live commercial, camera 1 will truck with Martha as she crosses into the patio.

Now it is eight minutes before air time. You know from experience that performers need a moment or two to catch their breath between rehearsal and performance. Even if you desperately need another three minutes to rehearse a move, you wisely call a break so that the lighting crew can make final adjustments and the cast can straighten seams, brush hair, and powder noses. You carefully set eight minutes on your stopwatch before you leave the control room and go to the stage. Now you reassure the cast, making performers feel comfortable and relaxed, giving them any final directorial points. With stopwatch in hand you know exactly how much time is left. A cast left too long alone on the stage begins to get nervous.

At a minute or so before air time (yes, you're cutting it awfully close), you wish them well and hurry to the control booth. As you enter, you remind your audio person that the program theme will be on tape. You check to make sure that camera 1 is in the patio and that camera 2 is positioned for the zoom. You take a deep breath and grin at the technical director (TD) seated beside you.

At two seconds before program time you roll your titles. A moment later, a red light blinks on, indicating that your studio has "channels." Simultaneously the program titles appear on a preview monitor. You have previously instructed your TD to matte the opening titles over camera 2—on the **effects bus**. Now you command quietly: "Cue Martha. Up on effects. Fade music under, cue announce!"

"The Martha McCuller Show" is on the air.

CHAPTER HIGHLIGHTS

— Interviews seldom stand alone; they are usually part of a larger program unit (news, sports, homemaker programs). Interviews provide information or entertainment from the guests' reservoir of knowledge.

— Hosts of interview programs fall into three categories: the Sincere Inquirer, the Personality, and the Provocateur. Sincere Inquirers take a background, supportive position, allowing the guest to star. Most allow a production assistant to conduct a preinterview, selecting facts and anecdotes as well as a theme for discussion. Thus, in the broadcast, the interview will be structured.

— With Sincere Inquirers, directors should favor the guest in camera treatment. To present full-face pictures, directors cross their cameras: the left camera photographing the performer on the right and vice versa. Directors search for stagings that will fit (approximately) the screen's aspect ratio. Two-person interview patterns fit this proportion nicely.

— Because of their celebrity status, the Personality interviewers attract as much or more attention than their guests do. Directors acknowledge their status by giving them equal or preferred camera treatment. With three cameras, the center holds a two-shot; cameras 1 and 3 favor host and guest in single or over-shoulder shots.

— The Provocateur takes off the kid gloves, probing into sensitive areas, provoking guests into statements they might not otherwise make. Directors must be sensitive to guests' reactions to avoid missing key moments. All good directors concentrate deeply on action before the camera, listening intently, taking their cues from performers.

— In shot selection, wide angles provide orientation. The closer the angle, the more impact it affords. Over-shoulder shots suggest relationship. High angles diminish; low angles make their subjects appear strong, dominant.

— Each program type (news, sports, panel show) presents different visual opportunities for directors. Sports programs provide a colorful backdrop for interviews. The playing field, marching band, cheerleaders, and players all add richness.

— In a one-camera interview, the camera should favor the guest, who usually attracts greater viewer interest than the host. When the two stand in profile, neither is favored. But if the host takes a slightly downstage position, the guest will gain visual impact by presenting nearly a full face to the camera.

— On panel shows, a variation on the interview pattern, guests exchange views or provide information. In staging panel shows, directors place together panelists with a similar attitude and in oposing positions (across a table from one another) those having conflicting points of view. To prevent audience confusion, placards or superimposed titles should identify panelists from time to time.

— Concealment of cameras presents a major problem because they frequently must photograph panelists seated directly across from one another. Lighting and the set may provide concealment.

— Directors of homemaker shows—as with most other program forms—perform a wide variety of functions. Their sphere of action is not limited to a control room.

A PROJECT FOR ASPIRING DIRECTORS

You are the director of a Provocateur interview. What camera angles would you use throughout the following interview? Remember to save your closeups for moments of greatest importance and impact. Also remember that your camera doesn't have to be on the face of the person speaking. It must be where the greatest *drama* is taking place, from moment to moment. Write your choice of camera angles in the video column.

PROVOCATEUR INTERVIEW

VIDEO	AUDIO
FADE IN:	PROV: (TO CAMERA) Hello. Welcome to "Personalities in the News." Our guest tonight is Brad "Dutch" Fleming, head coach of the Los Angeles Eagles. Brad was accused by the press last week of providing his players with certain steroids in order to increase their skills on the playing field. He has volunteered to give us the true story. (TO BRAD) Good evening, Brad. How are you feeling?
	BRAD: A little nervous, I guess. And anxious to set the record straight.
	PROV: First, I want to congratulate you on a great season. Only one loss! Best season the Eagles have had in nine years.
	BRAD: Well, the guys have worked hard. They deserve it.
	PROV: According to the <u>Los Angeles Times</u>, they got a little help from their friends.
	BRAD: That ... that's just not true. We're as clean as any team in pro ball. Cleaner than most!
	PROV: You say that with a lot of conviction.
	BRAD: Well, sure. We're plenty steamed up about those articles. Our lawyers say we got a damned good

PROVOCATEUR INTERVIEW

VIDEO   AUDIO

case if we want to sue. And we may. We may. The team is clean, Alan. Every one of our players has taken tests. And what did the league find? Nothing!

PROV: I talked to Blaisdell this afternoon, the Times reporter. He claimed to have testimony. From one of your players.

BRAD: He's never been near our players. He's full of it.

PROV: Then why would he make a statement like that? And why would the Times print the article if they didn't have proof? They don't want to be sued.

BRAD: Can I level with you? Blaisdell has had it in for the Eagles ever since I turned him down on some "behind the scenes" stories for his paper. He's just trying to make trouble for us.

PROV: And there's no truth to the story? No truth at all?

BRAD: None. It's baloney. All of it.

PROV: What if I told you I had a videotape of Blaisdell talking to one of your players? (NO ANSWER) What if that player confirmed the story, said that the owners were putting pressure on you, threatening to dump you if the team didn't get some wins.

BRAD: If you've . . . if you've got such a tape, I'd sure like to see it.

PROV: I've got it.

BRAD: You know, some people can rig these things, put words in people's mouths that they didn't really say . . .

PROV: Would you like to see the tape? We could run it now. Put it on the air.

PROVOCATEUR INTERVIEW

VIDEO	AUDIO
	BRAD: No, I . . . I'd . . . Tell you the truth, Alan, I'd rather take a look in private. I just don't believe that anybody on the team would . . .
	PROV: Maybe we should run the tape, give the folks at home a chance to make up their own minds. (CALLING OFFSTAGE) Jack, can we run that videotape now?
	BRAD: No! Look, look . . . we may have used some medicines to give the guys a little lift. Really gentle stuff. Legal stuff, you know? And it never happened more than once or twice. They were down, see? Depressed as hell. They needed help. Our trainer came up with these pills . . .
	PROV: Pills?
	BRAD: Hell, I don't think they were any stronger than well, just kind've pepper-uppers.
	PROV: So the L.A. Times story was true, then?
	BRAD: No, not really. Well, maybe kinda halfway. But it's not nearly as bad as they made it sound.
	PROV: We're nearly out of time. Good luck to you, Brad. (THEN) I do have to make a confession.
	BRAD: Confession?
	PROV: What I said about the videotape—I just made that up. There isn't any videotape.
	BRAD: What? What did you . . . ?
	PROV: (TO CAMERA) Good night and good luck. This has been "Personalities in the News."

FADE TO BLACK

CHAPTER

10

SHOW AND TELL: DEMONSTRATION

Demonstration programs inform, but you should recognize that almost all programs inform, even those whose only avowed purpose is entertainment. We learn about foreign life-styles from motion pictures filmed abroad. We gain insights into sociological or psychological problems in our culture from certain television shows or feature films. We even educate ourselves in new vocabulary, new styles of dress, new modes in home decoration, new hairstyles, and new emotional experiences from these same "noneducational" entertainment forms.

A primary difference between demonstration and entertainment programs lies in the program's goal. Demonstration programs—instructional, educational, or industrial—aim to teach rather than entertain, to provide information that may add to their viewers' reservoir of knowledge or instruction that may add to their skills. The spectrum of demonstration programs is wide, primarily because demonstration (like interviews) often appears as an element within other program types (such as homemaker shows).

As entertainment programs also educate, so do educational programs also (in varying degrees) entertain. As successful teachers understand, the informational pill often requires sugarcoating to stimulate or maintain student interest. In

When you are in a good working mood, images swarm through your busy imagination. Keeping up with them and catching them is very much like grappling with a run of herring.

Sergei Eisenstein[1]

directorial terms, sugarcoating simply means applying the principles described in Chapter 2: appealing to audiences through such classic elements as spectacle, humor, surprise, conflict, curiosity, sex appeal, or personal involvement through dramatization.

In this chapter we will examine the nature of the demonstration program, dividing our examination into three primary discussions:

— CREATING SPECTATOR INTEREST : a brief glimpse at the roles of audience and director

— ELEMENTS OF DEMONSTRATION : the director's techniques for creating vivid and understandable instructional shows

— TWO DEMONSTRATION PROGRAMS : a structural analysis of the cooking show and the fashion show

---

[1]*Film Form—Essays in Film Theory* (New York: Meridian Books, 1957), p. 261.

CREATING SPECTATOR INTEREST

The basic goal of television or film programs is communication—the director transmitting to an audience a message that is entertaining or instructive, or both. The success of the communication process depends on many factors. Two of the most significant are the skill of the Sender and the interest of the Receiver.

The Sender

Viewers will watch demonstration programs that deal with their favorite subject matter. If fine art intrigues them, they will watch a television program about art appreciation. If art lies outside their sphere of interests, they will avoid it—unless the director or the producer has managed to incorporate elements of such extraordinary appeal that they compel audience attention. If, for example, the art show is hosted by an internationally famous movie star, disinterested viewers might tune in. Although such an example is extreme—and outside the province of most directors—it demonstrates the importance of the mode of presentation.

If a program consists of a single lecturer in a monotone exhibiting little-known paintings and demonstrating their compositions on a blackboard, casually interested viewers will certainly yawn and tune out. At the other extreme, if that program consists of "living paintings" from the famous Laguna Art Festival, enhanced by music and lively commentary from an engaging host, those same viewers will probably watch in fascination.

The point is obvious. Viewer interest varies in proportion to the Sender's showmanship and creativity. When talented directors use the classic entertainment elements effectively, minimally interesting subject matter is often warmly received. Warning: The director has the responsibility to shape and interpret the message, but he or she is frequently tempted to highlight only the color, to allow significant content to become minimized or lost. Elements of entertainment provide more fun; they are titillating. But, like frosting without the cake, such programs are unsubstantial and ultimately unsatisfying.

The Receiver

Of the elements of audience appeal discussed in Chapter 2, the one most applicable to demonstration programs is *curiosity*. Most of us lead hectic lives. Our days bulge with the pressures of professional, educational, or personal activities. We watch entertainment programs for an escape valve. When we take the time to watch educational or instructional programs, our purpose is more specific, more goal oriented. We label it "curiosity," but in truth our curiosity is frequently mixed with self-interest.

Most of us seek to improve our knowledge of quasars or Jungian psychology or classical music to gain respect from peers, to be accepted into the world of sophisticated, well-educated people, or to win admiration from members of the opposite sex. We seek new or improved skills to gratify personal or avocational interests or else to improve our professional statuses, to help us make more money. If a television program helps us to achieve those goals, we watch with a sense of inner pride.

Understandably, we tend to select programs that require less effort and that provide more entertainment. However, when the reward is high and the motivation sufficiently strong, we will dig in, despite a lack of sugarcoating; the accomplishment outweighs the effort.

ELEMENTS OF DEMONSTRATION

Most of us learn best by doing. When programs can stimulate viewer participation—creating projects that require a hands-on experience for spectators—they will usually succeed in achieving their broadcast goals. Shows aimed at young children often use this principle beautifully, presenting projects that require paper and scissors or crayons, for viewers to use in "games" that teach skills or inform. Because the art of instruction—especially as it relates to student participation—belongs more appropriately in a teaching text than in one devoted to directing, I will not pursue it further.

A recurring motif in this book is the director's need to present program material in the most visual manner possible. The principle applies more to demonstration programs than to almost any other directorial category. Other than viewer participation, visualization creates the greatest audience impact and provides the most effective instructional tool. Creating appropriate visuals becomes the motivating factor for directors in programs that demonstrate objects or a skill or concept.

As in all programs, both fiction and nonfiction, the director's goal is twofold: (1) to show the audience what it wants to see and (2) to show what the director wants viewers to see. If a chef carves carrots into fancy curlicues, most audiences would want to see exactly how he or she does it; they would want to get closer. Accordingly, most directors intuitively would cut to a closeup of the action. If the chef has prepared a list of ingredients that go into a specific recipe, it is the director's decision when to show it to the audience and how often—perhaps once at the beginning of the demonstration and again at the conclusion.

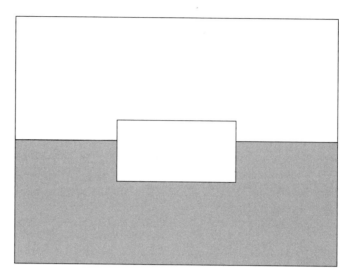

FIGURE 10.1   When objects are photographed from a flat, front angle, they become difficult to recognize.

Demonstrating an Object

When audiences can see an object for themselves—a gyroscope, for example—together with a demonstration of how the object can be used, they gain an understanding that would be impossible from mere "talking heads." Seeing the gyroscope spin at an impossible angle or on the point of a needle, virtually defying gravity, creates a vivid mental picture. Enlarging that understanding with film depicting how gyroscopes are used in the navigation of ships or airplanes, holding the vessel to course despite ocean currents and wind, demonstrates effectively because it demonstrates visually.

Closeness   As discussed in Chapters 6 and 7, close angles intensify dramatic values. They create emotional closeness or familiarity and provide impact. When introducing an audience to an unfamiliar object, all of these values are helpful.

When placing the object before your closeup camera, look at it as if you had never seen it before. When an object is presented at a flat angle, its shape and dimensions are difficult to ascertain (see Figure 10.1). When the object is turned slightly or if the camera is raised so that the viewer can recognize three dimensions, the image will be more clearly delineated (see Figure 10.2).

Lighting, too, can provide dimension and recognizability. If an object is a significant part of your demonstration, ask your lighting director to key it from a 45-degree angle and then to check it on a monitor (or from the control room) to make certain that the object is crisply defined.

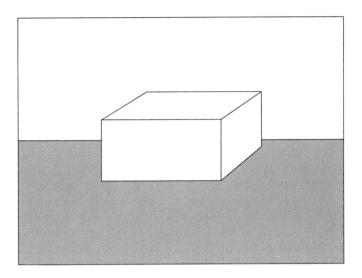

FIGURE 10.2 When photographed from a 45-degree angle and lighted to provide dimension, an object's shape and character may be easily discerned.

Separation Sometimes an object in a close angle tends to become involved or lost in its background. Separation may be achieved through color, lighting, or texture.

When the color of an object is too close to the color of its background, separation becomes difficult. Changing the color or tone of one or the other is an obvious answer. Also, by taking some (or all) light off the background, it will appear darker, providing better separation.

Backlighting the object separates it from its background, providing a rim of highlight and creating a three-dimensional image. **Backlight** also throws a shadow pattern on the floor (in foreground) that may help to define the object's shape.

When the background is "**busy**"—that is, if its pattern is small, convoluted, or intricate and its colors varied—the object is easily lost in the complexity of the background (see Figure 10.3). Generally, the key for easy recognition is simplicity.

Motion The simple act of picking up an object and putting it to use automatically solves many recognition factors. The object then changes position, lighting, and background. The audience also may determine its size in relation to a performer; in limbo, it could be any size.

Motion also demonstrates the object's practical aspects. The gyroscope, for example, reveals more about itself when spinning on a suspended thread than through a hundred beautifully composed camera angles on the still object. A stationary automobile is little better on television than in a magazine advertisement. When that same car races up a scenic mountain road, it creates more than

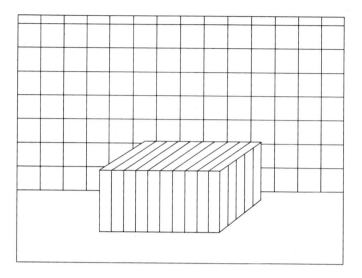

FIGURE 10.3   Intricate backgrounds create problems in recognition.

static beauty; it suggests speed, control, power. Viewers understand the object as much by what it *does* as by how it *appears*.

TV commercials utilize the demonstration process as a major sales tool. By seeing a product in use, and the happy reactions of users, spectators learn (are sold on) the remarkable effectiveness of the product. The reactions of others are usually what sell the product, not the specific qualities of the product itself. When a homemaker is thrilled at the effectiveness of a detergent or the rich taste of a new coffee, viewers tend to identify with her, sometimes in spite of themselves.

Demonstrating a Skill or Concept

Nowhere is the need for visuals more apparent than in demonstrating something that is itself nonvisual—an idea, a method, a concept, or a skill. If a program deals with the art of interior decoration, for example, words alone are virtually worthless. The director of such a program must insist on videotape, film, or photographs—or televise the program directly from the site to be refurbished— so that viewers can study the decoration problems firsthand. Creative planning could add greatly to such a program. With a photograph of the bare room, for example, the director could matte possible groupings of furniture into the room so that spectators could actually see what worked and what didn't. The decorator might then explain the specific reasons why one furniture grouping was effective and the other was not. But explanation without visual demonstration would fail to make its point, no matter how skilled the explainer.

Demonstrating ideas often calls for some degree of dramatization. Many industrial films teach employees such concepts as merchandising, salesmanship, how to resolve office conflicts, and stress management. Rather than put together films that simply lecture employees, producers sometimes build story situations in which new principles are dramatized, creating characters with whom an audience of employees can identify. I once directed an industrial film in which a girl accepted a job as a magician's assistant. When the magician trained her, she made every mistake in the book. As she learned how to learn, so did the audience.

Classroom methods of demonstrating a concept often work well on television. Experienced teachers use visual aids such as videotape, film, or still photographs in educating their students. Still photographs, when used with imagination, provide unusual opportunities for visual demonstration.

When using photographs, many directors try to keep their cameras moving to help create an illusion of reality. They either start wide and dolly in slowly, or start close on some feature of the photograph and dolly out. Sometimes, if the photograph is large enough, cameras can pan slowly from one feature of the photograph to another, the movement bringing life to the inanimate. Such treatment works far better when there are no human figures frozen in position in the photograph. On some occasions, music or sound effects can help create the illusion of three-dimensional reality in a still photograph. Bringing in music as a director dissolves to a photograph creates dramatic impact. Because the camera is focusing on a flat plane rather than dimensional subject matter, dollies in or out become somewhat difficult; be sure that your camera operator can handle the chore.

Other classroom devices also work on television. A blackboard, easel, or any surface on which pictures may be sketched helps to turn a verbal description into a visual one. Audiences, like students, need help in reinforcing ideas. A picture adds another dimension; it redefines an intangible concept, even if it is nothing more sophisticated than chalk on a blackboard.

Whenever possible, a skill should be performed *from the viewer's point of view.* Students of instructional programming have labeled such a shot a "zero-degree camera angle." The shot creates far greater learning impact than an objective angle because it seems to make the viewer a participant. A zero-degree camera angle is also the point of view of the demonstrator.

For demonstrating a skill, all of the described principles for demonstrating an object hold true. When an expert demonstrates how to prune roses, for example, viewers must see each of the pruning shears' cuts. When the expert explains that each cut should be diagonal, slightly above the bud that will later become a branch, it is critically important for the audience to *see* the bud and to *see* the diagonal cut. The branches must stand out clearly from the background. Backlighting, a contrast between foreground and background, and simplicity of background will all help to make the expert's demonstration visually effective.

The Director as Audience

In addition to staging action and calling camera shots, the director of a demonstration program plays another essential role, that of audience. In preparation and rehearsal, the director watches the demonstration with a critical eye. If the expert does not present material clearly and understandably, the director must ask for changes.

Because the director is a relatively disinterested observer, he or she can be objective in assessing the manner of presentation, asking such questions as: Does the expert go too fast, not giving viewers time to assimilate the material? Does the expert lead the audience carefully from the known to the unknown, from the simple to the complex? Are his or her explanations confusing? Do they involve too many esoteric words or phrases? Are the explanations visual? What can the expert do to make the demonstration *more* visible? By asking the questions an audience would ask, the director prevents a breakdown in communication.

The Expert

When demonstrating where and how to prune a rosebush, the gardening expert must be as aware of your problems as you are of his or hers. The expert must know, for example, which camera will take close shots so that action can be angled to that camera. You can help the expert (and yourself) by placing a monitor near the work area. Now your expert can see when a hand or shadow covers the significant action or when it is angled incorrectly to the camera. During rehearsal, allow your expert to experiment in showing action to the camera, becoming comfortable with the physical setup, even a little proud of his or her ability to accommodate your needs.

Although experts are highly skilled in their own fields, they are often unskilled (and uneasy) when appearing before television cameras. Nervousness can mar the effectiveness of their contributions. Remember, directors are responsible for the overall impact of a program. They do not simply "direct traffic" on stage. Their function is not limited to calling camera shots. Therefore, as director, you must make the expert feel welcome and secure. A cup of coffee helps. So does a smile.

TWO DEMONSTRATION PROGRAMS

The element of demonstration appears in programs covering a variety of **formats**: from cooking shows to documentaries, from homemaker shows to industrial films. We will examine in some detail a typical demonstration show and an atypical one. Whether or not you have any interest in these two specific program types, they will nevertheless provide insights to a broad range of other demonstration programs.

A Typical Demonstration Program: The Cooking Show

Literally and figuratively, the cooking show has become the meat and potatoes of many local television stations. It has also appeared on daytime commercial networks ("The Frugal Gourmet") and repeatedly on PBS ("The Julia Child Show"). TV stations may produce such a show inexpensively on a daily basis, featuring local epicurean personalities.

The director may bring visual appeal to cooking shows in three areas, all of which may be described by the word *appetizing*: the food, the host(ess), and the setting.

The Food A key concern in the preparation of food is that the audience be able to see it clearly, vividly, and up close. The obvious answer is to zoom the camera in to an extremely close shot. Unfortunately, the solution isn't that simple. Because most normal work areas are at waist height, a camera, even at the top of its pedestal, cannot gain a steep enough down angle to provide an effective view of the food. The angle does get steeper as the camera moves closer, and it is possible for work/display areas to be lower than normal (although low counters cause problems for a performer's back). These, however, are generally unsatisfactory, second-best solutions.

Many cooking shows use a large, double-mirror "periscope" for one of their cameras (see Figure 10.4). Through such a device, the camera is able to gain an acute high angle, providing viewers with an excellent perspective on the food. Such a large piece of equipment somewhat limits camera movement, although the camera may pedestal up to shoot between the two mirrors or may move to either side. Without a double mirror, cameras must pedestal up as high as possible and move as close to food as possible to get an effective shot.

Most cooking shows use two cameras. One generally stays wide, covering the host and providing orientation; the other provides close shots of the food in preparation. Thus, it is seldom necessary for the closeup camera to move from its mirror position except when shooting live commercials or titles. Such an arrangement is comfortable for the host in that he or she only has to play to one camera, which represents the viewer.

Most professional cooks and chefs are aware that an attractive appearance is as necessary to a dish as its taste. As a director, you should be equally aware. In all steps of preparation, stove, sink, and counters should be immaculate. When fat has been trimmed from meat, for example, or vegetables peeled, debris should be immediately cleared so that work areas remain spotless.

Of course, food should be arranged on plates or platters as artistically as possible. Garnish often adds eye appeal. Parsley, carrot curls, or fruit baskets may

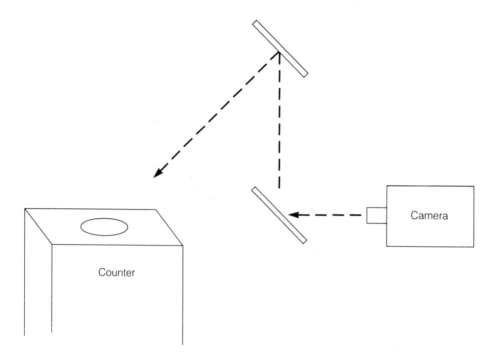

FIGURE 10.4   To provide effective down angles on food preparation, many cooking shows use a system of double mirrors.

not be utilitarian, but they effectively dress up the completed dish. Even the choice of china is important in creating eye appeal. If the host needs prompting in this area, don't be shy.

Commercial photographers learned many years ago that one of the secrets of making food appear appealing was to use plenty of light. Backlight, in particular, creates the glisten and sparkle that most of us associate with taste appeal. Glance through any magazine that features food advertisements and note the glisten that "sells" the food. If your lighting director isn't aware of such fine points, pass along the word. Most cooking shows feature a special downstage area where the completed dish is proudly displayed. More than any other, this is an area for special care and attention by your lighting director.

Many recipes require considerable preparation time, so it is often advisable to "cheat" in a live telecast by having on hand examples in various stages of progress. Thus, the host can prepare cake batter on camera and then avoid an hour's baking time by producing a second cake, which he or she has prepared prior to airtime. No need to apologize for taking shortcuts. Audiences understand the exigencies of a half-hour live telecast. Now, the host can decorate the cake, concluding the recipe with a grande finale of fancy frills and trimmings instead of a deadly wait or the promise to finish tomorrow.

The Host   In my years in live television I directed three different cooking shows. Two of the three had hosts who were truly marvelous cooks. The third was a mediocre cook but a delightful personality. Which of the three shows had the highest audience ratings? Beyond any question, it was the third.

People relate to people. True, we watch a cooking show primarily to learn. But learning need not be tedious. As Julie Andrews sang in *Mary Poppins*, "A spoonful of sugar helps the medicine go down!" When the learning process includes elements of entertainment, the medicine goes down easily. When a host brings humor, a personal zest, or style to a show, it lifts the show beyond the ordinary. Caution: A sense of humor cannot replace cooking ability. Audiences still tune in primarily to learn kitchen skills. But if five cooks have approximately equal talents, the one with a twinkle will steal the Nielsen points.

One of the delights of "The Julia Child Show" was her ability to laugh at herself when she misread a recipe or dropped a soufflé. The error made her human; her self-deprecating remarks endeared her to us. But those qualities would not have been so meaningful had she not been recognized initially as a superb cook.

The kitchen's immaculate appearance should, of course, be reflected in the appearance of your host. In dress, hairstyle, and work habits, he or she should create an appetizing image. Studio lights tend to work against that image; they are hot and cause perspiration. If your show occasionally breaks for filmed or taped commercials, those breaks are ideal opportunities for your host to blot that perspiration.

Because you're aware that a star's popularity and a show's success often go hand in hand, you should try hard to make your host look good. Make certain that the lighting is flattering. Try to create an ambience on the set that is warm and supportive. If the star is nervous or insecure, ask your crew for help in making him or her comfortable. Build confidence through sincere compliments; create humor to reduce tension; generate a feeling of competency. Your host will come to understand that the crew are professionals who know exactly what they're doing; they're not going to let anything go wrong.

The Setting   For a cooking show the setting should be cheerful and utilitarian. Whether modern or period, most designer kitchens use plenty of brick and wood with bright colors (paint or paper) for accent. As in all well-designed sets, your kitchen should not be too "busy"; avoid too many small, fussy details that obstruct easy readability. Thus, ceramic tiles are seldom a good choice. Most kitchen sets place oven and refrigerator to the side or in the background so that the primary work area can be in the foreground, usually a countertop and adjoining gas stove top.

One of the financial fringe benefits of cooking shows is the opportunity they present for merchandising tie-ins with local grocery markets. Major food distributors always are seeking display space in markets, so the cooking show is often able to attract national advertisers through the promise of such space. Local grocers usually are happy to provide displays in return for being mentioned in the advertiser's cooking show commercials. Thus all parties benefit. Merchandising sometimes includes appearances by the host at the market and on-the-air contests involving market and product.

The director must respect the host's involvement with advertisers and merchandising (stars frequently receive a percentage of advertising dollars), but commercial efforts should not impinge upon the content of the show. When this happens, audiences begin to feel that the program has become one long advertising message. To prevent such a reaction, directors should separate the work area from the commercial area so that the host must physically leave the work area to begin a commercial pitch. Such an area usually adjoins the downstage countertop and is of the same decor as the kitchen.

An Atypical Demonstration Program: The Fashion Show

Fashion shows, like other demonstration elements, usually are presented as part of a larger program format. They appear occasionally in newscasts in coverage of couturier showings. Because their appeal is primarily to women, they appear frequently in homemaker shows, sometimes as a part and sometimes as the entire program.

Fashion shows are unlike other demonstration programs in that the ambience becomes almost as significant as the product demonstrated. In this respect they are much like magazine advertisements and TV commercials. In truth, their goal is often as much to sell as to educate. The drama surrounding the presentation, the character of the models, the setting, the music, and the commentator create an atmosphere that shapes the spectator's receptivity.

The Models   A popular fallacy: Successful models are always beautiful. Not so. They are *sometimes* beautiful. More importantly, they wear clothes with style. By their demeanor, they make fashions appear attractive, important, necessary. Part of a model's effectiveness relates to her figure, usually lean and strikingly tall. The most sought-after fashion models are five feet, nine inches or taller. Part of their effectiveness relates to facial expressions.

When directing a fashion show, notice that the finest models create ambience as much from their expressions as from their physical demeanor. They wear clothes proudly; their faces proclaim, "I feel marvelous wearing this gown; I feel rich and sensuous; I love myself." Models also express these thoughts in their movements, in the way they walk, in their gestures such as pulling a collar about the throat or pointing out specific apparel features as they are described by a commentator. By their movements, facial expressions, and attitude, they create an unmistakable sense of beauty.

Directors sometimes are presented with an inexperienced or lackluster model. Although most TV directors don't have the time to give lessons, they can suggest what a model should be feeling in the hope that the model will have the dramatic capabilities to bring it off. Inexperienced models often are so concerned with the physical movements necessary to the staging of a fashion show that they cannot give thought to much else. Wise directors therefore keep their staging simple. When possible, they voice their opinions in the selection of models.

The Staging and Setting   Logically, decor should complement the fashions. Thus, if the gowns are formal, the setting should echo that formality, perhaps with "marble" columns, arches, or a circular staircase. The elaborateness of the setting, of course, depends on budget, the resourcefulness of the art director, and the set pieces available in the storage barn. Less formal attire might utilize a garden setting with lattice or a gazebo and greenery as space (or money) will allow.

Male models seem to function more effectively when using props. Tennis rackets or golf clubs, for example, give them something to do with their hands while modeling sports attire. When modeling suits or formal attire, they appear more natural carrying a briefcase, overcoat, or a corsage box. When the budget allows, such handsome props as a Rolls Royce or classic old car will not only add elegance to the setting but will also give the models something to relate to.

Imagination sometimes provides answers to budget problems. I once directed a fashion show featuring bathing suits. We tried to put together a beach setting. Because of time and budget limitations, it was impossible. Instead, we directed a spotlight into a pan of water, throwing shimmering reflections on a bare **cyclorama**. We introduced each model in a master angle, shooting through a small fish tank (borrowed from a neighborhood pet store), with multicolored fish and seaweed in sharp focus in the foreground, the model blurred and indistinct in the background. The camera then racked focus to the model. After a moment, we cut to the second camera, which provided a closer angle, eliminating the fish tank. But the underwater illusion was maintained with the ripple effect on the cyclorama and a few rock cutouts in the foreground.

Traditionally, the staging of fashion shows divides into two parts: (1) the revelation of a new model in a new outfit and (2) a closer inspection of the clothes and their features. The revelation can be as simple or as elegant as the setting allows. For example, if the fashions are formal, the model might appear at the top of a staircase. Camera 1 would follow her in a full shot as she descends the staircase and moves downstage (toward camera) to a spot that you have carefully marked on the studio floor. In this second, foreground position, you cut to a closer angle, perhaps a waist shot. As the model turns, revealing features of the gown, camera 2 follows her movement. Camera 2 may tilt down to the model's feet, revealing features at the hem, and tilt back up to her head and shoulders.

While camera 2 photographs this closer angle, your floor manager cues the next model to take her place at the top of the staircase where she waits, unseen by camera 2. Camera 1 meanwhile is lining up on the waiting model. When the first model has completed her stint, you cut to camera 1 as the floor manager cues the

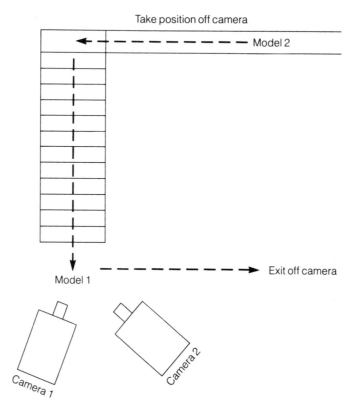

Take position off camera

Model 2

Model 1

Exit off camera

Camera 1

Camera 2

FIGURE 10.5   In this staircase pattern, camera 1 pans with the model down the stairs in
a full shot. Camera 2 provides closer coverage.

second model to descend. The first model now exits, unseen by camera 1. And
the pattern is repeated. (See Figure 10.5.) A principle that serves directors well,
in fashion shows as in other forms of staging: Entrances are interesting; they
create anticipation. Exits are boring. The conspicuous exception is the final shot
in many dramatic films in which an exit has traditionally marked the bittersweet
fadeout. (Recall the final shots in many Charlie Chaplin comedies and in most
Westerns.)

In maintaining the mood generated by the gowns, the formal setting, and the
music, it is often effective in a fashion show to dissolve rather than cut between
camera angles. A dissolve seems less abrupt, more fluid.

The Commentator   The commentator in a fashion show does far more than provide descriptions of various garments; he or she sets the pace for the program,
tying together all of the elements. Commentators work far more effectively off

camera than on because they must refer to notes and their presence often interferes with the mood generated by setting and models.

Careful coordination between commentator and director is essential. The director usually has no idea how much time the commentator will take in describing each gown. Therefore the commentator must provide a cue when to cut or dissolve to the next model. Usually a polite "thank you" to the on-camera model signals completion. Generally, the fashion show coordinator or commentator provides the director with a list of outfits to be exhibited and specific features of each so that the director may anticipate the commentator's words. Thus, when the commentator describes lace at the neckline, the director will be waiting and ready to cut to a closeup. Providing the commentator with a monitor also helps ensure effective coordination. By watching the screen, the commentator knows exactly what is being photographed at each moment, enabling him or her to adjust commentary to match.

The Music   As with the setting, music must complement the mood that the fashions generate. Easy rock 'n' roll or muted jazz might be appropriate for sports attire. Formal fashions might require something more stately: violins or harps or popular classics. Remember, the music must fit comfortably behind the commentator's voice; it cannot be strident. It is a background to the show, not the show itself.

CHAPTER HIGHLIGHTS

— Entertainment programs educate; educational films entertain. The latter create interest through such classic elements of audience appeal as humor, spectacle, conflict, curiosity, sex appeal, or personal involvement through dramatization.

— When demonstrating objects, directors must photograph them for maximum recognition, using such factors as closeness, appropriate angle, separation from background, and movement. Separation may be achieved through a differentiation in color and by backlighting. Movement that indicates how an object is used most clearly reveals its nature.

— In demonstrating a skill or concept, the director must create a presentation that is essentially visual through the use of videotape or film, still photographs, physical demonstration of a procedure (such as pruning roses), dramatization through actors, and even a chalk board.

— Experts bring knowledge and skill to a program but often lack familiarity with television procedures. Sensitive directors try to make them feel secure. Directors also try to react as a typical audience, making certain that the expert's presentation is not too fast, too esoteric, or too complex for easy assimilation.

— A cooking show is a typical demonstration program. Directors sometimes use mirrors for down angle shots of food. They ensure that food presentations appear attractive and appealing, in both preparation and lighting. The host's personality is a major factor in the success of a cooking show.

— A fashion show represents an atypical demonstration program—atypical in that ambience becomes almost as important as the fashions themselves. The setting should reflect the character of the clothing. Thus, formal gowns require a formal setting. Models "sell" their apparel as much through facial expressions as through physical movement.

— A fashion commentator usually works off camera, timing commentary to the director's coverage (or vice versa). A studio monitor helps the commentator synchronize words with picture.

PROJECTS FOR ASPIRING DIRECTORS

1. Create a cooking show. Try to find a concept that makes it different from any cooking show you have seen. Keep in mind the following questions:
   a. Who will be your host/star? Will there be more than one? What will be their working relationship?
   b. Design the set so that you as director will be able to cover the action effectively. Will there be a commercial area? Who will handle the live commercials?
   c. Will your show have audience participation? Contests? Recipes sent to viewers? Recipes received from viewers and demonstrated on the show?
   d. How will you handle titles? Music? Preparation and rehearsal time?

2. Critique a demonstration program. If you were the director, how would you have made the demonstration elements more visually effective? Be as specific as possible.

3. A sample project: The host of a show you are directing wants to demonstrate the game of Scrabble. Describe on paper (and diagram) how you would stage such a demonstration.

4. Another directing project: Your show's host wants to demonstrate a device that reproduces music digitally. Note that this will be primarily a demonstration of sound. How will you handle it?

5. You are the host of a demonstration program. Make a list of qualities you hope to find in your director.

CHAPTER

11

THE NEWS PROGRAM

News has come into its own. A decade ago many TV stations aired two daily quarter-hour newscasts, one at the dinner hour and another in late evening. Now those same stations broadcast two or three hours of news each day. To be sure, much of it fails to qualify as "hard" news. Many stations boast feature reporters, restaurant authorities, consumer advocates, medical or health advisers, entertainment reporters, and political commentators. The newscasts also include sports and weather experts, chosen as much for audience appeal as for their expertise. When you add to this potpourri the personal magnetism of anchor persons and the colorful computer graphics that enhance these "news" programs, you must conclude that, in most cases, they have become a new form of variety hour—news-oriented entertainment specials.

Because of the multiplicity of elements and the complexity of their visual structure, these programs have become tests of concentration for their directors. Although *concentration* is an essential characteristic of all directors, the term is particularly appropriate for news directors because they must coordinate script, local cameras, remote cameras, videotape, timing of sound elements, computer graphics, and personalities.

A second characteristic marks successful news directors: *awareness*. Once a newscast begins, a dozen disparate elements hang in the air. A weather reporter

*News, or orientation to reality, is essential to the happiness
and emotional balance of democratic man. Disorientation
is one of the things we have most to fear.*

Harry Skornia[1]

on remote at the county fair is having trouble with crowd control. A newsroom executive hurries in with revised script pages. The producer is on the phone to a remote, advising a performer to stand by. The technical director has punched up the wrong graphic. The aware director has a dozen eyes, tracking on-the-air elements as well as those in the control room. The latter are significant because they will affect or become on-the-air elements in minutes or seconds.

Aware directors are conscious of the problems of other members of the production team. If the audio engineer has difficulty with a microphone, that difficulty automatically becomes the director's problem. If the producer has miscalculated, cramming too many elements into a script, that timing miscalculation also becomes the director's problem.

A key part of awareness is knowledge. Successful directors work to build a knowledge of the functions, goals, and problems of other members of their news teams. Such knowledge provides security. If something goes wrong, they are better able to adjust or to compensate. Knowledge transforms technicians into directors.

---

[1]*Television and the News* (Palo Alto, Calif.: Pacific Books, 1974), p. 2.

This chapter includes three primary discussions:

— THE ELEMENTS: exploring the assembling and synthesis of written and visual news material within a narrow time frame

— THE DIRECTOR AT WORK: an examination of the news director's activities— in preparation, on the air, in the field, and in the studio—plus sophisticated technologies that are part of today's newscasts

— THE NEWS DIRECTOR—A COMPOSITE PORTRAIT: a glimpse into the backgrounds, characteristics, and philosophies of four news directors, network and local

THE ELEMENTS

The material in this chapter is based upon personal observation of the news operations at KHJ-TV, a Los Angeles independent station owned by RKO General, KNBC (a major Southern California network affiliate), and "Entertainment Tonight," a daily half-hour, syndicated, entertainment news program (multiple-camera) produced at Paramount Studios.

The differences among the three operations lie primarily in scope and size. KHJ-TV has a staff of just over forty; "Entertainment Tonight" (called "ET" by its staff) has approximately 160; KNBC has about 260, which is large for a local station. KHJ-TV has four field units; KNBC has ten to twelve.

Most viewers would be surprised at the depth and complexity of the typical news-gathering operation. The discussion here divides it into three parts: visual content, preparation, and paperwork.

Visual Content

Television stations recognize the acute need for visuals as a way of dramatizing news stories. An anchorperson simply telling (reading) a story fails to realize the promise of a visual medium; the program loses much of its entertainment value. Watching such "talking heads" is little better than hearing the same story on radio. Therefore, most TV stations search for ways to add a visual dimension. When such visualization is not available, stories frequently are dropped from the lineup.

The ideal way to dramatize a story is through **videotape**, creating impact because it places viewers at the event.

Videotape   How can a local TV station offer its viewers videotape of major news events happening thousands of miles away? Director Jay Roper with NBC News explains that a station affiliated with a network receives each day "an affiliate news feed from a network **satellite**: usually sports, some hard news, and many softer news items. The affiliate feed will sometimes include a promo for the local station to air, promoting that evening's network news." The NBC satellite is called Sky Com.

In addition to receiving visual material, network affiliates may also contribute. If a newsworthy event happens locally, the local station covers the event with its field units, recording the material on videotape. The station then ships the videotaped news material to its New York headquarters for possible distribution (via affiliate news feed) to other stations. Satellite use is expensive; if an event is not sufficiently newsworthy, networks use other means of distribution. If the event has major national significance, however, the network usually sends its own director and crew to handle coverage. In the latter case, the material would be transmitted to New York via a satellite **uplink**, for later airing on network news programs and satellite distribution to affiliates.

Independent stations also exchange video material via satellite. They form a network of their own called **INDX** (pronounced "index"), an acronym for Independent Exchange. Member stations both receive and contribute material. WPIX in New York functions as the origin for daily INDX transmissions.

About the middle of the afternoon, independent stations receive (via telex) a list of stories that INDX will transmit that day. Such advance notice helps stations plan the pattern of their newscasts. An hour or two later (usually about five o'clock), the satellite material begins coming in. The station records the material on two separate videotape machines, for protection in case one breaks down and to expedite editing. Each item carries a slate describing the story and its exact length. On the average, approximately forty minutes of material is transmitted each day. According to KHJ-TV executive producer, Bill Northrup, only about 10 percent of that material is actually used by the station.

A local TV station tries to serve the interests of its community; it therefore places a mix of local and regional news in the foreground, allowing world and national news to take a background position. To provide visuals for such stories, the station must rely on live or videotape coverage from field units, material borrowed from other local or regional stations, still photographs, or artwork and computer-generated titles.

Boxes Stations use still photographs, artwork, or lettering—or some combination of all three—to add visual interest to shots of anchorpersons. Most directors refer to these digital video effects as **boxes**. They usually appear in a corner of the frame above an anchor's shoulder. Photographs may come from a wire service or from the station's **frame storer** (such as an **Adda**), a computer that contains within its memory thousands of frames of visual information. Frame storers also preserve artwork, usually generic **graphics** such as the sun radiating heat waves to be used on the next blistering summer day.

When the director or assistant director calls for this material during a newscast, the computer feeds it to one of the control room monitors. The technical director then mattes the appropriate box into the frame with the anchorperson.

Sometimes videotape elements of a news story will originate within a box and then will zoom out to fill the screen. To move the boxed videotape to full screen, the director simply says "Move it." To change from one box to another, the command is "Change Chyron." Coming out of a videotape segment: "Ready 1, a box," meaning that camera 1 should stand by and that the technical director should be ready to matte the appropriate box over camera 1's picture.

These commands vary somewhat from station to station. There is no industry rulebook stating that commands must always be presented in the identical way. Because of the increasing number of visual effects available to today's directors, new commands are constantly appearing. There is some variation among directors, but most commands follow a fairly standard pattern. (See notes at the beginning of Part Three.) They must be *clear* and *terse*.

Lettering often appears in boxes. It is usually computer generated. The **character generator (CG)** most used for such purposes is a **Chyron** (pronounced Kyron), which can generate lettering in a variety of sizes, fonts, colors, or formats, stationary or moving. Such lettering usually is matted over a videotape segment near the bottom of the screen, identifying either a locale or person.

Computer generators are also used extensively in sports broadcasts, stating names of players and statistics or listing scores of completed or ongoing ball games. As the sportscaster ticks off the scores, the director cues changes from one graphic to next. When used during sports events, Chyrons have an operator either in the control truck or in another truck nearby, providing the director with a constant flow of statistical information. On "Entertainment Tonight," the Chyron operator sits in the **control room** at a console behind the director. Because the show is videotaped in a start/stop operation, stopping tape when something goes wrong and picking up after corrections have been made, the Chyron operator sits close at hand, ready to make any changes that the producer or director finds necessary.

Most news operations employ graphic artists to create the colorful visuals that appear in boxes. Sometimes the director has a voice in the selection of these visuals, sometimes not. The director of "Entertainment Tonight" meets with graphic artists each morning to suggest visuals that might be appropriate for specific stories. For a story about Malibu, for example, he might suggest a photograph (or artwork) of surfers with the word *Malibu* across the frame.

Sometimes graphic elements will move. Examples include clouds moving— or rain appearing to fall—on a weather map. Such an effect is generated either from videotape or from a computer called a **Dubner**.

Still photographs in full color sometimes are extracted from videotape stories when the videotape itself is not used. The single frame appears in a box for a shortened version of the original story.

Local TV Cameras Newsworthy local events usually are covered by three-person ENG crews: camera operator, sound technician, and reporter. On a day-to-day basis, independent stations do not employ the services of directors to stage such coverage; the reporter in consultation with the camera operator generally handles such chores.

The single-camera, three-person unit has the advantages of being compact, lightweight, self-contained (usually in a van), and able to move quickly from one locale to another. Such units maintain constant radio contact with the newsroom so that they can keep the producer informed of their progress and any unexpected problems. Also, after completing an assignment, the field unit may be dispatched by radio to cover a story breaking suddenly in another location. Most such units are constantly on the move, videotaping interviews, fires, award ceremonies, accidents, political speeches, and arrivals and departures of celebrities. Usually they leave the TV station in the morning and do not return until night.

Field units dispatch their stories to the newsroom in two ways: by **microwave link**, transmitting picture and sound as an event is actually occurring, and by recording the event on videotape. In the second circumstance, the tape is subsequently microwaved to the station, "bulk feeding" it, along with other material recorded that day, or it is picked up by a courier who will deliver it to the TV station. When the signal is transmitted by microwave, it is recorded at the newsroom. In both cases, the coverage is evaluated by the news producer and edited per his or her instructions.

Sometimes, when an event occurs at the time of the news program, it is transmitted live (via microwave link) to the TV station, where it is incorporated into the newscast. Such a practice is common during local elections, with sports or weather personnel on location, and with reporters providing updates on major catastrophes and other events.

The balance of the news program is supplied by studio cameras covering live performers in the news set. One practice currently in vogue ties together studio anchor personnel with reporters on location. The reporter's picture is fed into a monitor within the news set. (These monitors usually are on hydraulic lifts so they may be raised into view or lowered out of sight during a program.) With the monitor raised, anchorpersons may address or question the distant reporter directly. Sometimes no monitor is used; the reporter on location is chroma keyed onto a wall behind the anchorperson or wiped in using **digital video effects (DVE)**. Once the questioning is finished, the director usually cuts directly to the reporter so that he or she becomes the sole focus of attention.

Weather Reports

Just as the location reporter was chroma keyed onto the wall behind the anchorperson (or inserted via DVE), so are maps and graphics frequently displayed directly behind weather reporters. When a weather map is keyed behind a reporter, he or she must glance occasionally at an off-camera monitor in order to know where to point—to orient himself or herself in relation to the map. Some weather reporters hold a concealed switch in one hand to cue changes in the graphics. Others use word cues to signal the director when to make changes.

When backgrounds contain movement—as sometimes happens with satellite weather tapes—the director rolls the videotape full screen with the weather reporter keyed over it.

Preparation

The various elements within a news program come together slowly. Some major events such as elections or political conventions are planned weeks before the date of the broadcast.

Feature stories and local news events for Tuesday usually are discussed on Monday afternoon. The stories to be covered by field camera units are also determined at this meeting. The news director, the executive producer, and producer(s) discuss and evaluate ideas for future feature stories. Sometimes the concepts for these features originate with the personalities who will be reporting them.

On "Entertainment Tonight," major features are planned weeks in advance so that the producers may gain time for advance publicity and appearance in *TV Guide*. (Usually the second "act" of "ET" consists of feature stories; the first act presents entertainment news.)

Let's examine in depth a local, one-hour news program that airs at 9:00 P.M. Beginning early in the day, its structure slowly takes shape. By late morning, the producer and the staff have read copy from a number of Associated Press (AP) or United Press International (UPI) news wires and identified those three or four national or international stories that will probably be included in the broadcast. A telex will later confirm that INDX will provide video material for these stories. Aware of local events being covered by field units, the producer is able now to assemble a list of known "availables" for the night's broadcast: stories covered by each of the field units plus features in work. To this list will be added INDX material for national or international items, once they have been specified (mid-afternoon). The pattern will change a dozen times between now and nine o'clock as some stories fall by the wayside and others suddenly break, sometimes only minutes before the broadcast.

By late morning, material from the field units begins to filter in. The producer screens each piece of incoming material, evaluating its impact and probable running time. Writers are assigned to provide copy; editors are assigned to cut the videotape footage down to size, eliminating dull material, heightening drama. The length of the videotape will be tailored to fit the news copy rather than the other way around.

The newsroom atmosphere early in the day is relaxed and casual; writers joke with one another about events in the news. As the day progresses, the relaxed atmosphere gradually changes. By late afternoon, it has become quiet, business-like. By early evening, as the staff makes final changes in the rundown and hurries to assemble a final script, the casual atmosphere has been replaced by mild tension. No panic, no loud voices, simply a sense of urgency, a need to get a lot of work completed in a limited time.

At 3:30 the executives assemble for a production meeting. At some stations the director attends this meeting, a first glimpse at the news pattern with which he or she will deal. At other stations, the director will not arrive for another hour or two.

At this production meeting (and there may be others depending on late-breaking stories), the "final" assemblage of stories is discussed, evaluated, and arranged into a broadcast sequence. The newscast is divided arbitrarily into seven or eight sections, depending on the number of commercial breaks. Certain fixed, continuing patterns (which vary from station to station, depending on the programming philosophy of each) determine how elements are structured. For example, at KHJ-TV sports generally concludes the third section and is updated at the end of the newscast. The weather forecast begins the next-to-last section. As with all news programs, the headline story, the news item of greatest significance to viewers, begins the lineup.

Paperwork

After a sequence is determined at the production meeting, the producer distributes a rundown sheet listing each story and its position in the sequence, plus the writer, talent, and certain necessary technical information. (For a portion of a **rundown sheet**, see Figure 11.1.) When late-breaking stories rupture the pattern, revised rundown sheets are distributed.

To some, the newsroom would appear to be a paper factory. Paper spews forth from a dozen sources. Six or eight wire service teletypes provide coverage of sports, hard news, soft news, feature stories, weather, and human interest items. A dozen typewriters and word processors grind out script material, contributing to the avalanche of paper. In addition, papers lie inches deep in baskets and bins, on desks and on the floor, pinned to bulletin boards and taped to the walls above desks. Without paper, a news program could not happen.

An hour or two after the rundown sheet first appears, script pages begin circulating—about seven or eight o'clock. Although the majority of the stories were written earlier in the day, many were later revised or polished due to time limitations, an updating of news facts, or a change in the talent who would read the material. When the producer adds late-breaking stories to the lineup, others are necessarily deleted.

Writers feed news material into computers that print scripts with large type for distribution to the various departments. The **TelePrompter** later displays the material on video screens mounted on each camera and in the control room. A sheet of glass in front of the lens, angled at 45 degrees, enables performers to read the reflected news copy but still direct their gaze to the lens, apparently speaking directly to the home viewer. (Note that TelePrompter is a trademark; TV stations across the country use other similar devices.)

When script pages are first distributed, many are incomplete. These mostly blank pages, which some stations call **slugs**, indicate what material will later

```
PAGE   EJ  SLUG              PRODUCTION  RR TAL  v  GRAPHICS      TRT  BACK WRT
=================================================================================
            NUMBER ONE! THE CHANNEL 4 NEWS
            AT 4:00-WEDNESDAY JUNE 22,1988
            JOHN BEARD AND LINDA ALVAREZ
            ***   TEASE  3:59:13   ***

4-01  F<A> HEAD:LAKERS DAY   2-SHOT/VO>>>   J-L
4-01+ F<B> HEAD:WASH. HEAT   VO
4-01+ F<C> HEAD:EXERCISE     2-SHOT/VO>>>
4-01+ F<D> HEAD:SUCCESS STOR VO/2-SHOT

4-02  ---- BOC CX ------- 1:30 ----------- ------------ ------------- 1:30        ---
4-03  <A>  4P OPEN "A"      SOT                                       :15

4-04       INTRO FRED       BOX/3-SHOT   LA       LAKERS DAY              SS
4-04+ <B>  LAKERS DAY       BOX/SOT/CU/3 RGGIN    LAKERS DAY     2:00
4-05       TOSS NR-POP      BOX/MON      JB       MURDER CHGS.           YB
4-05+ <C>  POP ARRAIGNMENT  NR/SOT/NR/MN ERKSN                   2:00
4-06  <D>  PEYER            BOX/VO/CU    JB       PEYER          :45    EH
4-07       RAMIREZ          BOX          LA       RAMIREZ        :20    DK
4-08  <A>  PALISADES BODY   BOX/VO/CU    LA  >>>  PALISADES MR   :45    WC
4-09  <B>  FREMONT SHOOTING BOX/VO/CU    JB       SCHOOL SHTNG   :45    WC
4-09A      NUKE TESTS       BOX/2-SHOT   JB                      :20    AC
4-10  <C>  NN-DEFENSE       BOX/SOT      LA       DEFENSE SCDL   2:00   AC
4-11  <D>  SKY-WASH HEAT    BOX/VO>>>>>> LA       HEAT WAVE      :30    RH
4-12  F<A> SKY-CHICAGO HEAT VO           LA                      :30    RH

4-13  F<B> TSE #1-SKATEBOARD 2-SHOT/VO   J-L                     :15
                             NEWS BUMP   ATR
4-14  ---- CX #1--------- 2:10 ----------- ------------ ------------- 2:10        ---

4-21       NEWSTALK         BOX/2-SHOT   JB       FOR TH FILES   :25

4-22       NEWSTALK         BOX          LA       FOR TH FILES   :25

4-23       INTRO WX         3-SHOT       J-L                     :15

4-24       WX               CU           CHRIS                   3:30
4-24+ F<C> WX CASS          CHY^VO
4-24+      NAT'L MAPS       KEY: WX MACH
4-24+ F<D> RIVER            KEY: VO
4-24+      NAT'L TEMPS      KEY: GRAPHIC          <X> *NAT'L TEMPS
4-24+ F<A> SO CAL TEMPS     KEY: VO
4-24+      FORECAST         CHY^FG/3-SHT          <V> *4P FORECAST

4-25  <B>  SKY-SKATEBOARDS  BOX/VO       LA       SKTBRD PRTST   :45    RH

4-26  F<C> TSE #2-SUCCESS ST 2-SHOT/VO   J-L                     :15
                             NEWS BUMP   ATR
4-27  ---- CX #2--------- 2:10 ----------- ------------ ------------- 2:10        ---

4-31       AIDS             BOX          LA       AIDS           :20    WC
4-32       INTRO GENDEL     3-SHOT       JB                      :15    FL
4-32+ <D>  DRS. & AIDS FEAR BOX/SOT/CU/3 GNDEL    GENDEL HLTH    3:00
4-33       SKIN CANCER      BOX          JB       CANCER TRTMN   :20    SK
4-34  <A>  SUCCESS STORY    CU/SOT       LA                      2:00   WC

4-35  F<B> TSE #3-CARL EXTOR 2-SHOT/VO                           :30
                             NEWS BUMP   ATR
```

FIGURE 11.1 Portion of a rundown sheet for KNBC's 4:00 o'clock news: a blueprint of program material. Used by permission of KNBC.

appear there. As the copy for each story is completed, pages are distributed to replace the slugs. Seldom are all script pages completed by airtime. Professional directors are used to final pages (or replacement pages) being distributed during the broadcast itself.

THE DIRECTOR AT WORK

In directing on-the-air news programs, the role of directors is more integrative than creative. They must combine a dozen disparate elements into a crisply cohesive whole. Because these visual and aural elements appear so quickly, offering little opportunity for leisurely examination, and because each element directly affects others, news directors' concentration must be intense and coordination keen. Directors' prebroadcast preparation must be so meticulous that it will eliminate most of the possibilities for error.

News directors also work in the field, covering special events.

In the Field

A psychotic robber has slain half a dozen customers and holds others as hostages in a neighborhood market. A local television station has sent a director, camera crew, and reporter to cover the grisly event. Police have surrounded the market. From time to time, a police lieutenant calls through a bullhorn. The fear-crazed robber does not answer.

Upon arriving at the scene with a remote crew, the director's first function is to assess: Where is the drama? How can that drama best be transmitted to an audience? What subject matter will best convey the sense of fear that has descended on the scene?

A second responsibility falls on the director's shoulders: staying out of the way of police so that they can do their job. After a quick appraisal, the director places one camera on top of the remote truck, the high angle providing excellent orientation. From this vantage point, viewers will be able to see the market, police cars, and police personnel, plus curious and frightened bystanders. Cameras at ground level can focus on the reporter and provide close angles on the market, police lieutenant, and bystanders. If in an exchange of gunfire police officers are killed or wounded, the director has another responsibility: treating the mayhem with taste, avoiding gruesome shots of blood and dead bodies that would offend home viewers.

After the director and the reporter have created a visual and word picture of what transpired earlier, a long wait begins. The director suggests interviews with witnesses. Such interviews personalize an event; they reduce major catastrophes such as fires and floods to the level of people with whom spectators can identify. Most TV stations and networks caution reporters against naming victims until after next-of-kin have been notified.

The reporter talks to an elderly woman who was actually in the market at the time of the robbery. She's extremely nervous and upset, almost inarticulate, trying to describe events but breaking down and running off in the middle of the interview. Next the reporter interviews a teenage boy who arrived on the scene minutes after the carnage took place. He saw little or nothing but he's bright, full of words, eager to expand his moment of importance. Finally, the reporter grabs a brief moment with the police lieutenant, who is reluctant to talk; he merely states his concern for the hostages and that he has radioed headquarters to send a SWAT team.

All three of the interviews proclaim immediacy. While the teenage boy had little factual information to contribute, his words still provided viewers with a sense of being there. Weak as his interview was, it still proved superior to no interview at all.

The interview with the elderly woman was the most effective of the three. It contained immediacy and veracity of such crushing emotional force that viewers would be shaken, would feel the terror and panic that such an event could evoke. Her words communicated far more than facts; they created emotional involvement, transforming viewers into participants.

The interview with the police lieutenant provided a note of authority. It gave viewers a sense of being on the inside, of having special knowledge that even on-scene spectators did not have.

Experienced directors try to stage on-site interviews with as much location atmosphere in the background as possible, thus visually enriching the interview. On occasion, overly zealous directors manufacture background color, restaging events for their cameras, trying to add richness when the interview or location seems lacking. Most responsible executives prohibit such practices.

Most field interviews are staged or recorded under circumstances less calamitous than the market massacre. These day-to-day interviews usually have no director assigned; the cost factor would become prohibitive. Because single-camera interviews are a necessary part of today's news spectrum, we will discuss them briefly here, despite the fact that they rarely touch on a director's world. You may find yourself working as a camera operator, as talent, or as a field producer—and such essentially directorial knowledge may prove helpful.

When videotaping interviews with a single camera, it is essential to provide some degree of editorial flexibility. In any interview, certain words, phrases, or speeches are unimportant or confusing; they get in the way of more important thoughts. It's necessary to be able to delete such clutter, but this is possible only if someone has anticipated the editor's needs.

The primary angle, of course, favors the person being interviewed. A second camera angle, usually favoring the interviewer, provides the editor with something to cut to, thereby eliminating unwanted material. Accordingly, after an interview has concluded, camera operators usually shoot a reverse angle: an over-shoulder shot favoring the interviewer. If the person interviewed cannot (or will not) remain for the reverse angle shot, the camera operator generally shoots a closeup of the interviewer alone, faking it, reacting appropriately to the thoughts expressed earlier. This gives the editor something to cut to in order to clean up or shorten the interview. There is nothing unethical about manufacturing such a shot because it does not change either the message or its implications.

When speeches by newsworthy figures are covered, the need for editorial flexibility remains; newscasts usually have time only for a few sentences. Therefore, experienced camera crews shoot some relevant cutaway material, frequently shots of the audience listening or applauding or of the press photographing the speaker. Sometimes an extremely wide shot serves the purpose; from such a distance we cannot distinguish lip movements, so the editor may cheat and place any necessary words into the speaker's mouth.

In the Studio

While specific practices vary from station to station, the patterns of directorial preparation remain much the same. Usually, directors receive rundown sheets two or three hours prior to the newscast (see Figure 11.1). The rundown is an early glimpse of newscast elements, a blueprint that contains an incredible amount of information. The KNBC rundown, for instance, indicates:

1. Page number for each story
2. Whether or not the story uses videotape plus the video channel (A, B, C, or D)
3. Subject matter
4. Production elements (for example, Box/2-shot)
5. The talent
6. The graphics plus the graphics channel (X or Y)
7. Approximate running time of each segment
8. Each story's writer

Once directors have received rundown sheets, they may plan their camera coverage, based on knowledge of the news set and where each member of the news team sits. Most jot down their camera shots on the rundown and later transfer those markings to their scripts.

As you discovered in Chapter 9, directors usually cross their cameras. Thus, camera 1 (on the left) photographs the performer on the right—and vice versa. Camera 2, in the center, usually holds a two-shot. When other personalities enter the set—weather or sports personnel, for instance—camera 2 will widen to accommodate them and cameras 1 and 3 will provide closer angles.

Let's assume that the weather forecaster (we'll call him Mario) enters from camera right and sits beside the news team. Camera 2 has widened to a three-shot and camera 1 will provide a close up of Mario. Perhaps the director will ask camera 3 to frame a two-shot left (the two news regulars) and then cut between a two-shot of the news team and a single shot of Mario. (See Figure 11.2.)

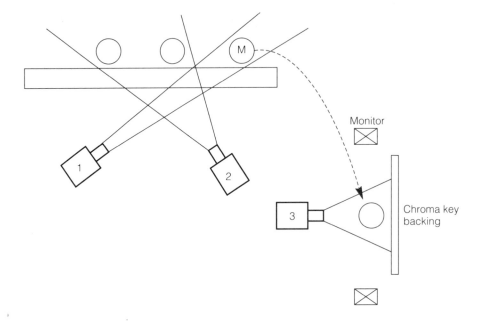

FIGURE 11.2   Cameras 1 and 2 cover the interchange between Mario (M) and the news
team before he moves to the weather area. Note TV monitors just off camera so that
Mario can see the graphics keyed behind him.

One additional factor may complicate the coverage. At the end of a short
interchange with the news team, Mario will move to his weather area. The so-
called weather area is nothing more than a wall painted to match the station's
chroma key specifications. Many stations use blue; KNBC uses green, primarily
because so many performers have blue in their wardrobes or in their eyes. If you
have taken production courses, then you will understand that with chroma key
the background color may be replaced with another picture, perhaps provided by
another camera, by videotape or graphics.

In the weather portion of news programs, graphics such as weather maps
usually are keyed into the area behind the personality. But the weather personality
cannot see the map on the wall behind him or her. The wall is bare. Therefore,
TV **monitors** are usually placed off camera-right and -left, and the weatherperson
orients himself or herself to the map by viewing the monitors. The director
generally changes the graphics behind the weatherperson from word cues. If the
weatherperson is consuming too much time, the director speeds things along by
changing the graphics so that the personality must hurry to catch up.

Let's assume that the weather area is located at extreme camera-right. If
camera 3 will photograph Mario in the weather area, then cameras 1 and 2 must

cover the preceding action. The director must allow a few seconds for camera 3 to move into the new area. Therefore, the director may plan to cover any chitchat between Mario and the news team by using only cameras 1 and 2. Because camera 3 cannot *pan* Mario into the area (the wall is bare, remember), the director usually cuts to the graphic keyed over the bare wall and lets Mario walk into it.

When script pages begin to arrive, the director transfers markings from the rundown sheet to the script. As indicated earlier, some of the pages are slugs, to be replaced later when the copy has been written. Slugs usually contain the page number and subject and running time of the story. Although directors seldom see the completed script prior to airtime, they're usually not concerned: the rundown has already provided most of the necessary information—and they have plotted all camera moves.

In addition to camera shots and camera instructions, the director also marks **roll cues** in the script. Years ago, when most news stories were recorded on film, directors needed five seconds for their roll cues. That is, they had to roll the film five seconds before a commentator finished speaking so that the film would be up to speed by then and the director could cut to it. Directors used to rehearse all news copy with commentators, timing their speech with a stopwatch to find the exact word on which to roll film. Today, with videotape players capable of attaining the necessary speed almost instantly, no such rehearsal is necessary. The director may roll tape and then cut to it an instant later.

Editors routinely place **digital leader** at the head of each videotape story. Digital leader displays a succession of numbers indicating how many seconds remain before the picture appears. We have all seen digital leader on television when the director mistakenly cut to a videotape too early.

For news programs, directors ask that videotape be stopped at the two-second mark for material containing sound and at the first frame of picture for a story without sound. The tape is held at those marks until the director cues the technical director to roll.

Today most digital leader omits numbers 1 and 2, replacing them with black to protect the director. Thus, if a director or technical director inadvertently cuts to the tape early, viewers will see no numbers, just a brief flash of black.

In marking scripts, directors gauge their roll cues either from experience (familiarity with the dialog speed of their news team) or they time the words with a stopwatch. They then circle the roll cues in their scripts.

In the Control Room

Control rooms are impressive. Even for television veterans, they stimulate the senses, inspire awe, create a sense of excitement waiting to happen.

In KNBC's news control room, the director faces a bank of fifty-six monitors, fourteen screens across, four high. Four of these screens represent feeds from the four studio cameras. Cameras 1, 2, and 3 cover the news team; camera 4 usually photographs graphics (for example, still photos).

The four videotape channels are labeled Alpha, Baker, Charlie, and Delta (A, B, C, and D), deliberately avoiding numbers to prevent confusion with camera numbers. Graphics come in from the graphics room (which processes the boxes) on channels X and Y. Other screens represent remote channels, on-air programming, and preview monitors.

To the director's right sits the technical director. To the left sits the associate director (AD). The audio booth is off to the far left, separated from control room clamor by glass doors. The producer sits in another glass booth to the rear. In the studio, a floor manager provides the connection between director and talent. The floor manager throws cues to the news team, provides them with timings, and passes along comments from the director and associate director (AD) when the news team is off camera.

In recent years the AD has taken some of the load from the director's shoulders. Today, he or she oversees the TelePrompter operation, tells switching control when to roll commercials, times the tape segments, counts them down and gives the talent timing cues through the stage manager, gives the rundown to operators in the graphics room, cues a reporter on location, and is responsible for timing the show. A tremendous load! Most of the characteristics that I have suggested for news directors are equally necessary for an effective AD.

Videotape   Sometimes videotape material is silent with the talent providing commentary. Sometimes the videotape contains sound: music, effects, or speech. In most news scripts, sound on tape is abbreviated **SOT**. (See Figure 11.3.) Sometimes a videotape story contains sound from beginning to end. Sometimes the sound appears at a point within the story and continues to its conclusion. And sometimes sound appears, continues for a brief period and then cuts out, the videotape continuing silently as the news team resumes its commentary.

For the director and AD, such intermittent sound elements demand the utmost in concentration and coordination. If sound appears at eighteen seconds into a story, for example, the AD or director must start a timer when the picture first appears and must alert the sound engineer to be ready for the SOT eighteen seconds later. (Directors start their own timers when they have no AD—or for protection, like the man who wears a belt *and* suspenders.) A sharp audio engineer won't need such warning; the script states exactly when the SOT appears. But meticulous directors leave nothing to chance.

The associate director recites a countdown to the floor manager so that the latter can signal the talent. The talent, in turn, must time its speech so that he or she finishes before the SOT occurs. Such a recitation by the AD also prepares the audio engineer to open his or her **pot** (sound channel). When such sound elements appear many times in a single program (which is usually the pattern), they can tax the crew's coordination.

Experienced directors always punch a timer (or stopwatch) when tape appears, whether the tape contains sound or not, whether they have an associate

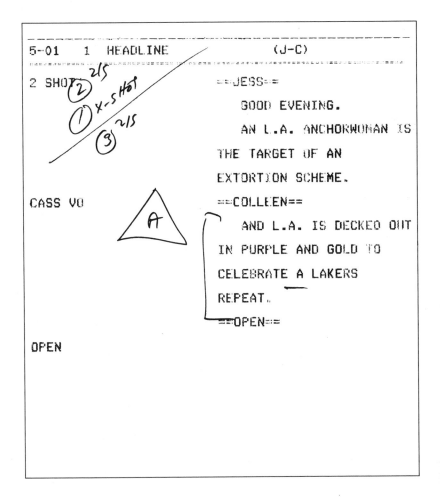

FIGURE 11.3   Portions of director Gene Leong's script for KNBC's 5:00 o'clock news program reveal various directorial cues and markings. Used by permission of KNBC.

sitting beside them or not. The script always indicates **total running time (TRT)**, so the director is forewarned. He or she must cut away from videotape before it runs out or look like an amateur. The only way that directors can *know* when the tape will run out is if they have started a stopwatch.

The script (Figure 11.3) provides SOT **out cues**, sometimes called **end cues**. They state the ending phrase in an SOT segment, usually the last two or three

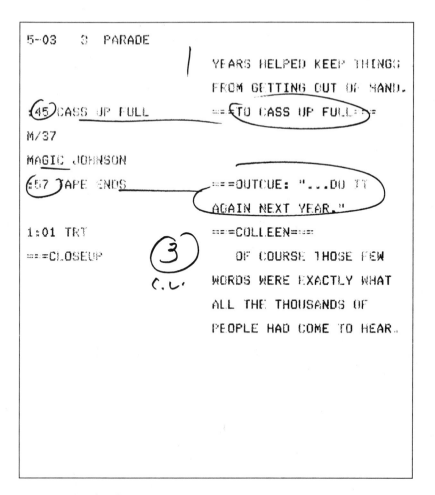

FIGURE 11.3   (*continued*)

words, so that the director will know exactly when to cut away from the videotape or to cue another sound source, such as studio talent.

Timing   Although the timing of many live programs often rests on the director's shoulders, such is not the case with news. Either the associate director or, in rare cases, the producer must get the show off the air on time. ("Entertainment Tonight" has a script supervisor who is responsible for timing.) He or she has backtimed the script as carefully as possible, considering the late arrival of some pages. Television inherits the term **back timing** from radio. When ADs back time a script, they work with known and unknown factors. They know the exact running time of each videotape segment, of opening and closing titles, of com-

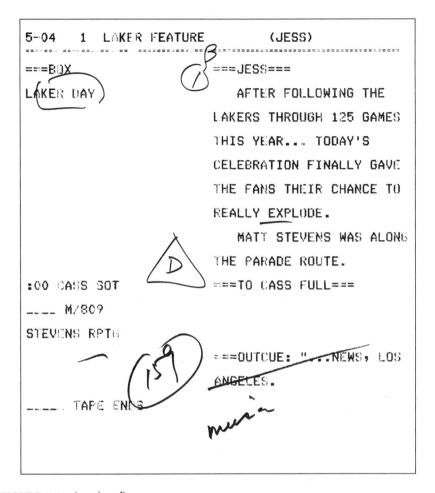

FIGURE 11.3   (*continued*)

mercials, and, when scheduled, the time consumed by a station break. But the time for commentary and ad libs by performers must be estimated.

Starting at the end and working forward, ADs calculate exactly when the director must hit closing credits. Prior to that, and still working forward, they estimate as closely as possible the time at which each segment of the newscast must finish. They mark these approximate timings in their scripts. Knowing that every show contains some elasticity, they cue performers (through the floor manager) to stretch or tighten their commentary, so that show timings will correspond to the back timings in their scripts.

Usually the last element in a news program contains some flexibility, expanding or contracting to accommodate the clock. At KHJ-TV, a sports update usually provides such a buffer. When the program must be stretched, the talent asks questions of the sports authority. Other shows contain long and short versions of closing credits, which provide further time control.

THE NEWS DIRECTOR—A COMPOSITE PORTRAIT

One of the questions aspiring directors most often ask about professionals is "How did he or she get started?" Such interest is understandable; these hopefuls are looking for clues that may help them find their own professional toeholds.

With this interest in mind, I will try to provide insights by examining the careers of four news professionals: two directors for local TV stations, one for a network, and one for a major news–entertainment program.

Gene Leong directs the 5:00 o'clock and the 11:00 o'clock news for KNBC in Los Angeles. His interest in entertainment began at San Francisco State University where he studied television and film courses. During a stint in the army's psychological warfare division, he had the opportunity to take additional television courses. After his army service, Gene obtained a job at a small San Francisco film company, American Zoetrope (Francis Ford Coppola's production company), where he functioned as a gofer, running errands, doing odd jobs, studying the operation, asking questions, absorbing as many details of film production as he could cram into his head.

With this professional experience behind him, he was able to land a job at KGO-TV in San Francisco, first as a stage manager and then as an AD. Later, when staff directors took their summer vacations, Gene filled in, directing a wide variety of programming, from cooking shows to children's programs.

He moved to KQED, a PBS station in San Francisco, as a producer-director and eventually to Los Angeles where, because of the larger market, he felt he would have greater opportunity for advancement. Gene worked as free-lance director in sports and even directed a few **sitcoms**. Eventually he found his way to KNBC, where he now concentrates on local news shows. Occasionally he has the opportunity to direct news-oriented programming for the network.

Friendly, low-key, and extremely bright, Gene Leong is meticulous about his work. He double- and triple-checks every aspect of his newscast, knowing that in such a time-intensive operation it's easy for mistakes to happen. When I asked him the requisites for a first-rate news director, he replied, "the ability to think fast—and to ad lib when necessary."

Chris Stegner directs the 9:00 P.M. news at KHJ-TV in Los Angeles. A handsome and articulate man in his middle thirties, he began his orientation toward show business by majoring in television and film at the University of Wisconsin. Soon after graduation, he managed to find a job at a small UHF station in the Chicago area. Later, with professional experience under his belt, Chris came to Los Angeles, where he worked primarily as a sports director. He points out that the requirements for directing sports are remarkably similar to those required for news—foremost, of course, the ability to think on your feet. In time Chris Stegner made the transition to news. Like Gene Leong, his control room demeanor is low key: his commands are quiet, crisp, precise.

Jay Roper directs news for the NBC network. He handles feature stories and newsbreaks that originate on the West Coast. Jay studied telecommunications at UCLA. While he was a student, he got a job as a page at NBC. A page works on the guest relations staff, showing visitors around, handling audiences for quiz or game shows, and serving the needs of various network departments. It's an entry-level job that affords the opportunity to meet people in a cross section of network activities. Many of NBC's top executives once started as pages.

Jay makes the point that it's important for students to get jobs in the entertainment industry before they graduate. Thus, after leaving college, they will already have a foot in the door.

He worked as a page for two years and then was offered a job in NBC's operations department, where he suffered through the drudgery of typing routine sheets. Then he worked in NBC's license renewal department. But Jay was unhappy. He felt he belonged in production. He kept hounding production executives, but the prospects seemed hopeless. Eventually he was successful in landing a job as stage manager for the local station (then KRCA). Jay worked for five years as stage manager and assistant director before being promoted to director.

A year or two later, President Nixon was slated to have his first West Coast press conference. The network news director was on vacation. An NBC exec asked Jay if he wanted to direct the program. Jay picked himself off the floor and replied that he'd like to. He confided in an interview that the project "scared hell out of me!" In spite of his fears, Jay directed the Nixon news conference—and it went beautifully. The next morning NBC asked him if he'd like to direct for the network full time.

Jay Roper has been directing network news for twelve years. When I asked him what qualities are necessary in a director, he replied, "All of us are frustrated actors. If we can't be in front of the camera, we do the next-best thing!"

Wayne Parsons is the Emmy-winning technical director of "Entertainment Tonight," a syndicated half-hour news-type program that airs six days per week. Wayne is included here because he directs the program from time to time, when the regular director is on vacation or busy on other projects.

While a student at Roanoke College in Virginia, Wayne got a job at the local TV station (WRFT), "pushing a camera." He worked ten hours a week for the minimum wage. Because of his keen interest in television, Wayne hung around the station a lot, picking up information. He learned switching and eventually became the best switcher on the staff. After graduation, he started working full time. A few months later the management moved him to the night shift, a promotion because that was where the important programming was.

Wayne stayed at WRFT three and a half years and then worked in advertising. But he missed the excitement, the immediacy of live television. So he wrote a letter to every program director in the state of Virginia saying, "I heard about a recent opening on your staff for a producer-director, and I would like to apply for the job." Wayne received one answer, was interviewed by a station executive, and got the job! His advice to aspirants: "In order to get lucky, you have to do something that will bring you luck, that will *make* your luck. You have to take risks." He echoes Jay Roper's feelings that students must get started in the business *before* they graduate.

Wayne Parsons feels that it's important for news directors to understand TV technology. Libraries are filled with books describing the electronic wonders that fill control rooms. Aspiring directors should talk to engineers, read manuals, write to equipment manufacturers. If students have a background of some mathematics and a little physics, they can understand 70 percent of the technical material.

Wayne tells the story of Marty Passetta, a top director of musical specials and awards shows, interviewing a young man for the position of assistant director on one of the Academy Award programs. Passetta asked him casually, "Do you know what a time bit user code is?" The hopeful shook his head in bewilderment; he had never heard the term. Alas, Passetta turned elsewhere for his AD. He felt that the time bit user code is an essential technological ingredient in staging musical numbers. Without that specialized knowledge, an AD would be handicapped.

Wayne Parson feels that editing is probably the best avenue to directing. The best directors need keen perception and a heightened awareness, an ability to retain images in their minds.

CHAPTER HIGHLIGHTS

— The most effective news programs use as many visual elements as possible. For national or international stories, network affiliates receive videotape coverage from a network satellite. When major events occur in their territory, stations also contribute stories via the satellite.

— Independent stations rely on INDX (Independent Exchange) for videotape coverage of distant news events. Stations receive approximately forty minutes of material each day, but use only about 10 percent.

— The bulk of local stations' news coverage is local or regional, using videotape borrowed from nearby stations or from their own three-member ENG units. These field units either transmit stories directly to the TV station or record material for later use. If no videotape is available, stations use still photographs as inserts, sharing the frame with talent.

— A production meeting determines the final news sequence, subject to change by later-breaking stories. This sequence is distributed in the form of a rundown sheet describing each story and noting the talent and visual elements involved. Stories are grouped into sections, divided by commercial or station breaks.

— The director's work divides into two spheres: the field and the studio.

— When a newsbreak occurs, the field director assesses the nature of the event and how best to transmit its drama to an audience. Interviews often personalize an event. Because most interviews are staged with a single camera, directors usually photograph a second angle featuring the reporter. The second angle allows the editor to eliminate unnecessary verbiage.

— News directors usually begin their preparation with the rundown sheet. From the information contained therein, they are able to mark most camera angles.

— When script pages arrive, the director transfers camera markings to the script and adds roll cues. When videotape is called for, it must be "rolled" by the director a moment before the picture appears. A digital leader at the head of each story displays a succession of numbers indicating how many seconds remain before the picture appears. Most directors set the leader at two seconds for sound and on the first frame for silent footage, which allows the tape to attain speed before the picture appears.

— Some videotapes contain sound elements. Sound on tape (or SOT) sometimes begins a story and sometimes occurs in the middle of a story. Associate directors always start a timer (or stopwatch) when videotape appears. Because scripts always state the exact number of seconds before SOT occurs, the timer enables the director to cue performers on the set. Scripts also contain end cues, or out cues, stating the last two or three words in an SOT element. When directors hear these words, they automatically cue the upcoming sound source.

PROJECTS FOR ASPIRING DIRECTORS

1. Write a letter to the executive producer of a local news program (or to the station's program director) stating that you are a directing student and would like to watch the airing of a newscast. Before writing, call the station and find out the name and exact title of the person you're addressing. Follow up your letter with a phone call.

2. If you strike out with one station, try another. When you attend a news program, introduce yourself to the director, if at all possible. If you are a student, invite him or her to speak to your class. If that is impossible, arrange to interview him or her and write a paper describing in as much detail as possible the director's career, its problems and satisfactions, and advice for newcomers.

3. Gain insight into the structure of a news program by creating a rundown sheet. If you have a VCR, record a one-hour news program that you admire. Prepare a blank rundown sheet similar to the one in Figure 11.1. Now study your recorded news program, stopping the VCR when necessary to fill in the blanks. Do not skip any of the categories except writer. When you have finished, ask yourself: Could a director block a news program solely from the information you have inserted?

4. On the same news program or another, time a videotape story that contains both silent footage and sound. With a stopwatch (or a watch with a second hand) time the silent and SOT elements within the news story. If there is commentary over the silent portion, note how well the commentator's words fill the soundless gap.

CHAPTER

12

"AND NOW A WORD FROM OUR SPONSOR": COMMERCIALS

Television commercials reach out and touch almost everyone. Their slogans and jingles have become part of the American vernacular. They are repeated by youngsters barely able to walk, who become aware of such social calamities as ring around the collar well before kindergarten.

So stunning are the production values in some commercials, so meticulous is their preparation, so powerful is their motivating force that they have become "an American art form . . . ahead of most TV programs in being in tune with the United States."[1] As shaped by some of the country's most creative directors, they provide viewers with such diverse entertainment elements as humor, music, spectacle, drama, famous celebrities, and extraordinary visual style. A case in point: the Pepsi-Cola musical commercial with Grammy winner Michael Jackson, which cost approximately $2 million, a sum once considered high for a feature motion picture.

---

[1]Albert Book and Norman Carey, *The Television Commercial: Creativity and Craftsmanship* (Chicago: Crain Books, 1970).

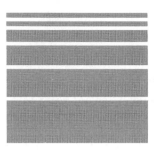

I don't mind the commercials. It's just the programs I can't stand.

TV critic John Crosby[2]

Presented with script and storyboard, directors must express their message creatively, with optimum impact and sales effectiveness, pleasing themselves, advertising agency, and client. The director's challenge is compounded by the need to cram a message into thirty or sixty impossibly short seconds. Of necessity, they must be high-density seconds; directors must distill the actions and emotions of performers. Each gesture must count. Each word must contribute. Each facial expression must advance the premise.

Because many commercials demonstrate the newest styles, technologies, and cinema techniques, they often become a stimulus for directors in other entertainment arenas. On occasion, they propel their own directors into more prestigious projects such as feature films. (I do not mean to denigrate the status of commercials; successful commercial directors receive respect from their industry peers as well as substantial financial rewards.)

---

[2]Quoted by Robert Hilliard, *Writing for Television and Radio*, 4th ed. (Belmont, Calif.: Wadsworth, 1984), p. 41.

The world of TV commercials offers its directors both opportunity and frustration: the opportunity to achieve near perfection in terms of realizing a script's promise and frustration in terms of limited creative control. A storyboard, client, or agency executive, for the most part, will hold directors to preconceived patterns and parameters.

This chapter will discuss:

— WHAT DO COMMERCIALS SELL?: more than just a product

— THE PRODUCTION PATTERN: how directors are hired, how they function in preparation, production, and postproduction, the director–agency relationship, and local commercial patterns

— PRODUCTION TECHNIQUES: the art of cramming a big message into a small time frame

WHAT DO COMMERCIALS SELL?

To the casual viewer there is a single, obvious answer to the above question. Commercials sell a product. A narrator describes its advantages; we see it displayed handsomely; a playlet dramatizes how it can simplify our chores or enrich our lives.

Directors of commercials quickly discover that selling a product usually is secondary to selling a life-style, showing viewers how to make themselves enviable. John Berger, in his delightful *Ways of Seeing*, calls the state of being envied "glamour."[3] According to Berger, advertising provides viewers with an image of themselves as they might be, transformed by the product, thereby becoming enviable. Thus, commercials sell but one thing: the happiness that derives from being envied.

At first glance, Berger's assertion seems simplistic. Yes, it is certainly true for blue jeans that make us appear sexier. It is true for shampoos that make our hair lustrous and manageable. And perhaps for detergents that will give us whiter, brighter sheets, thus making us the envy of neighbors. But what about such products as underarm deodorants or headache remedies or medications for hemorrhoidal relief?

---

[3]John Berger, *Ways of Seeing* (New York: Penguin Books, 1977), p. 132.

If we are to believe commercials (and most of us do!), such remedies will make us feel like our old selves again, jovial, good-natured, ready to tackle the world. If the commercials don't make those statements in so many words, the attitudes of their participants do—suddenly released from the depression of congested sinuses, smiling, inhaling deeply, able now to win the love of beautiful men or women or to battle the dragons of Wall Street. In short: enviable.

In view of this attitude, directors in preparing TV commercials must give as much thought to the ambience of the world they are creating as to the specific needs described in script or storyboard. The ambience—created by camera, wardrobe, lighting, photography, music, choice of performer, acting, set, and set dressing—provides an upbeat dream world filled with delightful fantasy characters who have a single purpose: to be envied. By buying the product, the TV viewer can enter their luminous world and win the envy of friends and neighbors.

Soft drink commercials frequently depict a joyous, fun-filled never-never land in which attractive teenagers rollick and flirt, kiss, play ball, dance, wash cars, splash one another in the surf, or picnic on sun-drenched beaches—creating a life-style that viewers can achieve for fifty cents. Coffee commercials sometimes sell taste, sometimes price, sometimes quality. More often they sell ambience: a man and woman alone in a mountain cabin, snowed in, savoring the pleasure of coffee shared in love.

Despite the fact that they play in a few seconds rather than in minutes or hours, playlets such as the mountain cabin coffee commercial are literally drama and use principles discussed in earlier chapters on staging the actor and staging the camera. One of the three basic functions of camera is to create mood or atmosphere, achieved through lighting, filters, choice of angle, and camera movement. A low angle will make a truck appear strong and powerful in the same way it brings those dominant characteristics to an actor. A wide angle lens will make a new model car appear longer and sleeker in the same way it makes a modest soap opera corridor appear endless. Fog filters have become almost a cliché in commercials, creating a luminous, dreamlike quality that in the audience's minds ultimately becomes associated with the product itself.

For the director, achieving any kind of ambience, positive or negative, requires creative as well as technological know-how. Skilled cinematographers create daily miracles in adding sparkle to foods, beauty to fashion models, and glamour to kitchens. Use of gels in lights or fog filters on cameras can transform the mundane into the mythic. But directors cannot always rely on the expertise of others. From analysis of the script and discussions with agency personnel, together with their own personal "vision" of what the commercial must be, the director coordinates all of the production elements to create a specific mood or atmosphere. Sometimes such an atmosphere is defined in the script. Sometimes not. But a director's ability to find the specific ambience that will make characters glamorous (enviable) and their life-style desirable lies at the heart of his or her commercial success.

THE PRODUCTION PATTERN

Television commercials are produced and directed on two levels: national and local. National commercials are usually first aired on network television. The most lavish cost hundreds of thousands of dollars and include colorful locations, extravagant visual or graphic effects, accomplished actors, dancers, and original music or jingles. They are capsulized versions of a Broadway musical or Hollywood melodrama with a single major difference: Their goal is to sell a product. If it delights audiences but fails in motivating them to buy, the commercial is a flop.

On the local level, commercials usually are filmed or taped by local advertising agencies or by TV station staff directors. Typically, such commercials advertise car dealers, local markets, tire dealers, department stores, and loan companies. Because audiences are necessarily smaller on a local level, dollars expended by advertisers are considerably fewer. The production scope of local commercials is thus considerably more modest than for national ones. Directors must work with relatively simple scripts and within relatively tight production schedules.

The National Commercial

Major commercial campaigns are prepared by advertising agencies with the consent and approval of their clients. Before a script or storyboard reaches a director's hands, its concepts have been pondered by agency and advertiser, evaluated and reevaluated, written and rewritten, tested and retested for style (do they properly reflect the advertiser's image?), for sales impact, and for their ability to capture and hold a viewer.

Most directors of national commercials either work for a production house or for themselves; sometimes a director *is* the production house. Usually when agencies solicit bids from production houses, it is because those houses provide the services of directors in whom they are interested.

The Sample Reel When advertising agencies consider which director to hire for a specific campaign, they frequently study sample reels submitted by various directors, assessing their past work, searching for directorial styles and backgrounds that seem appropriate for the upcoming campaign. As in other fields, directors tend to be typecast. Those who are experienced in food commercials tend to be hired for other food commercials. Those experienced in high fashion will inevitably be offered other high fashion commercials. Agencies take specialization so much for granted that they often joke about it. Typical is the story in which a director is called in for an interview and the producer asks, "Have you ever worked on breakfast cereals?" When the director answers "Yes," the producer asks, "Hot or cold?"

Because hundreds of thousands of dollars frequently hang in the balance—as well as the security of a valuable advertising account—agencies often use directors who have worked for them before, whose style, creativity, and dependability are known quantities.

Aspiring commercial directors inevitably ask how it is possible to assemble a sample reel without having directed commercials. This, of course, represents the classic chicken-and-egg paradox. You cannot get a job until you have experience. You cannot gain experience without getting a job. Catch-22.

Getting Started Seldom if ever do commercial directors begin their professional careers as directors. Usually they gain some measure of success in a related field and then move into directing. According to Tony Asher, who has been a producer for both advertising agencies and production companies, directors of national commercials usually have backgrounds in art direction, cinematography (or still photography), and editing. (Asher has worked on the Mattel Toys account for Ogilvy & Mather and for Paisley Productions.) Because word-oriented people such as writers tend to rise in an advertising agency and dominate its hierarchy, the visually oriented art directors often veer into other realms, specifically directing. Broadcast art directors (as opposed to print art directors) are basically "picture people" and therefore able to translate a commercial message into effective visual language.

Many directors are actually director-cinematographers. Many began as cinematographers and then, because of their unique abilities, expanded their craft to include directing. Actually, says Asher:

> A sizable number of commercials don't require the direction of any talent [actors]. They're made up of animated stills—close shots of food, or cars driving along a road while the voice-over describes the product. These directors don't have to say to actors: emote. Most of them are essentially skilled still photographers. They don't direct talent.They're storytellers in the visual language. They're painters.

Like cinematographers, film or videotape editors develop a keen visual sense. They come to understand the theory and practical application of fitting angles together to build effective dramatic sequences. As Asher says,

> Editors get such a good sense of how and why a film is working. Over and over again they find themselves in that situation where they simply can't make a sequence work and they complain: "if only the director had shot seven more frames of this or a reverse angle of that." They really understand why something works or what it would take to make it better.

Many, perhaps most, editors aspire to become directors. So do many, many members of the production world: writers, actors, producers, cinematographers.

Asher recalls a T-shirt he saw at a guild screening. It depicted Jojo the talking dog in conference with his agent. The dog was complaining, "But what I really want to do is direct!"

Stu Hagmann, one of the most successful commercial directors in the business, came from academia. He made his first short film with a $500 grant plus $1,000 of his own money, while he was waiting to join the teaching staff at Northwestern University. The film won first prize at the Venice Film Festival, first prize at Edinburgh, and first prize at the UNICEF Film Festival, as well as the Producers Guild Award for that year. He might still have gone on to teach but the Producers award included six months of employment at Universal—the award launched his career. Since then Hagmann has won virtually every award possible for a commercial director.

Hiring a Director Just as most people get several estimates before hiring someone to paint their homes, so do advertising agencies secure estimates or bids on what a contemplated set of commercials will cost. (Usually several commercials are filmed at one time; such a procedure is more cost effective.) Before agencies make any decision concerning which director to use, they assess the budgets submitted by producers at the various production companies.

Let's assume that you are a director associated with a major production house. Before your producer submits a bid, she automatically checks directly with you, asking that you read the script and study the storyboard. She asks for your estimate of how many hours you feel you would need to shoot the material, if you foresee any unusual production problems, and so forth. Based on your answers and the production elements called for in the script, the producer then submits an estimate to the advertising agency.

If the submitted budget figure is within the ball park, the agency then obtains the client's approval. Traditionally, the client first screams that the budget is far too high, but after some haggling eventually approves it. Once the client has approved both budget and director, the agency invites you to its offices for a first meeting with the agency producer. From this moment forth, you and the agency producer become a team, working closely through preparation and production. It is imperative that you both see eye to eye on the look, feel, and style of the commercial.

During the weeks that follow, you find it impossible to get the commercial out of your mind. You examine its content in the light of your own professional experience. You study the implications of the storyboard.

The Storyboard A storyboard is simply a series of drawings (frames) prepared by the agency's art department, illustrating various key moments in the commercial: camera angles, actions by performers, how the product itself should be displayed. Many believe that storyboards came into being because clients lacked imagination, unable to visualize how a commercial would look merely from reading a script.

> A storyboard has more value than making the commercial clear to the client. It is a blueprint for all the pertinent artists, artisans and mechanics to see, if not follow. This is the most articulate manner in which the advertising agency can present its thoughts to all concerned, both in the agency and outside. The board is a guide, a direction.[4]

During preparation you probably discover more effective ways of shooting certain elements than have been presented in the storyboard. When you have ideas for changes—no matter how slight—you must discuss them with the agency producer. Chances are, the producer will initially resist them. After all, the client has approved the storyboard as is. Changes can be disruptive. And yet you have been hired because of your creativity and expertise. Your reaction is a fresh one, providing agency executives with a valuable new perspective. If you are persuasive, the producer will take your suggested changes back to his or her creative staff—and to the client—for approval. Once (and if) they are approved, you should try to get such changes incorporated into the storyboard, so that everyone will clearly understand your intent and there will be no recriminations later.

Sometimes the producer will suggest that you shoot the questioned element (an action, line of dialog, piece of business) two ways: as presented originally in the storyboard and as you would like to change it. Such duplication is costly in terms of time, patience, and money. Yet it happens again and again. Because commercials must please executives with a variety of needs and pressures, it is not always possible to satisfy everyone with a single stroke. That is why directors sometimes print multiple takes—in order to keep everyone happy, accommodating different opinions regarding line readings or specific actions.

Although many directors rail against storyboards, claiming that they limit creativity, that they reduce the directional function to little more than "painting by the numbers," storyboards do provide subtle insights that might be difficult to obtain otherwise. Ad agency creators usually are bright. When you study their concepts as defined in storyboard drawings, you must try to recognize their goals as well as the reasons for stagings or camera angles that appear less than optimum.

> If the lawnmower is drawn way back in the frame with lots of green grass in the foreground, you might say that there is insufficient identification of the product and you would be right. However, in this case, the agency/client wants to emphasize the successful lawn, and this is a compromise.

---

[4]Ben Gradus, *Directing: The Television Commercial* (New York: Hastings House, 1981), p. 55. All quotations appearing in this chapter from *Directing*, copyright 1981 by Ben Gradus, are used with permission of Hastings House, Publishers, Inc.

The storyboard is quite useful in creating a communication with the agency to determine what the sketches mean. Understand each frame with all its nuances—our business is nuances. The nuance is the director's stock in trade; that's what he's hired for, since the storyboard has the facts.[5]

Producer Tony Asher describes two primary criteria that determine a director's latitude in relation to a storyboard: the director's reputation and the nature of the product. Directors who have won Clio awards (awarded yearly for excellence in TV commercials), whose services are in constant demand and whose talents are universally recognized, are allowed considerably more creative freedom than are relative newcomers. In some cases, they are allowed almost total freedom in shaping a commercial. Says Asher:

If you're dealing with certain kinds of products, almost every word of copy is scrutinized by lawyers and network continuity departments and government agencies. Anyone can complain, "That's an odd way of saying those words." The agency's answer is, "That's the only way they'll approve those words. They have to be in *exactly* that sequence and we have to say them *exactly* that way! This shot can't be any higher than that because it has to include the full table so the kids will know how big the toy is and we have to have a wide angle shot to show how many people are in the room." Almost every shot is required. So the freedom you have to be creative is limited.

Casting Although casting patterns differ depending on the agency and the amount of preparation time, a tentative selection of performers is often made before the director is hired. An agency may have its own casting director or it may have hired an independent. In either case, the casting director will have called in performers to read for the commercial's key roles. Usually the agency producer supervises such early sessions.

Some agencies videotape performers who seem physically right for a part and whose abilities make them likely candidates. (The largest agencies build libraries of such videotapes, constantly updating them, weeding out ineffective or undependable actors.) Often the director is asked to look at these preselected videotapes and render an opinion. For creative, conscientious directors, such a procedure can be frustrating. As director you must work closely with actors, under conditions of extreme pressure, and because you are responsible for obtaining the best possible performances, you will usually want as much voice as possible in selecting the performers. Often directors suggest performers with whom

[5]Gradus, *Directing*, p. 56.

they have worked, whom they believe superior to those previously selected. The agency must respect the director's experience and usually will call in such performers for a reading.

You are always consulted for the final decision. So is the client. Such decisions generally are made at a major preproduction meeting that takes place a week or two prior to shooting.

Set Design Final approval of set design also is given at the preproduction meeting. As with casting, set design sometimes begins before a director enters the picture. When you have the opportunity of conferring early with the set designer (or art director), you and the agency producer examine and approve preliminary sketches based on script or storyboard, discussing such important aspects as color, dressing, camera needs, and the staging of actors. Because time is critical, long crosses (say, from a kitchen cupboard to stove) must be shortened; each fractional second must be evaluated. Highly significant is the social level of the family inhabiting the set. Whether it is wealthy, middle class, or just beginning its climb up the economic ladder is of tremendous importance to client and agency. Usually the economic status is spelled out either in the script or in a preproduction agency booklet that defines the characters, their life-styles, attitudes, and economic status.

Many directors prefer to shoot on **location**, in actual buildings rather than in manufactured sets, feeling that the ambience will contribute significantly, especially to a slice-of-life commercial. Locations usually are more difficult to work in than a soundstage setting, because they require more production time and afford less flexibility, but many directors feel that the values gained more than outweigh the difficulty and expense. Once locations are selected, Polaroid pictures are taken for approval by agency executives and the client.

When a location is contemplated, you visit the locale to determine its appropriateness to the script/storyboard as well as any potential production problems. Is there space to park trucks? Where will the sun be morning and afternoon? Will billboards or advertisements need to be covered? On interiors, will there be room for camera and lights? Will the windows need gels (gelatin sheets placed over windows to reduce the brilliance of incoming light)? Will acoustics create sound problems? Will dialog have to be **looped** (recorded again, synchronized with lip movements)? Will the art director have to make structural changes? Will the location have to be dressed up or down in accordance with the economic status of its supposed occupants?

The Preproduction Meeting After all of the *pre*preproduction meetings comes the preproduction meeting itself at which most key decisions are made. The meeting usually is attended by representatives of the client and the agency as well as some production team members such as director, cinematographer, hairdresser, home economist, casting director, wardrobe person: production by committee.

Usually, it's the agency producer who conducts the meeting. He has an agenda, and nowadays those production meetings are sometimes very elaborate. You're given a book—you think it's a meeting of the U.N.—and the book has everything written out in it and people turn solemnly from page to page like they're in a religious ceremony. The simplest things are stated over and over again. But, there is an agenda and (the producer) usually makes that agenda and conducts the meeting.[6]

The meeting addresses itself to each detail and aspect of the commercial, from the color of a kitchen to the upholstery in a car. Should earrings be diamond or should they be gold? Should fingernails have polish or should they be natural? Should a blouse have long sleeves or short? It is at this meeting that you lay out your staging plans, frame by frame, setup by setup. If there are deviations from the storyboard, they will be discussed at great length. You try to sell your ideas for possible improvements. Final casting decisions are often made at this meeting as well as approval for locations and set plans.

At preproduction meetings some directors dominate; others sit quietly, absorbing viewpoints of agency and client, synthesizing ideas, clarifying items of confusion or disagreement. From this point on, you become captain of the ship, pulling all of the production elements together for the "shoot."

Production   Good directors who prepare meticulously visit their sets or locations a day or two before production begins to make certain that all details are as they have planned. (Some visit their sets or locations many times, soaking in atmosphere, visualizing the action, making preparatory sketches.) No director likes surprises. Discovering a mistake two days before the shoot is okay; there is time for the mistake to be rectified. Discovering that same mistake on the morning production begins can be disaster. No matter that it's not your mistake. It's going to cost precious time to fix it.

Directing commercials differs in one significant aspect from directing other fictional or nonfictional forms: the number of "bosses" on the set. At one point or another, the following keenly interested individuals may visit your set, studying, evaluating, and perhaps trying to correct or modify your procedures. From the advertising agency: creative director, senior copywriter and copywriter, art director, producer, account supervisor, and account executive. From the client: the advertising manager, the brand manager, and others.

All of these so-called bosses are nervous. Nervousness seems to be endemic to the advertising industry. Many fear for their jobs. Many fear that you will go over budget. Many fear that their promises to the client will not be realized. Many worry that you will change the character or style of their commercial. Many worry that the commercial itself is unsound. All look to you as either culprit or savior.

This apprehensive audience presents two problems for the director. The first

---

[6]A director quoted by Gradus, p. 113.

constitutes an invasion of your territory. Client or agency visitors sometimes register disapproval of a line reading or piece of action, voicing their disapproval either to you or to a member of cast or crew. Voicing comments directly to you would not be a problem except that opinions sometimes differ; an agency exec may suggest one thing, the client another. Moreover, it is disconcerting and time consuming to receive suggestions from half a dozen sources. Wise directors generally insist that all comments or suggestions be funneled through one person, usually the agency producer.

When client or agency visitors speak directly to cast or crew, it is a far more serious matter. As indicated earlier, everyone wants to be a director. When commercials are being filmed or taped, everyone wants to comment. Sometimes the comments are spurred by boredom, sometimes ego, sometimes the need to correct what appears to be a mistake.

Directors cannot allow guests to speak directly to cast or crew. All such comments must be funneled through them (the directors), quietly, away from the set. The cliché that "a ship can have only one captain" applies forcibly to any company filming commercials. Comments from offstage undermine the director's authority; actors begin to seek approval from multiple sources. Although directors are hired and paid by an agency, they must insist on autonomy on the set. In his excellent book *Directing: The Television Commercial*, Ben Gradus recalls a speech he often recites to the agency/client group.

> I'm here to make the best possible commercial and to please all of you in the process. I trust you will all have helpful suggestions and I promise to listen seriously and respond. However, there can be only one director on the set if the actors and crew are to perform their best. All ideas and objections should be channeled through one representative, in this case preferably the agency producer. Then the cast and crew will not become confused and neither will I. Henceforth, no one but I will address the cast and crew directly and, please, all suggestions should be presented to the agency producer, who will relay them to me. I will listen.[7]

The second problem the apprehensive agency/client audience presents is shock or consternation at how you are staging a scene or your choice of lighting or camera angle or whatever. Such a reaction should never happen.

More than anything else, a director is a communicator. Earlier I mentioned that directors don't like surprises on the set. Neither do agencies or clients. If directors expect to work again, they must spell out their intentions in great detail to the agency producer. If the storyboard indicates a closeup that the director doesn't feel is necessary, the director must communicate that feeling to the producer rather than have someone realize the omission the following day in the projection room. Perhaps the director can convince the producer of the validity

---

[7]Gradus, p. 100.

of the change. Perhaps. If the client has already approved the storyboard, chances are that the producer will insist on shooting the sequence as indicated. If the director feels strongly about such a departure, then he or she may elect to shoot the sequence two ways: as defined in the storyboard and as the director prefers. Then the ultimate decision will be made later in the editing room.

Postproduction   More and more often, TV commercials are being edited in videotape. Even those photographed on 35-millimeter film are now transferred to tape, because computerized editing is both effective and incredibly fast. In addition, optical effects may be tried, changed, and incorporated in the commercial without the wait (or expense) of sending film to an optical house.

Directors theoretically are allowed to work with the editor and put together a "first cut." After the first cut, the film is then turned over to agency personnel and the director will move on to other projects. Most directors shoot a commercial with a clear idea of how they would like the film (or tape) to be assembled, but the wise ones also respect the input of the editors. Editors bring to a project an objectivity that is impossible for a director who has spent weeks thinking of little else. The best editors have a sense of showmanship, an instinct for what plays and what doesn't, as well as a sense of dramatic rhythm in building sequences. Although directors don't always accept editors' suggestions, they would be foolish not to consider constructive ideas. They therefore listen respectfully and incorporate those suggestions that they feel will improve the spot.

> Neither the director nor producer nor any agency person can bring to the editing the unique perspective that the editor can. He alone has not been encumbered with the past, with the meetings and the dicta and the nonsenses and the difficulties of shooting and the preconceived ideas. He alone can truly measure good from bad in terms of audience acceptability and effectiveness with total disregard for politics and personalities and problems. His should be a strong voice and vote.[8]

The Local Commercial

Because commercials produced for a local or regional market reach a far smaller audience than those aired nationally, budgets are smaller and production schedules tighter. Only one element increases: pressure on the director.

The nature and scope of such a commercial depend, logically, on the size of the market. A town of 50,000 will see far more modest local commercials than will, say, New York or Denver. In the larger markets, directors may work with major advertising agencies, local agencies (large or small), or, on occasion, directly with the client. Sometimes these directors are employees of the agency or the client. More often, they work for the local television station or cable outlet.

---

[8]Gradus, pp. 187–188.

Most local advertising agencies are enterprising, responsible, and often extremely sophisticated, but a few are schlock artists, intent on squeezing three commercials into the time normally required to produce two, on badgering directors into taking shortcuts or giving more of their time than should be legitimately expected. Directors with self-respect either walk away from such exploitation or refuse to go beyond the expected norms.

The simplest local commercials consist of nothing more elaborate than a series of photographs and printed material with a staff announcer reading the copy. Others may use videotape of used-car lots, restaurants, or markets with local personalities making the pitch. On occasion, the local director gets the opportunity to be genuinely creative.

Let's imagine that you are a staff director for a TV station and an advertising agency approaches you either directly or through your station's management. Perhaps you have worked with members of the ad agency on certain local shows. They liked your work. When they decided to create a series of local commercials, the agency's creative director thought of you.

While the agency will be using some station facilities, you will be employed as an independent contractor. In other words, the commercials will not be done as a part of your normal staff work. You will be paid extra for them and will probably direct them on your own time. The amount of money you receive will not purchase a Mercedes 350 SL.

Your station is delighted to see you direct the commercials. Such cooperation with a local advertiser tends to strengthen relationships and encourage future purchases of airtime on your station. The program director will try to arrange your work schedule to accommodate agency needs.

The account executive hands you mimeographed scripts for three one-minute commercials. (Storyboards are rare on a local level.) A prominent mortgage loan company will be advertising on TV for the first time. The exec grins guiltily: The company's president will work on camera in all three spots. Two of the three commercials depict young couples anxious (but financially unable) to buy their first home; the president closes these commercials with a personal sales message. The third consists in its entirety of a heart-to-heart pitch from the president.

You realize that the effectiveness of these commercials will depend in large part on the effectiveness of the company president. You ask the essential question: Can he act?

The account executive shrugs. He *says* he can. It turns out that the loan company's decision to advertise on television was prompted by the assurance that the president himself would create the company image. Now it will be up to you to "get a performance."

The president's office will be built at the station. Some wood-paneled flats, greenery, a bookcase, and a handsome desk. When you talk to the station's art director you specify that the desk must be large and expensive. (Big executives have big desks—and you know that the man has an ego.) The office must also seem comfortable, not intimidating. Maybe pictures of the wife and kids.

The young couples are selected from the local university's drama department. They will receive small change for their efforts, but it's worthwhile for them. The commercial provides professional experience that will look good on their résumés. You and the agency exec select two couples (and a standby in case of illness), both wholesome and engaging.

The following week, using a minimal crew from the station, you tape the young couples looking wistfully at small but attractive new homes. (Local commercials usually are shot with videotape; it's cheaper than film, flexible, and can be edited quickly with a full range of optical effects.) To tape this location material the agency has purchased two hours of crew time. You asked for more. The agency couldn't afford it. It's a tight schedule, a fraction of the time you would have for a national commercial.

Aware of the pressure, you brought your actors here yesterday and meticulously rehearsed the sequences. (With professional actors you would have to pay for that privilege.) The taping goes well; viewers' hearts will go out to these attractive young kids aching for a home of their own. You cover the action in as many angles as time will allow, making certain the audience sees how much in love the couples are.

You have had two meetings with the president of the loan company, a white-haired gentleman with a warm, ingratiating manner. You discussed the commercial copy with him; he seemed comfortable with it. You explained that the material would be placed on a TelePrompter so that he could look directly into the camera lens but still have the security of reading his material. (TelePrompter material is reflected onto a glass surface fixed at a 45-degree angle directly in front of the camera lens.)

At these earlier meetings you were impressed with the president. He had authority. He spoke well. Most of your doubts disappeared.

They return in full force the minute rehearsals begin. The president seems nervous, apprehensive. You try to kid with him, to build his confidence, to put him at ease. But his mind is elsewhere. The scripted action is simple. He will begin his pitch seated at the desk. At one point he will rise, move around the desk and downstage where he will pick up the company logo.

He seems to improve during rehearsal, becoming comfortable with the action. But the minute you attempt a take, he tightens up. The material suddenly sounds memorized and phony.

You change a few words, making them more colloquial, clearing the changes with the agency exec, who tries to conceal his own apprehension. Again, when you try for a take, the president tightens up. The words sound hollow. You know you're facing disaster so you decide on a little ruse. You change the action slightly and announce that you want to rehearse the new pattern. You privately instruct your technical director to roll the videotape on a secret word cue from you. No, you won't be in the booth. You'll be out on the stage, watching the apparent rehearsal.

Without the pressure of a take, the president is relaxed and comfortable. Unaware that videotape is rolling, he speaks the words warmly and with assurance. He's marvelous. Later, when the agency people see the tape, they congratulate you on your ingenuity.

Although directors of national commercials most frequently begin as cinematographers, art directors, or editors, directing for a local TV station or cable outlet provides superb training. Also, superior local commercials may provide the beginning point for a sample reel. This is not to imply that directing on a local level is easily achieved. Most jobs in the highly competitive entertainment industry are difficult to obtain. Yet talented new people find their way into television every day of the week, every week of the year, so it can be done.

PRODUCTION TECHNIQUES

Although production techniques for commercials do not differ substantially from techniques for other dramatic or nondramatic forms, one area of special directorial concern is the product itself.

Displaying the Product

For most advertisers, understandably, every element in a commercial takes second position to the product. No matter how lavish the production values, how catchy the jingle, how famous the guest star, if the product is not displayed handsomely, in a manner that creates a positive audience reaction, the advertiser will be unhappy. Often the product or its logo is the last shot in a commercial, the image that will remain longest in an audience's mind. You can understand the advertiser's need for perfection.

Directors often display small products in **limbo**, that is, on a stand with a curved backing so that no floor line or horizon line is visible. At the station where I directed, such stands were called **gizmos**. Sometimes the product is displayed in the same manner as many high fashion models—upon a sheet of seamless paper that curves back behind the product and up out of the frame, again creating a bland, neutral background, generally of a pleasant pastel color, never taking attention away from the product.

Occasionally, other objects may be placed within the frame to enhance the product, to lend their own emotional colorations. An orchid, for example, beside a perfume bottle, might suggest exquisite beauty or delicacy. Bright red, polished apples with leaves and stems beside a can of applesauce might make the product seem more appealing. Such trimmings can add color and appeal to a product, but be sure always to clear your concept first with the agency executive. What seems to be a first-rate idea to you may be counterproductive, for legitimate merchandising reasons.

Backgrounds also add their own colorations to a product. I was once assigned to direct a car commercial in which the script called for a "Beauty Shot." I asked the agency representative exactly what the term meant. Apparently inexperienced, he hemmed and hawed, telling me just to make the car look great. The car had a sleek, classic look, so I asked our art director to bring in Greek columns (cutouts) which we arranged to form a colonnade. With the car brilliantly lighted in front of a cyclorama, the colonnade created a simple, yet classically rich setting, thereby lending richness to the product itself. The agency rep nodded when he saw it. Yes, that was what he had in mind all along.

As you watch TV commercials, be aware of how much backgrounds contribute to the ambience. Note how often car commercials are photographed in front of expensive mansions or elegant clubs or restaurants. Even in relatively confined studio commercials, an inventive background can enrich a product. A tweed or leather background, for example, might enhance an aftershave lotion. An artificial grass mat creates an outdoorsy background for a package of golf balls. Because the product shot is of such concern to advertisers, most directors and agency producers give it extraordinary attention. I recall a beer commercial in which the agency producer spent fully half an hour patiently coaching the gentleman who was going to pour the beer. The beer glass was illuminated from beneath— through a hole in the base of the gizmo—to highlight the bubbles. All that would be seen of the gentleman was his hand. But because the head on the beer had to be perfect, the manner of pouring became critical.

When the commercial features food, a home economist usually is hired (or on the company payroll) to make certain that its presentation is appetizing. The economist usually supervises the entire food background: tablecloth, place settings, silver, candles, and flowers. When the food is to be served hot, multiple items usually are kept in ovens just off camera so that a dish may be whisked from oven to camera, in time to capture on film the heat radiating from it. Directors who specialize in food commercials understand the importance of highlights. Sometimes they spend literally hours lighting a closeup. One director I know used to brush certain food products with oil, causing them to glisten attractively.

The trade secrets for making food appear attractive are usually carefully guarded by economists skilled in the commercial field. But here are a few hints from Maggie Kilgore:

— Mortician's wax will plump a deflating chicken.

— Dry ice will create the illusion of frost or smoke.

— Beef is rarely cooked through because it tends to shrink.

— Egg whites and liquid soap can create foam on coffee, beer, or soda.

— Glue or corn starch will thicken gravy or sauce.

— If ice cream isn't the primary product, mashed potatoes will make effective fake ice cream to serve as a stand-in.

Of course, there are limits to what directors or home economists may do to enhance a product's attractiveness:

> Economists are also hired for their integrity in guaranteeing truth in advertising. Such concern grew out of the landmark 1970 court case in which the Federal Trade Commission ruled that a Campbell's Soup commercial was deceptive because it used marbles at the bottom of soup bowls to make the vegetables protrude above the soup.
>
> Now food stylists cannot change or adulterate a product being advertised—if it is foreground food. The FTC also forbids such practises as substituting frozen green beans, which look more appetizing, for canned.[9]

Packaging of the product often demands special attention. Standard cartons as they are displayed in stores frequently contain so much copy that they appear "busy" when photographed in their usual state, so crowded with visual material that the impact is diminished. Most advertising agencies provide special packaging for commercials, closely resembling the product as routinely packaged and sold in stores and yet simplified for greater readability and impact. When photographing them, directors make certain the packages are perfect: clean, dust free, without flaws in design or lettering, no matter how minute.

Foil packaging or products that contain metallic or mirrored surfaces pose special photographic problems. Because their surfaces may reflect lights, the camera, or even the director himself, they must be surrounded by "tenting," a light cloth that will be invisible when reflected and will eliminate distracting reflections. The cloth must have a small opening through which the camera can poke its nose to photograph the product. When photographing metallic surfaces such as watches, jewelry, or silverware, the glisten and sparkle become part of the sales appeal, so plenty of high intensity lighting is essential.

> There are times when the package has to be placed in closeup into an exact spot during a scene. Often a good way to do this is to have the fingers *remove* the package from its correct final placement, running the camera in reverse. . . . However, keep one piece of rhythm in mind. If you study the movement carefully, you will note that the hand will slow

[9]Adapted from Maggie Kilgore's "The Unsung Hero," *EMMY, The Magazine of the Academy of Television Arts & Sciences*, April 1987, pp. 48–51.

up toward the end of placing it down. Therefore, when playing the action in reverse and picking it up instead of placing it down, start the pickup slowly and then, after it is off the table a couple of inches, move it out faster. If you don't do it this way, there will be an unnatural jump of the product into place in the final straight-on version.[10]

Reversing camera action is a useful technique for commercial directors because it ensures that action will conclude exactly where needed and lighted to perfection. When the storyboard calls for an automobile to drive toward camera, for example, stopping with the radiator grill in a tight closeup, such a shot can best be achieved by reversing action. Thus, the car would start in camera closeup, tight on the gleaming grill, and would back away.

Directorial Techniques

Tony Asher has produced literally hundreds of commercials and worked with dozens of directors. Here are his words concerning common directorial problems in producing commercials:

Q. Is there any special talent that directors need in order to tell a whole story in thirty seconds?

A. A sense of proportion, obviously. A sense of *reduced* proportion. More specifically, they must know how much material they can get into a commercial. Very often agencies try to include too much.

Q. But agencies have stopwatches. Don't they time their material?

A. What happens is, you write yourself a nice commercial. And it runs thirty-four seconds and you say, "Hey, that's no problem. We can certainly cut that down." And then somebody puts in four lines of mandatory copy. And "Member F.D.I.C." And somebody suggests a short musical intro. Now, people do put stopwatches on these things but they read the copy in their heads, silently—and not leaving the right pauses for this and that.

By the time it gets to the director, it's been sold to the client that way and the client's satisfied that he's included his nine essential points. If the director can't get all of those points in, he's going to be the one who has failed. Well, the director puts his own stopwatch on the material—visualizing the way it will probably play—and he always comes out with something like forty-four seconds.

Q. And that's when the agency trims the material?

A. Well, the director says, "What's going to go?" And the producer

---

[10]Gradus, p. 47.

says, "Nothing can go. It's been sold to the client. It's got to be delivered just that way." And the director will sometimes say: "Hey, guys, it's not going to work. We're not going to take the job on that basis. We can't deliver what's on this storyboard in thirty seconds." What happens, of course, is that eventually a whole scene is taken out—or if four points are supposed to be made, one of them gets lost and only three are made. So, for a director, having the ability to know in advance how material will play—that's crucial.

Examine the best commercials on TV. You'll find that usually they make a single point. By taking the full thirty seconds to make it, advertisers are able to dramatize their message with impact. Trying to cover too much territory in a limited time creates only confusion.

Telling a story in a minute or thirty seconds creates other directorial problems. Every moment has to count. Let's return to my interview with Tony Asher:

Q. How do you cram it all in?

A. You have to work with shorthand. You have to understand how to get from a raised eyebrow or a gesture, what might take thirty seconds or a minute in a feature film. Certain things happen simultaneously that might not happen simultaneously in real life. Or you're able to use a tiny gesture to communicate something that you didn't see nine-tenths of—but you understand what happened earlier because you saw the final moment. The really good directors understand how little you can get away with and still successfully communicate an idea.

Q. How do you learn that?

A. By doing it—or by studying others who know how to do it. You ask yourself, "Why do I know he just got through washing a car?" Well, because his pants are wet and he's got a sponge in his hand. To find ways to make those things clear in nonverbal, non-time-consuming ways is really tricky. Sharp directors do that. They create impressions without actually showing a piece of action. It's *economy*, really. Economy of words, economy of motion, economy of visual image.

Q. Does a director plan all of those economies before walking on the stage?

A. Absolutely. Frame by frame. Line by line. Action by action.

Director Stu Hagmann claims that the essence of directing prize-winning commercials is style:

What interests me is style, not flair. An awful lot of commercial production is done strictly with flair. It's like a lot of hollandaise on an egg without the muffin under it. It looks OK on the plate, but you can't pick it up. . . .

I guess it's whether you can or can't tell a story. Some people simply can't tell a story. They haven't read any literature, to begin with. All they are interested in doing is creating an effect rather than guiding a person's mind.[11]

CHAPTER HIGHLIGHTS

— Commercials sell more than a product; they sell a life-style. They demonstrate how viewers can achieve glamour, the state of being envied.

— Directors of national commercials frequently come from three related fields: art direction, cinematography, and editing. Local TV stations or advertising agencies often provide invaluable training.

— Advertising agencies usually present directors of national commercials with storyboards, graphic depictions of the essential commercial action. Although many directors feel that storyboards limit creativity, they also recognize that storyboards provide subtle insights and clearcut avenues of communication.

— In preproduction, directors often discover that preliminary casting has already been done. Many large agencies maintain videotape libraries of actors from which selections sometimes are made. Agencies consult both directors and their clients in determining final casting. During preproduction, directors select locations, work with art directors on set design, and plan their staging.

— Final approval of casting, set design, storyboard changes, wardrobe, and other details are made at a preproduction meeting attended by advertiser, agency, and director, plus key members of the production team. The director communicates his or her vision of the commercial, move by move, detail by detail.

— Because both agency and advertiser usually attend the shoot, the director must maintain control of cast and crew, preventing interference from visi-

[11]*EMMY, The Magazine of the Academy of Television Arts & Sciences*, July/August 1985, p. 22.

tors. Suggestions regarding staging or line readings should be funneled through the agency producer, who will present them to the director in private.

— In postproduction, directors work closely with editors, preparing a director's cut. Because seasoned editors bring both skill and objectivity to a project, wise directors welcome their input.

— In the pressurized field of local commercials, directors face the dual problem of limited time and limited money. Scope depends upon the size of the market. Many local directors work for television stations or local agencies.

— When assessing values, directors must give primary importance to the product. Enhancement can be achieved through appropriate backgrounds, accompanying props, effective lighting, and use of simplified-for-camera packaging.

— Experienced directors understand the inflexibility of time and refuse to shoot commercials containing excessive material. To cram a story into limited time, directors use a kind of visual shorthand, condensing actions that cannot be dramatized in full.

PROJECTS FOR ASPIRING DIRECTORS

1. To win audiences quickly, commercials use the elements of entertainment (specifically, *pleasure*) discussed in Chapter 2. Study a few commercials on TV. Make a list of those that use each of the following elements. Some commercials use more than one element. Can you find a single commercial that uses *all* of these elements?
    a. Spectacle
    b. Conflict
    c. Male-female chemistry
    d. Order/symmetry
    e. Surprise and humor

2. You are the director of the following thirty-second commercial. You must first conceive subject matter that will effectively dramatize the commercial's message. Then you must plan the shots you would use in photographing that subject. Describe any special use you would make of camera, lighting, filters, music, or actors to give the commercial special appeal. (Note: I am not providing a storyboard because you will create your own camera treatment.)

VIDEO	AUDIO
	ANNOUNCER (wryly) You don't need Chapman's Frozen Fruit Bar to get you through the summer heat. What's a little discomfort anyway? Try to ignore that furnace out there, those simmering sidewalks and your sandpaper throat. Try not to think about Chapman's icy sweet bars, filled with generous chunks of real fruit: papaya, orange, pineapple, mango, banana. Put 'em out of your mind because you're strong. Sure you are. And because you're strong, you've probably even forgotten the name: Chapman's Frozen Fruit Bar.

3. Now that you have conceived the imagery and the camera shots for your commercial, prepare a simple storyboard of not more than ten frames that illustrates your vision. See the description of storyboarding in the Projects at the end of Chapter 6.

CHAPTER

13

MUSIC

Music evokes emotion; it is probably more evocative than any other art form. Rock concerts make our pulses pound. Movie love themes sometimes make us cry. Beethoven symphonies exalt us. In any program in which music is a major component, the director faces the obligation of communicating not just the music itself but also something of the emotion generated by it. As we shall see, it is usually unsatisfactory merely to photograph the instrument that produces the music. A notable exception is when that instrument is the human voice and the performer is exquisitely talented. Emotion may be expressed through technical means (camera, lighting, lenses, editing); through scenery, props, and wardrobe; through imaginative use of related imagery; and through the skill and sensitivity of the artists.

Musical programs present an additional challenge. Because music is written primarily to be heard, its target is the ear. This book has stressed (and will continue to stress) that one of a director's continuing goals is to express concepts in visual terms, to pictorialize. Thus, when directing programs in which the primary element is music, directors face a paradox. How can they effectively depict the instrument, whether it be an electric guitar or a human voice? Do they wander far afield, creating visual imagery evoked by the mood of the music? Or

Since the beginning of time—1956—rock 'n' roll and TV have never really hit it off. But suddenly it's like they've gotten married and can't leave each other alone.

Keith Richard (Rolling Stones)[1]

do they simply photograph the sound-producing apparatus? Do they cut to video-taped silhouettes of lovers walking on a fog-shrouded beach or lightning splitting an ominous sky? Do they train their cameras on a row of violinists or on a spotlighted vocalist belting out a rock favorite in front of a backdrop?

But musical and musical variety programs are a minor source of music on television. Audiences hear far more in the music that provides a background to dramatic action. Such music, called a **score**, is composed solely to enrich dramatic action. Most of the time, audiences remain unaware of its presence. Scoring dramatic material in TV and motion pictures can be enormously creative; it has almost become an art form.

This chapter divides into five discussions. The first four concern the musical program; the fifth studies the process of scoring.

— PREPARATION: KNOWING THE SCORE: examining artist, script, and musical score

[1]"Sing a Song of Seeing," *Time*, 26 December 1983, pp. 54–56.

— CAMERA AND EDITING: analyzing a director's techniques in covering classical, pop, country, and rock music

— DRAMATIZING THE MUSIC: creating visual elements that reflect music's emotional content

— STUDIO VERSUS LOCATION: contrasting the advantages and disadvantages of each

— SCORING DRAMA: insights into the purpose and functions of music as a background to dramatic action

PREPARATION: KNOWING THE SCORE

For a director, there is no feeling worse than walking into a control room inadequately prepared. Even for the most resourceful, most seasoned directors capable of bulldozing their way through any perilous situation, insufficient preparation inevitably undermines the security of a project. It destroys confidence. It consumes precious rehearsal time in missteps that must be corrected. It communicates itself to cast and crew, creating tension, sabotaging their best efforts, and vastly increasing the flow of stomach acids.

Preparation of a music show depends in part on the director's familiarity with (and understanding of) music. Obviously, a director lacking knowledge of classical music requires more time preparing a symphony orchestra broadcast than does a musicologist. When I directed in live television I had absolutely no background and little knowledge of the music world. I was able (with help from friends) to hum a tune on key. Management, therefore, with assailable logic, assigned me to direct almost all of the station's musical programs, from symphony orchestra broadcasts to big band concerts to western music variety hours. It was challenging.

Whether you have musical expertise or not, you will need to take certain steps in the preparation of any musical program. These steps concern artist, script, and score.

Artist

When possible, study the artists during an actual performance, whether they be vocalists, rock musicians, or concert pianists. Failing that, try to find videotape or film of programs they have previously presented. Videotape or film is actually preferable to a live performance in that you can run a particular sequence over and over again, familiarizing yourself with the artists, their music, and their style of presentation. Make notes. Ask yourself such questions as:

— Are they colorful or drab? If drab, how will you create audience interest? If colorful, where does the color lie? How will you emphasize it?

— Are there specific personalities that shine? How will you capture their spark?

— Does their physical setup lend itself to effective camera treatment? How will you change it—without destroying their musical effectiveness?

— If a vocalist, is the vocal style intimate or physical? (How will this affect your camera treatment?) Attractive or unattractive? Are there "good" angles or "bad" angles?

— Are any of the numbers the same ones that will be performed in your show? If so, you are, in effect, watching a rehearsal and can plan your treatment more specifically: camera angles, lighting, and setting.

When possible, meet with your star performers well in advance of the broadcast. Let them know how delighted you are to be working with them and share some of your concepts for the format. At this point, you or a station administrator must discuss certain key issues. Will the show be performed in the studio or at a location? If a studio show, what kind of setting will be most appropriate? Will there be an audience? Who will be the host? Will the show be scripted or off the cuff? (Who will write the script?) How will costs be handled? If the program is to be sponsored, how will commercials be integrated? If your performers have reservations about any of your program ideas, it's better to find out early. Welcome their creative suggestions. Make them feel they are part of the team.

Script

In many cable and local TV station operations, the director is also the producer and therefore responsible for a program's content as well as its form and style. The preliminary meeting with artists or their manager should prompt them to deliver a suggested list of musical selections as early as possible. You will probably want to discuss an appropriate sequencing of these selections. This sequence provides the skeletal basis for a script.

Many directors believe that their work begins only after a script has been written. Although some working directors remain aloof from the writing process, the good ones understand that quality shows begin in the scripting. They know from bitter experience that it's almost impossible to create a fine program from a mediocre script.

If a script is to be prepared for your music program, don't be shy. Contribute ideas before the concrete has hardened. If the writer's concepts are stodgy, impractical, or old fashioned, be tactful in suggesting changes. Some directors have the golden ability of making a writer believe that their ideas actually originated with the writer. Because ego is so inextricably involved in the creative processes, most writers tend to function best when working with their own concepts rather than concepts someone has foisted on them.

Score

It is important that you become familiar with musical selections prior to rehearsal. (When you begin rehearsal, you should already have mapped out a pattern of camera angles.) Listening to cassettes or to compact disks will help immeasurably in your preparation. If possible, get photocopies of the actual scores.

Counting bars is essential in order to cover a musical program effectively. If you cannot read music or count bars, it is imperative that you have someone—an associate director or musician—sitting beside you who can. Lacking musical expertise and denied an associate director, I frequently relied on a stopwatch for timing musical passages. It was helpful but not always dependable, because musical tempos vary. With singers the problem is less acute. Neatly typed lyrics (as they are actually sung) provide an effective directorial road map.

The reason for familiarizing yourself with the music is so that you will be able to direct your cameras to the right place at the right time. It's embarrassing to cut to a closeup of a piano keyboard at the exact moment a solist begins to sing.

CAMERA AND EDITING

In photographing a ballet such as *Swan Lake*, some directors prefer to dissolve rather than cut between camera angles. By such transitions from one camera to another, they attempt to enhance the mood generated by music and dancers. All camera and editing techniques carry certain implications; they influence audience reactions; they create a sense of rhythm, mood, or atmosphere. Audiences remain largely unaware of such subtle influences; they occur below the level of full consciousness. It is impossible to provide a comprehensive list of optical implications because they vary from circumstance to circumstance and from program to program. For example, in drama a dissolve often implies the passage of time or marks a change of scene. In a ballet pas de deux, it is merely a graceful change of angle.

Transitions between camera angles are a director's most elementary technique in dramatizing any message. Although digital video effects have revitalized musical transitions (especially in MTV), the cut and the dissolve are still the most often used.

The Cut

A cut is clean, crisp, sharp, and incisive. In an uptempo musical number, a straight cut seems to fit the pace and character of the music. In such selections, most directors prefer to cut on the beat so that a sequence of cuts (every two bars, every four bars) almost becomes part of the music; it has a rhythm of its own, reflecting the musical pattern.

A properly executed cut is invisible, whether the musical tempo is fast or slow. When cuts are made on movement, they are virtually undetectable. If a singer whirls and moves downstage during a musical bridge, for example, a cut to a wider angle will go unseen if executed during the whirling motion. Similarly, when a cut is executed as a singer begins a word or musical phrase, the cut will seem invisible.

The theory behind invisible cutting is similar to a technique stage magicians use. When a magician removes a rabbit from an inner pocket with his left hand, his right hand simultaneously makes a flourish high in the air. The viewer's eye follows the larger action. When a singer or dancer whirls flamboyantly, the observer's eye, drawn to the larger action, is oblivious to the change of camera angle. Similarly, when a singer begins a new phrase, audience attention goes to the new thought he or she is expressing and thus is unaware of a change in angle.

Another key principle is also at work. Good editing always follows audience interest. As interest in the performer heightens, spectators want to move closer. Effective editing parallels that increased psychological need for closeness by providing closer camera angles. When cuts are synchronized with audience desires, the audience simply does not see them.

The Dissolve

A **dissolve** from one angle to another is more fluid, less incisive than a cut. It presents a blending or merging between two angles in which the first, for a moment or two, is actually superimposed over the second. In early motion picture scripts, it was called a lap dissolve, suggesting the overlapping of images. Early directors in live television used the command "lap" for such a transition, perhaps because the word was easier to say than "dissolve" (for example, "lap to one," "lap to film").

A dissolve can be any length: almost as quick as a cut or extending as long as a director desires. Television dissolves usually last about two seconds. In drama, extended dissolves often imply a connection between the first image and the second as the first image remains briefly superimposed over the second. For example, when a scene ends on a burning car and the picture dissolves slowly to a hospital bed, to a close shot of the car's driver (the wreck superimposed over her face for a moment, ghostlike), the director perhaps implies that the driver carries the memory of the burning car in her mind; she cannot forget it.

In ballet, a dissolve echoes and reinforces the fluid motion of the dancers, helping to maintain mood. Very slow dissolves during a ballet sequence (especially when photographed against a dark background) provide multiple images of the dancers, enhancing the poetry of their motion, suggesting figures reflected in a pool.

All ballet, of course, is not *Swan Lake*. Modern ballets are often stylized or bizarre in flavor. With such ballets, dissolves might be totally inappropriate, suggesting a fluidity that does not exist in either the action or the music. Whether in music, sports, drama, or special events, camera and editorial treatment must reflect the character and mood of the production.

Shots and Angles

In general, the impact of a shot varies directly with its closeness. A closeup intensifies the significance of subject matter; a wide (or long) shot tends to diminish it. Close shots gain impact by contrast. Thus, in drama, directors and editors try to save close shots for critical moments, when they are dramatically motivated, aware that if an entire scene is played in close angles, nothing will have impact.

With instrumental numbers, this rule must be modified. It is virtually impossible to assess one portion of a selection as being more "significant" than another. Use wide shots sparingly. Their primary purpose is orientation. Close and medium angles usually comprise most of the camera coverage. A wide shot of a symphony orchestra or even a rock group is visually unexciting, although a rock group of three or four members allows the camera to be considerably closer than a seventy-piece orchestra and therefore provides more interest.

Classical Music Close shots of drummers, violinists, or flautists playing instruments can provide colorful material—but a closeup of a tuba player with inflated cheeks and eyeballs bulging can prove ludicrous, destroying mood and provoking laughter. If laughter does not conflict with the musical mood, however, then such a shot might prove refreshing. For example, if the Boston Pops Orchestra were playing a rollicking polka, humor would not be out of place, and such a tuba closeup might add zest to the proceedings. The point, again, is that the shots must contribute to the overall character or flavor of the music.

If close shots provide interest, extreme closeups often add more. Fingers on piano keyboards, violin frets, harp strings, or valves on wind instruments are visually dynamic. So are drumsticks rolling on a timpani, bows sawing on a string bass, Chinese cymbals clanging, and even toes tapping out a beat. Don't be afraid to be *bold* with your closeups. Bold angles energize musical selections that otherwise might be visually bland.

Close shots do not always have to include instruments. With a virtuoso pianist or violinist, for example, closeups of the face often provide tremendous visual emphasis, enriching the mood and creating psychological closeness between audience and performer. If you have watched violinist Itzhak Perlman during a concert, you will recall facial expressions of ecstasy or grief, eyebrows occasionally furrowing in the intensity of concentration or lips smiling suddenly in enjoyment of the musical moment—expressions that vastly deepen audience participation and help to communicate the music's emotion.

Shots of the orchestra conductor are effective for three reasons. First, they often communicate emotionality. Second, they provide relief from incessant shots of musical instruments. Third, they provide directors with a convenient "escape hatch" if they get in trouble—if you've miscounted bars and find yourself temporarily lost, cutting to the conductor affords instant salvation. Some conductors, of course, are naturally flamboyant; they dramatize their music with great histrionic flair, providing directors with an unending source of exciting material. Stoic conductors usually end up with considerably less screen time.

To photograph conductors effectively, the director must conceal a camera at the rear of the orchestra. Such a hidden camera may also provide effective shots of the audience. Boston Pops programs on PBS often feature group or individual audience reaction shots as a part of their camera coverage, especially if the music is light in flavor and audience members contribute to the ambience through smiling expressions or fingers tapping rhythmically on tables.

Many of the principles discussed in Chapter 3 will be helpful in planning shots of instruments or musicians. The dynamic diagonal line, for instance, occurs repeatedly in side angle shots (such as shooting down a row of violins or horns). These side angles, sometimes called **raking shots**, derive much of their visual energy from the rhythmic repetition of form as both musician and instrument are repeated in diminishing perspective. It is not always necessary to include an entire row of musicians. Sometimes close raking two-shots or three-shots of instruments (such as string bass or cello) provide bold, effective angles: bows sawing on strings or fingers on frets.

One of my favorite uses of the diagonal line involves the piano keyboard. In most programs featuring a pianist, directors use three or four relatively standard angles: a full profile shot of musician and piano, a close shot of fingers on keyboard, and angles through the raised piano lid, featuring musician but eliminating keyboard. I discovered that two cameras shooting down and close on the piano keyboard (from opposite sides) when superimposed would create crossed diagonal lines of considerable beauty: two keyboards, four hands in motion. Caution: Such an angle contains directorial traps because neither camera may pull back when on the air without revealing the presence of the other. Solution: Either the on-the-air camera must remain briefly on the keyboard to allow the opposing camera to escape or the director must cut to a third camera shooting through the raised piano lid, close on the pianist. In this latter angle, both keyboard cameras can move away unseen.

Thus far I have discussed fixed shots from a stationary camera. Movement may also help to dramatize music. Dollying or zooming in (or out) creates a visual transition from wide to medium to close. As indicated earlier, such a move parallels the viewer's increasing interest as it goes from the ensemble orchestra to an individual section or instrument. Conversely, as interest changes from a specific instrument or section to the entire orchestra, dollying or zooming out becomes appropriate.

What is the purpose or reason for such a move when a simple cut achieves the same result, only faster? For one thing, a forward movement does not pretend to be invisible. With dollying in, the audience becomes aware, with some anticipation, that they are being carried slowly forward, from the general to the specific, to a new focus of attention. As much as possible, such movement should be synchronous with the musical tempo. Conversely, dollying out takes viewers from the specific to the general, reorienting and releasing them from whatever tensions may have been generated by closer, more dynamic angles. Dollies or slow zooms outward are often used at the conclusion of a number.

Because panning shots are seldom used in classical music programs, they remain fresh and visually interesting. They are simply another way to transfer attention from one orchestra section to another. As with a dolly shot, a slow pan heightens audience interest in its destination. Panning shots become more dynamic when panning, say, from an entire woodwind section to a single violinist, huge in the foreground, with the element of surprise adding to the visual boldness.

When photographing a full orchestra, the trucking shot remains the most expressive of all camera moves. When used properly, its lateral movement reflects the music's emotional content far more powerfully than either the dolly or the pan. Imagine a raking shot down a row of violinists. Now, as the music begins to swell and soar, the camera begins to glide past row after row of violinists, its sweeping movement propelled by the music, echoing its emotionality. A trucking shot magnifies audience interest because it constantly unveils new imagery, entering camera frame from the side (and thus, at apparently great speed), sweeping dizzyingly past in lateral motion and then disappearing to make way for new forms, new faces, new instruments.

Popular, Country, and Rock Music   Because these forms of music are less tradition bound than classical music, directors may be more imaginative in their treatment of them. If camera style becomes a reflection of subject matter, classical music will be relatively conventional, punk or acid rock will be bizarre and uninhibited, with popular and country music somewhere in between.

I will discuss rock 'n' roll as an opposite pole to classical music, setting out a few guidelines for its direction. Popular and country music will not receive special mention here; you can infer the camera styles for these, centered in the musical spectrum between classical and rock, from the discussion.

MTV, the current cable music television channel, daily offers a virtual textbook of camera gimmickry, distortion or multiple-image lenses, bizarre angles, off-center framing, negative or polarized images, and color enhancement or distortion as visual accompaniment to its music. Watching an hour or two of such often imaginative visual treatment of rock music will stimulate your own imagination and provide potential directors with an ever-enlarging bag of tricks.

Many of the camera and editing rules discussed earlier in the chapter apply equally well to rock. Cutting on the beat, for instance, is good practice for any uptempo selection. With rock, generally more frenetic in tone and tempo than classical or pop, you may want to cut more frequently.

If boldness in closeups seemed essential with classical music, it is imperative in photographing most rock selections. Whereas head, shoulders, and violin would constitute an effective closeup in classical, a head alone is often its rock equivalent. Cameras are frequently hand held, which allows tremendous freedom of movement. Boldness is sometimes achieved with wide angle lenses, which distort the images for dramatic, occasionally comedic effect. Such lenses magnify foreground and diminish background, creating an illusion of extraordinary depth. The wider the angle, the greater the degree of distortion.

Boldness is also achieved with bizarre angles such as shooting the drummer from extreme low angles to frame past an assortment of drums in the foreground. Such bizarre angles frequently shoot up into studio lights, allowing, almost seeking, lens flares, which contribute to the frenetic mood.

Both of these so-called rules, however, must be subordinate to the dramatic principle of progression. *Progression* simply means *growth* or *development*. In drama it is a climbing action in which events gradually build to a series of crises, each higher than the last. The highest dramatic peak of all usually occurs at the end of a play and is called the climax. The principle of progression occurs again and again in the entertainment world. Preliminary boxing (or wrestling) matches are always followed by the main event. Musical extravaganzas usually climax with a grand finale, the entire cast gathered on stage in an elaborate setting, lights flashing, umbrellas twirling, to sing the show's hit song. Oldtime vaudeville entrepreneurs used to place their headliners in the next-to-closing spot, because that was theoretically the climax.

When sequencing a music show, arranging the order of numbers, try to create a sense of growth in the program. If you're staging a half-hour show with a commercial in the middle, you are, in effect, staging a two-act show. Therefore, each act should end on a high musical peak—and the second peak should be higher than the first because it constitutes the climax of your program.

Chapter 7, Staging the Camera, discusses at some length the concept of camera progression, in which audiences experience a sense of building tension merely from increasingly close angles. Ideally, the increase in tension generated by camera angles should reflect the increased tension generated by actors and the dramatic situation. But the increasingly close angles alone create some illusion of progression.

The director can apply that principle to music. If you begin a number with extremely close shots, you may leave yourself with nowhere to go. If you want to end a number extremely close on hands, instruments, and faces, you must plot your camera progression in advance, calculating where you want to be at the halfway point and how wide your angles must be at the start. Of course, it is often effective to begin extremely close—fingers plucking guitar strings or tapping a bongo drum—so long as you drop back to wider shots for orientation and then build your progression from there.

In directing rock numbers, be adventuresome. Explore new directions. It's no crime to try something wild and fail; it *is* a crime not to try. Even if your sorties don't always work, they will at least add freshness to the look of your show. Some of that freshness is derived from lighting and scenery and use of "borrowed" imagery from videotape and film (which I will discuss shortly), but imaginative camera angles and crisp, clean, on-the-beat cutting will make your show professional and viewable.

Vocals   In the early days of television, Dinah Shore starred in a quarter-hour show twice weekly over NBC. For a few delightful months I was the stage manager of that show. Its director, a sensitive and talented man named Alan Handley, photographed most of Dinah's ballads in essentially the same camera pattern. At the start of each song, camera would be wide. As Dinah wove her spell, creating greater and greater involvement with her audience, Handley moved his camera closer. At the moment of greatest intimacy, of deepest emotionality, the camera was close, a head-and-shoulders shot. Then the song ended. As the audience descended from the mountaintop, the camera drifted back again. In a sense, the director's camera paralleled the movement of the viewer's consciousness, moving closer to the singer as she created her magic and moving away when she had finished.

With genuine talent, it is usually unnecessary for a director to resort to fancy camera angles and gimmicry. Whether it be an actor in a monolog, a comedian, or a singer, if the artist has the ability to capture and hold viewers with the intensity of his or her performance, then the director will only disturb audience involvement by arbitrarily changing angles.

Students in television production classes tend to become fascinated with hardware. They are dazzled by cameras and microphones and the control room's display of electronic magic. Many come to believe that a show's success depends on the skilled application of technique and technology—that the combination of imaginative lighting, optical effects, and a sweeping crane shot will spark audience applause. Understandable—but largely untrue. What appears in *front* of a camera contributes far more significantly to a show's success than the gaudiest display of electronic wizardry. Cameras, lighting, microphones, editing devices, and fancy lenses are only a means of communicating a message. But it's the message that counts, not the communicative devices.

When a vocalist is genuinely talented, don't get in the way. Let the artistry shine through. If the singers are less than talented, they may need some directorial help. Theatricality in the form of visual or audio effects often makes mediocre singers appear better than they are. Echo chambers add resonance to weak voices.

A background of dancers, flashy scenery, or lighting effects or constantly chang-ing camera angles sometimes dazzle audiences and distract their concentration from a lackluster singer. Spectacle (sound, motion, and color) is often a director's most valuable tool (see Chapter 2).

Enhancing a number through the use of spectacle, of course, is not limited to mediocre performers. Content dictates treatment. A simple, honest ballad suggests simple and honest treatment. A lively show tune or production number suggests a more theatrical staging. If the performer is genuinely talented, that talent inevitably will shine through, regardless of camera cuts, fancy angles, or production gimmicks—or the lack of them. If your budget will allow dancers and elaborate scenery, fine. If not, you may fall back on the hardware that once fascinated you in your TV production classes and utilize chroma key, multiple-image lenses, and fancy camera angles to create your own brand of spectacle.

Be kind to your vocalist. In drama it is often effective to cut to a bold, gritty closeup, revealing every pore and pimple in an actor's face. Not so with singers. During final rehearsal (with singer in makeup), zoom or dolly a camera to a closeup to determine just how close you can get before the shot becomes unflat-tering. When vocalists sing a love song, it obviously breaks the illusion to reveal gold fillings in their teeth or, on a sustained note, the color of their tonsils. Your singers' success is your success. Make them look good.

Vocalists sometimes sing to their studio audience rather than to television viewers, perhaps because it is easier to communicate with people they can see. If the singers are talented, TV viewers will accept such misplaced attention, but usually it is with the feeling that they have become observers rather than partici-pants in the singer–audience relationship. Some vocalists (usually those who play club or concert dates) occasionally leave the stage and sing to individual audience members. Such moves never fail to generate excited reaction from the live audi-ence. But the viewers at home are less enthusiastic. They are no longer recipients of the singer's favors. They feel, with understandable envy, that they are being ignored. Successful TV directors remain conscious of viewers' expectations, steering their performers' attention away from studio spectators. The camera lens should be the only audience.

DRAMATIZING THE MUSIC

Just as writers must understand the theme of stories they write, so must directors understand the essential mood, flavor, or motif of a musical program. The central concept may begin with performers, writer, producer, or director. Once it has crystallized in everyone's mind, then the creative wheels may begin to turn.

The developmental pattern (much abbreviated here) might go something like this. Music is selected to fit the concept. Performers, writer, or director may suggest a sequence. The production team (writer, director, producer) discuss production concepts. All understand that these concepts will inevitably change as the show develops, but you have to start someplace. The writer prepares a first-draft script. The production team tears it apart and puts it back together again, searching for better ways to dramatize the initial concept and to cement the program's elements. The writer rewrites and polishes the script. The director prepares. Musical rehearsal and timing. Camera rehearsal. Broadcast.

Creating visual imagery that enhances the emotional content of music is something like taking a Rorschach test; no two people react to a given stimulus in an identical fashion. A rock song or a concerto will stimulate a dozen different concepts in a dozen different minds. There will, of course, be thematic similarities, but specific actions, settings, and camera teatment will vary enormously.

During each developmental step, the director and other members of the production team build upon the original concept, enriching it, expanding it, bringing it to life in a number of practical ways.

Dramatizing Rock Music

Dramatizing rock music has become one of the most innovative and creative of popular art forms. Many directors create elaborate sets with dozens of extras to dramatize the mood or atmosphere of a particular song.

Treatment of rock music generally falls into two categories: photographing the musicians as they play and devising free-association imagery that either dramatizes the song's story content (sometimes literally, sometimes with tongue in cheek) or depicts its mood. The two techniques are not always separated. Often they are interwoven, cutting between shots of a drummer and water dripping from a faucet or a beautiful girl riding a white horse. Sometimes shots of musicians are intercut with as many as three or four divergent visual themes interpreting the song's musical or lyrical pattern.

Photographing Musicians Camera coverage of rock musicians was discussed at some length in the section on popular, country, and rock music. Directors sometimes employ additional techniques that fall generally into the category of dramatizing although, as stated before, the categories of depicting and dramatizing frequently overlap. Traditionally, directorial treatment must depend on the nature of the music. However, certain common stylistic approaches seem relevant to many rock 'n' roll selections, especially those that are bizarre in nature or that wander far from reality.

Directorial treatment of rock music—especially as seen in MTV—often recalls the psychedelic experience. Its style suggests the dream worlds of Dali, Magritte, and de Chirico. It derives its subject matter from the subconscious mind. This concept explains why distortion lenses, light flares, and bizarre camera angles seem to fit the music. Modern directors are limited only by their imaginations in achieving such a dreamlike state. Techniques include:

— Scenery (projected on a screen, matted via chroma key, or actually constructed) such as a jail cell, a mausoleum, an ice floe, or a whorehouse.

— Colorful locations such as a waterfront, field of daisies, abandoned factory, or ghost town.

— Smoke (generated by a smoke pot or dry ice) clinging to the floor and suggesting that the musicians are floating on clouds.

— Color distortion, achieved either with gels on lights, color correction filters on cameras, or electronic wizardry. Such distortion may be mild or extreme (faces turning green or lavender), the latter suggesting not so much the dream world as the nightmare.

— Split screen, achieved either with optical mattes or with multiple-image lenses. Screen may be divided into as many as sixty-four images.

— Lighting effects, from crisp silhouettes to harsh cross lighting or key lights placed extremely low, throwing huge shadows upon a background. Also, flashing of key lights to suggest lightning (perhaps accompanied by thunder).

— Superimposing effects or imagery over the musicians. Smoke, flames, ocean waves, storm clouds, rose petals, whatever fits the musical mood.

— Electronically reversing polarity, rendering blacks white and whites black: a negative image.

— Shooting through or past foreground objects, such as water dripping down glass beads, racking focus from beads in foreground to musicians in background and vice versa.

— Tilting, polarizing, or stretching the image electronically.

— Star filters to create stars out of highlights (a glittery, sophisticated look) or fog filters to create a soft, hazy, fantasy feeling.

— Computer-generated effects that may twist, bend, flip, or twirl any of the above.

These effects constitute only a fraction of the number available to the imaginative director to help achieve a dramatization of musical content in photographing rock musicians.

Creating Dramatic Imagery The treatment of dramatic imagery often suggests the grotesque or erotic world of the psychedelic trip. It begins with lyrics and the musical content, but then veers into strange and fascinating directions that, when skillfully conceived and performed, add new dimensions to the music.

No book can instruct in the creative processes. Each person possesses a unique reservoir of imagery and develops a distinctive and personal way of reaching his or her subconscious mind to tap that reservoir. It might be helpful, however, to trace the free-associative wanderings that the mind sometimes experiences in the creation of imagery for a rock song.

Our assignment is to create a visual treatment for a rock ballad, "Burn Me." The lyrics of this song of dubious origin state: "Burn me with your love:/Warm me, take me, bake me/Fill the night with fires that wake me/Though I run, they overtake me./Baby, burn me with your love."

In creating imagery, the director or writer would probably read and reflect on the lyrics, allowing the music to create a mood. The stream of consciousness elicits a random succession of words and images: A bedroom. Thunder, lightning. Girl awakening in terror. Too obvious, too literal. So is the idea of running from flames. Or from some nameless terror. A girl running—perhaps in slow motion, as if in a nightmare. Down long corridor, endless. She falls, can't get up. People in her way, blocking her. They keep pulling at her, trying to stop her. Hideous people. Smiling. Mocking. One of them has the face of Satan. She keeps looking behind her, but we never see what's chasing her. Still so damned literal. Yet the idea of running in a nightmare is universal; everyone has experienced it. Red is the dominant color. Wherever she turns, fire symbols confront her. Descent into hell. Advertising sign: picture of the devil, smoke coming out of mouth. Maybe we get some humor in here: kids popping smoke-filled balloons. A street beggar walking on hot coals. Statue in the park smoking a cigarette. Lightning strikes a tree; it erupts in flames. Hard to do. Cost a fortune. Always the Unseen Fear pursuing her. How to end it? Does she get caught? Down subway steps. (Descent into hell.) Trapped against subway wall. White, shiny, tiled. She screams. Busy passersby ignore her. Blank, cold faces—like mannequins. Camera becomes the moving point of view of the Chaser. And we see her illuminated by a red glow that gets brighter and brighter as we move closer. Might work. It's visual, plenty for a director to work with. Rather literal. But it's a beginning.

Dramatizing Classical Music

Unless a concert has an unusually strong thematic flavor, it seems unmotivated and out of character to cut away from the orchestra or soloist for other than audience reaction shots. Purists would frown even on those. When symphony orchestras celebrate a special or festive occasion (for example, Christmas), it is usually with the playing of traditional airs that evoke intense audience emotions. In these instances, occasional, tasteful use of imagery may be used to complement the music.

Let's use Christmas as an example. Shots of Madonna and Child (oil paintings) or the nativity scene might accompany "Adeste Fidelis"; shots of Dickensian carolers or church bells tolling might accompany "The First Noel." Such scenes might offer visual amplification of the program theme, but the focus of interest always must remain on the orchestra or the soloists.

The flavor of the occasion may be echoed in the setting. Again, such visual enhancement should be understated. An orchestra is visually "busy." That is, its elements are numerous and difficult to delineate, especially in a full shot. The "busyness" is compounded by television's 525-line screen that prevents crisp definition. Intricate scenery or fancy trimming exacerbate the problem. Simplicity is the key to the setting.

Because of the size of symphony orchestras, their broadcasts usually originate in a concert hall or amphitheater. Individual artists perform in both television studios and locations. Recall the piano concerts at the White House by Vladimir Horowitz and others.

Vocalists It was once fashionable on television to create little stories out of popular vocals, with scenery, props, and occasionally actors or other singers. The pattern gradually lost favor because many of the "stories" were self-consciously clever or cloyingly cute and took attention away from the singer.

Today scenery and lighting are used to dramatize the mood of the song rather than its story. A garden setting, for example, creates an atmosphere even before the singer walks on stage. The setting also provides the singer with action during the song, which contributes to a feeling of ease, of naturalness. Many numbers have orchestral bridges in which the vocalist has nothing to sing and therefore must stand, awkwardly waiting to continue. A setting provides a graceful answer—walking in the garden, smelling flowers, feeding a bird in a birdcage, or swinging gently in an old-fashioned swing.

If an entire show is built around a vocalist, locations provide enormous color and scope. One Perry Como television special, for example, opened with its star aboard a small boat in the New York harbor and featured as its background the Statue of Liberty.

STUDIO VERSUS LOCATION

There are two initial options for your show's setting: studio or location. Directing a show in the studio offers many practical advantages. You have far more control of all production elements. In a musical show, sound must be a major consideration. Studios are acoustically superior to most locations. You are not bothered by airplanes flying over, trucks rumbling by, or police sirens. Sound in open air shows tends to lose "liveness" because there are no reflective surfaces to create reverberations. Conversely, indoor locations are often filled with hard surfaces, bouncing sound back and forth unmercifully, creating disturbing echoes. Both problems may easily be solved by prerecording sound tracks and allowing vocalists to lip-sync their performance. In the Perry Como special just described, such prerecording allowed total camera and performer mobility. TV audiences seldom wonder where the orchestral background is coming from.

Lighting is less of a problem because modern lighting units are relatively small and easily transportable. Still, on location you probably won't have the flexibility or lighting quality that you have in a studio. The same holds true in other areas. If a prop or costume is missing, someone must return to the studio to get it. Or run to a store to purchase another. Everything seems to take longer on location simply because you are working in someone else's backyard.

On studio shows, directors sit in a control booth, separated from cast and crew. On location, separation becomes isolation. Miles of cable connect cameras, microphones, and production lines (usually) to a control truck. The isolation is so complete that directors must depend almost entirely upon their floor managers. They don't have the luxury of bellowing into a studio address system. (An abrasive habit, even in the studio!) If the floor managers are not alert and authoritative, a dark cloud of frustration and helplessness settles over the director. Like trying to pick up a needle with ice tongs.

But locations also offer major advantages. When the location sufficiently enriches a show, complementing it in mood and character, then advantages far outweigh discomfort or inconvenience. No matter how skillfully designed the setting, no matter how artful the lighting or sound, a studio show always looks like a studio show; it has a slickness, a glossiness that, with less than first-rate talent, seems synthetic.

Some years ago I directed "The Tex Williams Show," a western musical variety program that for a time originated at Knott's Berry Farm in Buena Park, California. The show first was staged in the streets of the western ghost town, surrounded by hundreds of ebullient fans. The ghost town background was ideal for a western variety show, and the fans lent tremendous color and excitement.

Later, because the show blocked traffic, interfering with sales in many of the Knott stores, they decided to move us. The Knott organization built us a small but handsome western street of our own in an open area with a grandstand large enough to accommodate triple the number of spectators. We broadcast the show every Sunday afternoon for over a year from that location. Ultimately, probably because of location costs, the show moved to studio E at NBC. It expired a couple of months later.

"The Tex Williams Show" started as a colorful, slam-bang Western, energized by the proximity of fans and by the earthy, good-humored atmosphere of the ghost town. You could almost smell the hot buttered popcorn. If a stagecoach came rattling down the street during a comedy routine, so much the better! If a baby cried (and one always did—right in the middle of a soulful ballad!), it simply contributed to the noisy, familial, cotton candy atmosphere.

When the show moved to the more sedate arena, separated now from its audience, the production values vastly improved; we had better control of sound and camera. Yet the show lacked excitement. It no longer rubbed elbows with its fans. Even though it originated at Knott's Berry Farm it had become a studio show, operating in front of painted sets in a traditional audience–stage pattern. Some of the warmth had disappeared. And with it, some of the show's appeal. The final move to a television studio completed the sterilization process.

When the right chemistry exists between a show and its location, magic happens. The show emerges with character and colorations never possible in a studio. Put Luciano Pavarotti in a cathedral singing Christmas songs and the TV screen lights up the room. Put Madonna in Madison Square Garden or Sting in the Los Angeles Coliseum, and you create vibrations impossible to achieve in any television studio.

SCORING DRAMA

Producers hire composers for the same reasons that they hire directors: respect for their work and the feeling that they are appropriate "casting" for a project—that is, that the composer's style fits the character of the producer's film.

The composer goes to work when most other members of the production staff pack up and go home: after the editing has been completed. Now the composer sits in a projection room with the producer and (sometimes) director and views the film for the first time. After the screening, the composer discusses the project with them, offering preliminary suggestions about musical treatment. Ideas at this meeting are usually tentative because most composers prefer to allow their ideas to simmer before putting notes on paper.

The composer will view the film again later on a **moviola**, this time with a music cutter (editor), determining exactly where music is needed and the general nature of each cue, trying to incorporate the producer's and director's ideas.

After composers have written a score, they hire an orchestra to record it. The size of the orchestra depends on the composer's musical needs as well as the size of the producer's budget. Previously, at various points during the writing, the composer has played motifs for the producer, seeking agreement and approval. Now, at the recording session the producer hears the full score for the first time. With the high-fidelity equipment in most recording studios, the score sounds magnificent; it will never again sound quite as rich and stirring.

Following the session, the music editor winds the taped music cues on reels and synchronizes them with the film. The editor spaces the cues by placing blank leader between them, so that they will occur at the appropriate spots when the film is run. This synchronization will be tested and refined during the dubbing session. If you remember, **dubbing** is the process of integrating dialog, sound effects, and music into a single composite sound track.

At the dubbing session everything comes together. The producer and director supervise as sound effects are balanced and blended with dialog and music. Now the picture springs vividly to life as the full audio perspective emerges. For the first time, the producer and director can assess and, when necessary, reshape the relationship between the picture and its sound components. In the process, music cues sometimes are shortened; sometimes they are repositioned; sometimes, if they fail to accomplish their dramatic purpose, they are totally eliminated.

Purposes of Scoring

Scoring adds dimension to a picture. It creates impact and it evokes emotion. Perhaps the reason for music's extraordinary power over our emotions is that it goes almost directly to the subconscious. We watch and participate in a dramatic scene, identifying with characters, experiencing their problems. We sit before a TV set or in a theater almost in an alpha state. If the show is well written, directed, and acted, we don't even hear the musical score, so involved are we in the dramatic situation. We don't hear the music *consciously*. It penetrates directly to our sub-conscious, where it can stir us on a fundamental level.

Because music's power to evoke emotion is so great, the relationship between picture and score becomes enormously important. That relationship is curiously symbiotic. Music affects the character of a scene; the scene affects the character of its music. Think of the popular song "As Time Goes By" as used in *Casablanca*. Before appearing in the film, it was just another fairly average romantic ballad. After it became the movie's love theme, audiences associated it with specific characters and imagery; it had acquired something of their identity; the song evoked bittersweet emotions arising directly from Rick's and Ilsa's star-crossed relationship. The song (which became part of the score) enriched the picture; the picture enriched the song.

Music can enhance—but it can also detract. If a score intrudes for any reason, it can undercut a scene's emotional impact. Once audiences become aware of a musical presence, if music is too loud or the wrong coloration, they will temporarily withdraw their participation in the story, suddenly aware that they are being manipulated.

Music can also detract if it overstates. In the teens and twenties, piano or organ accompaniment to silent films tended to be on the nose—that is, exactly the musical treatment that audiences expected; it was obvious and heavy handed. When cowboys pursued Indians, the pianist played galloping chase music. When the sweet, white-haired grandmother died, the organ moaned its variation of Hearts and Flowers.

These scores worked well when movies were relatively new and audiences were unsophisticated. Today's audiences are more demanding. Music that is on the nose, that mirrors screen action too exactly, swooping when the hero leaps from a tree and cascading when he slides down a rock face, is referred to as Mickey Mouse music. Such a label derives from cartoons in which the musical phrases exactly coincide with the actions of characters.

Sensitive composers often try to play against the action of a scene. An excellent example is the use of a zither playing a rather festive melody in Sir Carol Reed's *The Third Man*, an award-winning mystery that played in the darkened streets and sewers of postwar Vienna. The zither smacked of the old world and seemed entirely appropriate as a scoring instrument. And the gay melody provided a surprising and curiously chilling counterpoint to the mysterious proceedings.

Because of TV's quantities of police and private eye shows, composers face the challenge of scoring violent action without the music becoming clichéd and melodramatic. The answer often is to keep the score simple. I recall the climax of one action adventure program in which the composer effectively played against the overstated action by using a single instrument for scoring, a timpani drum in occasional, sporadic beats. Boom—boom, boom—boom—boom, boom. You get the idea.

Themes/Motifs

Most composers relate specific melodies or musical instruments to significant characters. When audiences hear a theme repeated as a background to scenes involving a specific character—especially in moments of emotional intensity—they associate the theme with the character. Thus, if a woman wanders on a mountaintop grieving for her lost lover, the composer might weave the lover's theme into the musical score, thereby evoking the lover's image and simultaneously providing the audience with insight as to the woman's mental processes; the music tells us what she's thinking about.

Themes are not limited to characters. They often define relationships. Those most frequently heard are love themes. The next time you watch a love story, listen to the music. Listen intently. Try to analyze how the composer is using themes. You will find that he or she is (1) identifying certain themes with characters, (2) changing those themes subtly from scene to scene so as to comment on the dramatic situation, and (3) stimulating your reactions by reprising those themes at emotional moments. If you remember *Love Story*, you may recall that the piano theme first used to underscore a delightful and moving love relationship was used later in the film to evoke audience tears. When the love theme in *Jagged Edge* was reprised, also played on a piano, it commented bitterly on a love that had turned to horror.

Music often provides hints about the nature of characters. A comical character frequently has a humorous theme; a fearsome character usually has a dark and foreboding theme. Recall the theme from *Jaws*. It instilled fear. Whenever we heard it we sank lower in our seats, nervously anticipating some stomach-wrenching depradation. Even when we didn't see the shark, the music told us that he was near at hand. Recall "Lara's Theme" from *Doctor Zhivago*, richly melodic and romantic in flavor. Years after we had seen the film it still evoked memories of Lara's tragic love affair.

Chapter 2's discussion of rhythm and repetition offers clues to a composer's use of themes. When a theme is repeated, it evokes memories of the scene first associated with it. If something cruel happened on screen the first time we heard a theme, the repetition will make us cringe, probably without realizing why. If another, even greater cruelty occurs the second time we hear the theme, we become slightly traumatized and the music becomes a symbol of that trauma. The third playing of the theme makes us anxious, apprehensive as we await another shocking manifestation of cruelty. (The process is somewhat similar to Pavlov's famous experiments in conditioned reflexes. Remember?) Repeated again and again, musical themes become cumulative, acquiring greater context, evoking new memories in audiences and new emotional colorations.

Source Music

Music that originates from a source within a scene is (logically) called **source music**. Another way of identifying source music: It is music that the characters in a scene can hear—as opposed to the musical score, which only the audience can hear. Examples of source music are a radio or TV set, a band in a nightclub, or a character playing a musical instrument.

I include source music in this discussion because music that begins as a source often finds its way into a movie's score. Returning to our old friend *Casablanca*, the love song was first played on the piano in a Paris bistro and again on the piano in Rick's Café Americain. The film's musical director, Max Steiner, then incorporated the melody into the film's score. Note: When it was played on the piano it was source music; the characters could hear it. When it was played by a full Warner Brothers orchestra as background music, working on our emotions and commenting on the dramatic content of scenes, it became a score; the characters could not hear it.

The practice of converting source music to score is far from new. It began in the early 1930s with the advent of musicals. Characters sang love songs or danced with their sweethearts in elaborate settings. Those songs later provided the underpinnings for an effective score.

Today suitable source music can often be found in commercial music libraries and need not be composed or recorded especially for a project.

Spotting Musical Cues

When composers view a completed film, one of the questions they ask themselves is, "Where will I insert music?" Where will it enhance the film? Where will it intrude?

Finding the places where music will be inserted is called **spotting**. Composers look for several traditional places in a film in which to spot their **music cues**.

Among these: beginnings and endings, transitions, silent action, and emotional moments.

Beginnings and endings Like the overture in a Broadway show, title music establishes the flavor of what is to come. It sets the mood. If a comedy, the opening music will probably be bright, bubbly, quixotic, telling the audience "Get ready to laugh." If a thriller, the title music will establish an entirely different mood, vaguely unnerving, warning of traumas to come. Listen to the themes of various TV programs. Notice how different in flavor is the title music of a police show from that of a sitcom. These themes become signatures, trademarks evoking images or memories of a program.

The closing musical passages of motion picture and TV stories (prior to closing titles) usually are positive in flavor, suggesting resolution, reflecting the positive outcome of the story itself. Often they are extremely brief—a **tag**, or final musical punctuation. Composers frequently use tags to punctuate individual scenes, especially in comedy. Such tags become exclamation points for comic moments. If a theme has been established for a character, the tag might become a whimsical variation of that theme.

Transitions A transition between (say) a lighthearted scene and one that carries ominous overtones places a burden on the audience, asking it to shift emotional gears quickly, often in a couple of seconds. Composers help audiences make this emotional adjustment by the use of musical transitions. By changing flavor within a musical cue, the composer leads the audience gracefully from one mood to another.

Transition cues also sustain or reinforce a mood. When a major shock will occur in the upcoming scene, the composer realizes the necessity of not only sustaining the ominous mood of the previous scene but also building on it, not allowing the audience to catch its breath, taking viewers to an even higher level of anxiety and apprehension. Therefore, the composer devises a cue filled with somber tones, hinting of horrors to come, preparing the audience for the monster's appearance or the corpse rising from its coffin.

Filling the Holes Nature abhors a vacuum. Producers abhor a vacuum in their sound tracks. When no characters are speaking dialog and sound effects are minimal or absent, some producers (and some composers) feel that audiences become uneasy. As a result, music often appears at such moments merely to sustain a mood, to transport audiences over the desert of silence to the next oasis of sound.

Such thinking is reminiscent of the early days of talking pictures when producers, excited over their new toy (sound), crammed every frame of film with dialog, songs, music, or effects.

It is refreshing to break rules, to thwart audience expectancy. And audiences enjoy variations from the norm. Nothing in the rulebook says that we have to fill silences. Nothing says that we have to play tags at the ends of scenes, the ends of acts, or the ends of shows. Clichés abound in television and film. Genuinely creative directors, writers, and composers work hard at finding ways to avoid them.

Emotional Moments   When producers want to make an audience cry, their most valuable tool is music. It worked for David Wark Griffith in 1909. It worked for David Selznick in 1939. It works today.

One component of music that often succeeds in evoking emotion is *melody*, which the dictionary defines as "the rhythmical succession of single tones producing a distinctive musical phrase or idea." The word *rhythmical* is significant; it implies repetition. Think of melodies from TV or film (or anywhere!) that have burned themselves into your mind. You will find they usually contain simple musical phrases that are repeated again and again. Do you remember the theme from the TV series "Hill Street Blues?" It consisted of three notes repeated over and over, with almost insignificant variations. This absurdly simple, enormously successful melody was recorded by a number of musical groups and played repeatedly on radio and TV.

Music that underscores emotion should carry within itself the seeds of that emotion. It should reflect the emotion, comment on the emotion, or somehow evoke the emotion, but not (as we have discussed) in a way that is heavy handed or on the nose. A difficult assignment. But that is why skilled composers earn handsome fees!

CHAPTER HIGHLIGHTS

— Because music evokes emotion, music programs pose a special challenge to directors: how to complement that emotionality in visual terms. Answers arise from use of technical elements—camera, editing, and lighting—and from appropriate visual imagery.

— As with other program forms, music shows require careful preparation. Directors should familiarize themselves with performers and the music selections they will play or sing. When scripts are used, directors should involve themselves in the creative processes early, working with the writer to ensure a top-notch presentation.

— In directing music shows, the ability to count bars is almost a requisite. Directors who lack such ability must rely on a musically oriented associate or a stopwatch.

— Use of camera and editing should match, as much as possible, the flavor and mood of the music. Cuts are crisp, sharp, and incisive, ideally suited to uptempo music. Because dissolves are more fluid, more languorous, they create visual poetry for certain dance numbers and romantic ballads.

— With symphony orchestras—as with most groups—wide angles usually are dull. Close angles create visual impact. Raking (side angle) shots down a row of musicians create an effective visual rhythm. The faces of some virtuosos communicate the music's emotion far more than shots of instruments do.

— With rock music, MTV provides a virtual textbook of techniques for directors. Because the music is often bizarre and adventuresome, directors are encouraged to echo that sense of adventure in their camera treatments. The dramatic element of progression is still at work, however: The program should build to a climax.

— With genuinely talented vocalists, directors should keep production elements simple, allowing the talent to shine through. With less-than-gifted performers, directors should try to fortify performances through addition of such theatrical elements as dancers, fancy settings, and background singers.

— Music shows may originate in a studio or on location. Both offer advantages and disadvantages. Studios provide greater security; the director has better control of all production elements, including sound, lighting, and effects. But locations sometimes offer excitement, an opportunity for original visual treatment and an ambience that is seldom possible in a studio.

— Composers score a motion picture or TV drama by adding musical cues that will reinforce story values and add emotional impact. Themes for the main characters or relationships help to identify characters and to evoke emotion.

— Music cues are usually spotted into a dramatic program at the beginning, to establish the flavor of the show, to smooth moments of transition, to help audiences change moods, to fill long silences, and to enrich emotional moments by reinforcing story values.

PROJECTS FOR ASPIRING DIRECTORS

1. You are the director of a musical show. Select a group that you would like to feature in your show. Pick five of their most successful songs. Now (a) arrange them in a sequence that you feel would make an effective program, and (b) decide how you would present each of the songs. Write a paragraph on each song detailing the staging, setting, lighting, camera treatment, or anything else that you feel would make an effective presentation.

2. Take one of the songs from your program in the preceding project, and decide how you would present it as a music video, with unlimited time and money. Let your imagination run wild. Type the lyrics in script form and make notes on how you would edit your MTV, line by line, phrase by phrase.

3. Run a feature-length motion picture of your choice on a videocassette recorder (VCR). Log the exact spots when music occurs. After you have finished, answer these questions:
   a. How many minutes of music were in the film? What percentage does this represent of the film's total running time?
   b. Did any of the music cues seem out of place, at odds with the dramatic material? Can you analyze why there seemed to be a mismatch?
   c. Try to figure the composer's purpose in spotting each of the first ten music cues. How were they used: As transitions? To fill a silent track? As background, to reinforce emotion? Can you find music cues that don't fall into any of these categories? Describe them on paper and try to figure out the composer's purpose.
   d. As an experiment, record the title music on a separate audiotape. Now play the title music from this film over the opening titles of a second film with its sound turned off. Analyze the effect. Did the incorrect title music enhance or detract from the second film? If it enhanced, can you figure out why? If it detracted, can you figure out why?

BOOKS THAT MAY BE HELPFUL

Armer, Alan A. *Writing the Screenplay—TV and Film*. Belmont, Calif.: Wadsworth, 1988.

Arnheim, Rudolf. *Visual Thinking*. Berkeley: University of California Press, 1969.

Arnheim, Rudolf. *Art and Visual Perception*. Berkeley: University of California Press, 1974.

Barr, Tony. *Acting for the Camera*. Boston: Allyn & Bacon, 1982.

Beaver, Frank E. *Dictionary of Film Terms*. New York: McGraw-Hill, 1983.

Berger, John. *Ways of Seeing*. New York: Penguin Books, 1977.

Blumenthal, Howard. *Television Producing and Directing*. New York: Harper & Row, 1987.

Brady, Ben, and Lee, Lance. *Understructure of Writing for Film & Television*. Austin: University of Texas Press, 1988.

Burrows, Tom, and Wood, Donald. *Television Production—Disciplines and Techniques*. Dubuque, Iowa: Wm. C. Brown, 1986.

Clurman, Harold. *On Directing*. New York: Macmillan, 1972.

Cole, Toby, and Chinoy, Helen Krich, eds. *Actors on Acting*. New York: Bobbs-Merrill, 1949.

Cole, Toby, and Chinoy, Helen Krich, eds. *Directors on Directing*. 2d ed. New York: Bobbs-Merrill, 1979.

Dmytryk, Edward. *On Screen Directing*. Woburn, Mass.: Focal Press-Butterworth, 1984.

Egri, Lajos. *The Art of Dramatic Writing*. New York: Simon & Schuster, 1946.

Elisophon, Eliot. *Color Photography*. New York: Viking Press, 1961.

Field, Syd. *Screenplay: The Foundations of Screenwriting*. Enlarged ed. New York: Dell, 1984.

Fischer, Heinz-Dietrich, and Melnik, Stefan Reinhard, eds. *Entertainment—A Cross Cultural Examination*. New York: Hastings House, 1979.

Funke, Lewis, and Booth, John E., eds. *Actors Talk about Acting*. New York: Avon Books, 1963.

Giannetti, Louis D. *Understanding Movies*. Englewood Cliffs, N.J.: Prentice-Hall, 1976.

Gradus, Ben. *Directing: The Television Commercial*. New York: Hastings House, 1981.

Henning, Fritz. *Concept and Composition*. Cincinnati, Ohio: North Light, 1983.

Lewis, Colby. *The TV Director/Interpreter*. New York: Hastings House, 1972.

Moore, Sonia. *The Stanislavski System*. New York: Penguin Books, 1984.

Pincus, Edward. *Guide to Filmmaking*. New York: Signet, 1969.

Reisz, Karel. *The Technique of Film Editing*. New York: Hastings House, 1968.

Reynertson, A. J. *The Work of the Film Director*. New York: Hastings House, 1970.

Rose, Tony. *How to Direct*. New York: Focal Press, 1954.

Schramm, Wilbur. *Men, Messages, and Media: A Look at Human Communication*. New York: Harper & Row, 1973.

Seger, Linda. *Making a Good Script Great*. New York: Dodd, Mead, 1987.

Sherman, Eric, ed. *Directing the Film—Film Directors on Their Art*. Boston: Little, Brown, 1976.

Sievers, David W., Stiver, Harry E., and Kahan, Stanley. *Directing for the Theatre*. Dubuque, Iowa: Wm. C. Brown, 1974.

Silver, Alain, and Ward, Elizabeth. *The Film Director's Team*. New York: Arco, 1983.

St. John Marner, Terrence (in association with London Film School). *Directing Motion Pictures*. New York: A. S. Barnes, 1972.

Stanislavski, Constantin. *An Actor Prepares*. Translated by Elizabeth Reynolds Hapgood. New York: Theater Arts Books, 1936.

Truffaut, François, and Scott, Helen G. *Hitchcock*. Rev. ed. New York: Simon & Schuster, 1984.

Wurtzel, Alan. *Television Production*. New York: McGraw-Hill, 1983.

Zettl, Herbert. *Sight, Sound, Motion: Applied Media Aesthetics*. Belmont, Calif.: Wadsworth, 1973.

Zettl, Herbert. *Television Production Handbook*. 4th ed. Belmont, Calif.: Wadsworth, 1984.

GLOSSARY

Above-the-line Phrase referring to so-called creative areas of production: producer, director, writer, actors, composer. Most commonly used in budgeting.

Action! The word most directors use to cue actors/performers to begin a scene or performance. When not a directorial command, the word applies to movement by actors within a scene.

Adda The commercial name for a computer that creates and stores video graphics.

Ad lib Dialog or action that is spontaneous and unrehearsed.

Angle (camera angle) The position from which a camera photographs action. For example, high angle, low angle.

Answer print A film that has been color balanced by a processing laboratory and that contains a composite sound track. Also called a *first trial composite*.

Antagonist The person, force, or element that opposes the protagonist, preventing him or her from reaching a goal, thereby creating drama's primary conflict.

Arc A curving camera movement. May be circular, semicircular, or very slight.

Art director The person who designs sets and coordinates visually related production materials.

Asides In drama, lines of dialog spoken directly to the audience.

Aspect ratio The screen proportion: three units high by four units wide in TV, wider for most feature films.

Assistant director (AD) In film, either a first or second assistant; people who assist the director in staging scenes and maintaining order and discipline on a set. The "second" handles paperwork. In live or videotape work, the AD usually times a show.

Audio   The sound portion of a televised program or commercial.

Authority   A quality in performers that suggests they are at ease, in control of themselves and of the audience.

Automated dialog replacement (ADR)   A computerized system for looping, recording new dialog to replace unsatisfactory dialog and to synchronize it with an actor's lip movements.

Backlight   A highly directional light that illuminates subject from the rear, adding highlights and separating subject from background.

Back story   A character's background or history that contains the most significant moments that shaped the character.

Back timing   A process for ensuring the correct program length: timing each element and then determining where the program should be, moment by moment as it progresses.

Balance   The comfortable state in pictorial compositions achieved by objects stabilizing each other in a frame, comparable to figures on a seesaw.

Beat   A smaller dramatic unit within a scene; a scene within a scene; a change in direction of scene content.

Below-the-line   Refers to technical and production personnel; term is most commonly used in budgets.

Blocking   The director's manipulation of performers within a set: their movement, relative positions, and business. Also, the manipulation of camera.

Boom (microphone boom)   A long metal arm holding a microphone, manipulated to follow performers about a set, picking up dialog.

Box   A rectangle containing visual material, usually keyed into the corner of a TV newscaster's frame.

Business   Personal actions by a performer in a dramatic scene (for example, combing hair, sewing, ironing, reading a magazine).

Busy   Term used to describe subject matter, often a background, that has an intricate pattern.

Canted angle   Sometimes called a "Dutch" angle in which the camera tilts to the side, placing the horizon at an angle, suggesting disequilibrium.

Cartridge ("cart")   An audiotape recording on a loop that rewinds as it plays.

Cassette   A plastic case containing either a videotape or audiotape recording.

Character generator (CG)   A computer device that reproduces letters or numbers electronically on a TV screen, affording a choice of size, font, and color.

Cheat   To make an adjustment of a camera's or actor's position that is so minor that the audience will not notice.

Cheated POV   A point of view shot that is higher, lower, or closer in order to make a dramatic point. For example, an angle cheated closer so that audience can read a newspaper headline.

Chroma key   A special effect that replaces a background color (often blue) with selected video material, as from another camera.

Chyron   Commercial name of a computer that generates and stores graphic material.

Cinematographer (director of photography)   Person in charge of a camera crew, responsible for lighting, composition, exposure.

Claustrophobic frame   An angle that is uncomfortably close to subject matter, restricting the audience's view.

Closeup (close shot)   A camera angle that is close to a performer or object, delineating detail and intensifying dramatic values.

Closure   The resolution of a screenplay, the coming together of story elements.

Color correction filters   Filters that change a scene's color values, usually subtly, for aesthetic effect.

Complementary angles   A pair of camera angles that are similarly framed but provide reverse perspectives on a subject. For example, over-shoulder shots, one favoring actor A, the other favoring actor B.

Composition (visual)   A harmonious arrangement of 2 or more elements, one of which dominates all others in interest.

Computerized graphics   Pictures, effects, or titles generated by a computer. For example, Adda, Chyron, Quantel.

Control room (booth)   A room from which a director and technicians control the rehearsal and broadcast (or videotaping) of a show.

Corner key (CK)   Usually graphics that are keyed into the upper corner of an anchor person's frame. *See* Box.

Coverage   The camera angles a director needs for dramatizing values in a scene and for effective editing. For example, a full shot, over-shoulder shots, closeups.

Covering   Photographing the action.

Crane   A metal boom arm with a camera mounted on one end, allowing vast sweeping upward or downward movements (craning).

Crawl   The movement or scrolling of titles on a screen, usually upward.

Cross   An actor's move across a stage or set.

Crossing the line   A jarring reversal of characters' positions within a frame caused by photographing characters from opposite sides of an imaginary line that bisects them.

Cue cards   Cards containing large, printed dialog, usually held just off camera for performers to read.

Cues, music   In scoring, musical phrases, figures, or passages composed for insertion into dramatic material.

Cut   A director's command to break off a performance, as at the end of a take. Also an instantaneous change of camera angle, accomplished by joining together separate film or videotape segments or by switching electronically between cameras. Also, the deletion of program or script material.

Cutaway   A cut to a person or action that is not the central focus of attention, perhaps to a spectator. Sometimes used by editors to delete unwanted footage.

Cyclorama (cyc)   A large cloth background for performances, often blue in color.

Dailies   A screening of film or videotape usually photographed the previous day.

Day for night   Photographing dramatic material outdoors in the daytime but, through exposure and filters, made to appear as night.

Deep focus   Photography in which the picture is in sharp focus from foreground to background. Usually accomplished by a small f-stop and correspondingly brighter illumination.

Depth of field   The range of a picture that is in sharp focus.

Desensitizing   Diminishing the impact of subject matter upon an audience because of repeated viewings.

Diagonal line   The most powerful force in visual compositions in terms of directing viewers' attention.

Diffusion filter   A filter placed in front of lens that softens the appearance of subject matter, often flattering or idealizing subject matter.

Digital leader   Numbers appearing at one-second intervals at the head of videotape. A visual countdown used for cueing.

Digital video effects (DVE)   An electronic device for inserting graphics or other material into a video picture.

Director   The person who shapes a program's message through the manipulation of artistic and technical elements, seeking to maximize their effect upon an audience.

Directors Guild of America (DGA)   The union to which professional directors belong.

Disequilibrium   The uncomfortable state resulting from story or visual elements that are askew, normalcy having been ruptured.

Disinvolving moves   Moves by one performer away from another. Such moves often suggest retreat, escape, or discomfort at physical proximity.

Dissolve An optical transition between scenes or camera angles; a blending or overlapping of images as one "melts" into the other.

Dolly (crab dolly) A four-wheeled camera vehicle that allows movement in all directions. A dolly shot moves forward or backward in relation to subject matter.

Dominance Achieving visual importance within a frame; attracting more attention from viewers than do other objects in the frame.

Downstage That part of a stage or setting that is nearest the audience or camera.

Dry Rehearsal without cameras. Also usually without set or props, as in a rehearsal room.

Dubbing The combining of dialog, music, and effects into a single sound track.

Dubner A computer that creates the illusion of movement in its graphics.

Editing The cutting together or splicing of film or video elements either to maximize dramatic effect or to achieve correct timing.

Effects bus That part of an electronic switcher that controls effects such as wipes or dissolves.

EFX Abbreviation for *effects*.

Electronic field production (EFP) Video recording of drama or feature stories away from the studio. EFP usually requires postproduction editing.

Electronic news gathering (ENG) Using portable camera, sound, and recording units to transmit or record news or feature stories.

End cue *See* Out cue.

Establishing shot A comprehensive angle that provides excellent orientation. Usually a wide or high angle.

Executional command The director's command to the technical director to carry out the function that was previously set up. For example: "Take it" or "Dissolve!"

Exposition Information that the audience needs to know to understand a story. Usually revealed through dialog or visual clues.

Expressionistic Screenplay, visual material, or directorial treatment that is intended primarily for emotional effect.

Fade in/out Transitions from black to full picture, from full picture to black. Fades often signify a curtain or commercial break.

Fax Abbreviation for *facilities*.

Feed lines Lines of dialog that serve no real purpose other than to cue another actor's response. For example: "And then what happened?"

Fill light Light that illuminates shadow areas, softening them.

Filter A glass or gelatinous material placed in front of a camera lens that changes the optical or color quality of an image. Also, an electronic instrument that eliminates certain sound frequencies, usually in postproduction, to achieve the effect of sound heard through a mechanical device such as a telephone.

First trial composite A film print in which color values have been balanced from scene to scene and which also contains a composite sound track.

Flare A halation or bright flash when camera is pointed at a light source.

Flashback A moment, episode, or sequence that is retrospective. It is usually recalled by a character and therefore colored by his or her memory.

Floor manager The director's assistant on the stage floor. When directors are confined to control booths, floor managers become their arms, legs, and voices on the floor.

Fog filter A camera filter that creates a soft, luminous image with varying degrees of diffusion. Used extensively in TV commercials.

Format The pattern or form of a TV program.

Frame The perimeter of a TV/film picture; a single photographic unit of film. Also a verb: to enclose or encompass subject matter.

Frame storer   A computer that holds enormous quantities of visual information. Examples: the Adda and the Chyron.

Freeze frame   An optical effect in which a moving subject suddenly becomes stationary, frozen in position. Also, the final frames of a commercial after the audio portion has concluded.

f-stop   Indicates the size of the lens aperture; the higher the f-stop number, the smaller the opening; the smaller the opening, the greater the depth of focus.

Full shot   *See* Establishing shot.

Gizmo   *See* Limbo.

Golden Mean   Believed by some to represent the aesthetically perfect spot within a frame in which to place a figure or object.

Graphics   Two-dimensional visuals, such as drawings, maps, and charts, prepared for TV by graphic artists or by computers.

Hand-held camera   Technique that directors use to evoke feeling of news or documentary coverage, to create appearance of gritty reality.

Heavy   A villain: evil or amoral, often sadistic.

High angle shot   Camera angle in which the camera looks down on a subject, often used for orientation. Also used to make characters appear inferior or weak.

Honey shot   Coverage of pretty young women—usually cheerleaders—sprinkled through a sports event to add sensual appeal.

Horizontal line   In visual compositions such a line tends to create feelings of stability, peace, serenity, and security.

Identification   The viewer's emotional involvement with (usually) the protagonist in drama; the viewer becomes the protagonist.

INDX   An acronym for Independent Exchange, an exchange of visual news material relayed to independent TV stations via satellite.

Involving moves   Moves in drama by one character toward another; moves that are seeking or aggressive in nature or arising from need.

Jump cut   A badly conceived cut between cameras that frame subject matter from almost identical angles. The subject appears to jump. Also, cuts between continuous action, as when frames have been deleted.

Keying   *See* Matte.

Leader   Videotape or film that precedes visual information. May be blank. *See* Digital leader.

Leitmotif   A motif or theme associated with specific person, situation, or idea; usually reprised for dramatic effect.

Limbo   An area used for shooting commercial inserts or small displays. Product is often displayed on a curved, neutral-colored, seamless backing called a *gizmo*.

Line of sight   The line created by a character's glance. Spectators tend to follow the line and glance in the same direction.

Location   An actual setting away from the studio. For example: a beach or factory. In TV a location is sometimes called a *remote* or a *nemo*.

Logo   A commercial name, insignia, or trademark, usually distinguished by unique artwork or lettering.

Long lens   A lens with telescopic properties, bringing distant objects close. Long lenses flatten space and diminish distance, reducing the speed of objects moving toward or away from camera.

Long shot   Photographing subject matter from a distance, providing orientation.

Looping   A postproduction technique for replacing dialog on tape or film, synchronizing new dialog with lip movements of actors. Used when original sound track is flawed.

Low angle   An angle created when a low camera shoots up at characters, making them appear powerful, dominant.

Magazine format   The structure or pattern of programs made up of various feature stories. For example: "60 Minutes" and "20/20."

Master angle A single camera angle in which the action of an entire scene is played out. Usually a full shot.

Matte To cut one picture or title into another electronically. Also called *keying*.

Medium shot A camera angle somewhere between a long shot and a close shot. Framing performers above the waist or hips.

Microwave link Using a parabolic dish to transmit a picture from a remote truck to a TV station or from a station to its transmitter.

Monitor A TV screen in a control booth or studio that reveals some aspect of a program's content. For example: a preview monitor.

Montage A term that originally referred to the editorial assembling of film segments. Montage today describes a rapid succession of images that convey a single concept. For example: a whirlwind courtship.

MOS Letters standing for "midout sound," signifying any sequence that is filmed silently. An affectionate carryover from early Hollywood and the German director who first spoke the words.

Moviola A motor-driven device for viewing and editing 16mm and 35mm film. The copyrighted name of a commercial product.

Multiplexer A system of prisms and mirrors that direct projected images into a television camera.

Music cues *See* Cues.

Music video Short filmed or taped musical numbers that convey the story, mood, or emotion suggested by a song. Often bizarre or highly imaginative in their imagery.

Nemo *See* Remote.

Neutral density (ND) filter A camera filter that reduces the quantity of light transmitted without affecting the quality of the image.

Night-for-night photography Actually shooting night scenes at night rather than using filters and exposure to create illu-

sion of night. More time consuming than *day-for-night*

Nose room Additional space in the direction in which a character is looking or moving, for pictorial balance.

Objective camera The customary camera treatment of dramatic material, in which the camera remains a neutral, objective, dispassionate viewer of the action.

On the nose A term used to describe dialog or action that is blunt, obvious, and heavy handed.

Out cue The final words in an SOT segment, alerting the director to cue the upcoming sound source.

Overcut In TV, when camera A is matted over camera B, overcutting replaces camera A (the foreground shot), allowing the background camera (B) to remain as is.

Overhead shot An angle from an overhead camera, creating a privileged, god-like perspective. Sometimes used in musicals to accentuate patterns formed by dancers.

Over-shoulder shot A camera angle framed past the foreground head and shoulders of Character A, with a frontal view of Character B.

Panning shot A shot created when the camera head turns horizontally with its base remaining stationary. The field moves horizontally. From the word *panorama*.

Pathetic fallacy When the forces of nature mirror the inner state of characters. From *pathos*.

Person As used in Chapter 4, the protagonist; a character with whom audiences can become emotionally involved.

Pickup shot A shot that continues the camera coverage of a scene from the point where it was interrupted. A time-saving technique that avoids the need to reshoot an entire master angle.

Plastic material Pudovkin's term for actions that help to define a character.

Playing against Dialog or actions that

take a direction different from what scene content seem to dictate, countering (say) tragedy by use of humor to prevent dramatic excesses.

Point of view (POV) shot   A subjective camera angle that becomes the perspective of a character. We look at the world through his or her eyes.

Polarizing filter   A filter that controls' glare or reflections or darkens skies for night effects.

Postproduction   The period following completion of principal photography, a time usually devoted to editing, scoring, and dubbing.

Pot   An abbreviation of *potentiometer*, a device that controls sound volume on an audio control panel.

Preparation   The emotional attitude or frame of mind that directors suggest to actors before they play a scene. Also, the period of preproduction in which directors plan their work.

Preparatory command   A director's command that alerts a crew member to be ready for a command of execution. For instance, "Ready [camera] 1" or "Prepare to fade out."

Preproduction   The preparatory period before a film or TV program begins photography, usually devoted to script polishing, casting, set design, and search for locations.

Preview monitor   A screen (monitor) in a control booth on which the TD previews the upcoming shot for the director.

Problem   As defined in Chapter 4, anything that prevents a protagonist from reaching his or her goal.

Producer   The person who is in charge of a production but who is subordinate to an executive producer.

Production board   A flat rectangular board prepared to facilitate scheduling and budget. Cardboard strips listing scenes, performers, extras, and other information are affixed to the board describing work to be accomplished each day.

Production values   Elements that add richness to a movie or TV production. For example: colorful locations, costumes, expensive sets.

Program line (PL)   The wire or radio link that allows communication between the control booth and the floor, usually between floor manager and the director or AD.

Progression   The traditional climbing action of drama, a growth in dramatic tension. Increasingly close camera angles represent camera progression.

Props (properties)   Items to be used or handled by characters in a dramatic scene.

Proscenium arch   In the theater, the arch that separates the stage from the auditorium.

Protagonist   The character with whom audiences become emotionally involved. The hero. The person with a problem.

Provocateur   An interviewer who goads guests, seeking revealing answers by asking provocative questions.

Pull focus   *See* Rack focus.

Rack focus   To change a camera's focus from foreground to background (or vice versa). Audience attention tends to follow the change in focus.

Radio lines   Dialog that comments on action that the viewer can see and is, therefore, unnecessary.

Raking shot   An angle in which objects or characters are photographed from the side so that viewers see them in depth.

Remote   A television program or segment that occurs away from the studio. Also called a *nemo*.

Rhythm   In visual composition, the pleasing repetition of images. In drama: repetition of phrases, actions, or musical themes for increased dramatic effect.

Riser   A raised platform.

Roll cue   The director's command or action to roll videotape or film.

Rough cut   An editor's first assembly of film or tape. The director's first opportunity to view and appraise the completed project.

Rundown sheet The paper that describes stories to be included in a news program; a sequence of events. Also called a fact sheet.

Satellite A communications device approximately 1,000 miles out in space that relays video and audio signals from one point to another.

Score Background music for a program that will intensify its dramatic or emotional values.

Screen direction The consistent pattern of movement from angle to angle: left to right or right to left.

Screenplay (script) A scenario that creates drama through plot, characters, and dialog—or that defines a sequence of events.

Selective focus Placing focus on point of greatest dramatic interest; viewer attention usually fastens on object in crispest focus.

Set dresser The person who provides furnishings for a set once it has been constructed. For example, tables, chairs, paintings, ashtrays.

Setup (camera setup) A camera position with its accompanying arrangement of lights.

Short lens (wide angle lens) A short focal length lens with a wide viewing angle. The wider the angle, the more distortion it creates, magnifying foreground objects and diminishing those in background.

Shot sheet A list of shots that each camera operator is expected to provide for a show.

Sitcom A situation comedy.

Slow motion A technique for slowing action so that movement may be studied. In drama, usually used to provide dramatic emphasis.

Slugs Pages inserted in early news scripts that contain only the barest facts about a news story, not the commentary.

Social function The element of companionship that TV and other entertainment provide.

SOF Sound on film.

SOT Sound on tape.

Source music Music that originates within a scene, that the characters can hear, as opposed to a *score*, which audiences can hear but characters cannot.

Special effects (SFX) Trick photography, optical effects, or those devised on the set. For example: smoke, explosions, bullets hitting a target.

Spectacle An element of audience appeal, generally using sound, motion, or color.

Spine A character's basic goals, needs, or drives. Also, a story theme.

Split screen A movie or TV screen that is split into two or more segments, each containing picture information.

Spotting Determining where music cues should be inserted into dramatic material.

Stage manager *See* Floor manager.

Star filter A filter that reproduces light sources or flares in a star pattern, creating an upbeat, glittery effect.

Storyboard A series of frames or sketches indicating key dramatic moments in a scene, script, or commercial. Often includes dialog beneath sketches.

Studio The room, building, or group of buildings devoted to TV or film production.

Style A director's personal pattern of treating material, including staging of camera and performers, script elements, and music.

Subjective camera Implies that camera has become a participant rather than an observer. POV shot is an example. Suggests a character's mood or emotions through a more generalized treatment (lighting, music, lenses).

Suspense Usually associated with films generating anxiety or apprehension; actually a component of all well-conceived drama. Concern or worry over a story's outcome.

Switcher The electronic device a technical director uses to cut between video cameras or to create optical effects.

Tag A final line of dialog, action, or musical cue that serves as punctuation at the end of a scene.

Take An instantaneous change from one video source to another. A verb or a noun.

Tally light A red light on a video camera that indicates when it is on the air.

Technical director (TD) Person in charge of crew in live or videotape operations. Edits and creates video effects through an electronic switcher.

TelePrompter A device that displays news or other copy for performers to read. Reflected in glass angled before lens so that performers may look directly into camera. A registered trademark.

Tilt Moving the camera head upwards or down, the camera base remaining fixed.

Timing cut A cut that eliminates or shortens tedious or unnecessary dramatic material, either by cutting away briefly or by cutting ahead of the action.

Total running time (TRT) The minutes and seconds of (usually) a videotape segment.

Tracking shot A shot that moves with performers, usually in their tracks, ahead of them or behind them.

Trucking shot Camera moving with performers, usually from a side angle. Term sometimes used interchangeably with *tracking shot*.

Turnaround time The time between work periods, after work on Monday and before work on Tuesday. Penalties usually given for *short turnaround* (less then ten hours off duty).

Two-shot An angle that includes two performers.

Undercut When one camera is matted over another, to change the background to another video source.

Unit manager The person responsible for schedule and budget. Also, assembles below-the-line personnel.

Uplink Used to transmit a video signal from an earth station to a satellite.

Upstage That part of a stage farthest from the audience or camera.

Vertical line Tends to create feelings of strength, power, spirituality in a visual composition.

Video Relating to television or to videotape. The visual portion of a TV production.

Videotape A plastic tape that can magnetically record various audio, video, and control track information.

Videotape recorder (VTR) An electronic device that records audio, video, and control signals on videotape.

Whip pan (also *swish pan* or *blur pan*) A sudden fast, sometimes blurred pan from one subject or scene to another.

Wide angle lens *See* Short lens.

Wipe A transition effect in which one image moves into the frame, replacing another. Move may be horizontal, vertical, diagonal, or whatever the imagination may invent.

Wrap To conclude production activities for the day or for the picture. "That's a wrap": words that a cast and crew are always happy to hear.

Zero-degree camera angle An angle that exhibits a device or skill from the audience's perspective.

Zoom shot A change of angle that moves slowly or quickly from a wide to a closer angle or vice versa. Executed through a lens adjustment.

INDEX